Lecture Notes in Bioinformatics 8701

Subseries of Lecture Notes in Computer Science

Dan Brown   Burkhard Morgenstern (Eds.)

Algorithms in Bioinformatics

14th International Workshop, WABI 2014
Wroclaw, Poland, September 8-10, 2014
Proceedings

 Springer

Volume Editors

Dan Brown
University of Waterloo
David R. Cheriton School of Computer Science
Waterloo, ON, Canada
E-mail: dan.brown@uwaterloo.ca

Burkhard Morgenstern
University of Göttingen
Institute of Microbiology and Genetics
Department of Bioinformatics
Göttingen, Germany
E-mail: bmorgen@gwdg.de

ISSN 0302-9743   e-ISSN 1611-3349
ISBN 978-3-662-44752-9   e-ISBN 978-3-662-44753-6
DOI 10.1007/978-3-662-44753-6
Springer Heidelberg New York Dordrecht London

Library of Congress Control Number: 2014947978

LNCS Sublibrary: SL 8 – Bioinformatics

Typesetting: Camera-ready by author, data conversion by Scientific Publishing Services, Chennai, India

Printed on acid-free paper

Springer is part of Springer Science+Business Media (www.springer.com)

Preface

This proceedings volume contains papers presented at the Workshop on Algorithms in Bioinformatics 2014 (WABI 2014) that was held at the University of Wrocław, Poland, Institute of Computer Science, during September 8–12, 2014. WABI 2014 was one of seven conferences that were organized as part of ALGO 2014. WABI is an annual conference series on all aspects of algorithms and data structure in molecular biology, genomics, and phylogeny data analysis that was first held in 2001. WABI 2014 was sponsored by the European Association for Theoretical Computer Science (EATCS) and the International Society for Computational Biology (ISCB).

In 2014, a total of 61 manuscripts were submitted to WABI from which 27 were selected for presentation at the conference, 26 of them full papers not previously published in journals, and one short abstract of a paper that was published simultaneously in a journal. Papers were selected based on a thorough reviewing and discussion process by the WABI Program Committee, usually involving three reviewers per submitted paper. The selected papers cover a wide range of topics from sequence and genome analysis through phylogeny reconstruction and networks to mass spectrometry data analysis. As in previous years, extended versions of selected WABI papers will be published in a Thematic Series in the journal *Algorithms for Molecular Biology* (AMB), published by BioMed Central.

We thank all the authors of submitted papers and the members of the Program Committee for their efforts that made this conference possible and the WABI Steering Committee for help and advice. In particular, we are indebted to the keynote speaker of the conference, Hélène Touzet, for her presentation. Above all, we are most grateful to Marcin Bieńkowski and the local Organizing Committee for their very professional work.

September 2014

Dan Brown
Burkhard Morgenstern

Organization

Program Chairs

Dan Brown   University of Waterloo, Canada
Burkhard Morgenstern   University of Göttingen, Germany

Program Committee

Mohamed Abouelhoda   Cairo University, Egypt
Tatsuya Akutsu   Kyoto University, Japan
Anne Bergeron   Universite du Quebec a Montreal, Canada
Sebastian Böcker   Friedrich Schiller University Jena, Germany
Paola Bonizzoni   Università di Milano-Bicocca
Marilia Braga   Inmetro - Ditel
Broňa Brejová   Comenius University in Bratislava, Slovakia
C.Titus Brown   Michigan State University, USA
Philipp Bucher   Swiss Institute for Experimental Cancer
                             Research, Switzerland
Rita Casadio   UNIBO
Cedric Chauve   Simon Fraser University, Canada
Matteo Comin   University of Padova, Italy
Lenore Cowen   Tufts University, USA
Keith Crandall   George Washington University, USA
Nadia El-Mabrouk   University of Montreal, Canada
David Fernández-Baca   Iowa State University, USA
Anna Gambin   Institute of Informatics, Warsaw University,
                             Poland
Olivier Gascuel   LIRMM, CNRS - Université Montpellier 2,
                             France
Raffaele Giancarlo   Università di Palermo, Italy
Nicholas Hamilton   The University of Queensland, Australia
Barbara Holland   University of Tasmania, Australia
Katharina Huber   University of East Anglia, UK
Steven Kelk   University of Maastricht, The Netherlands
Carl Kingsford   Carnegie Mellon University, USA
Gregory Kucherov   CNRS/LIGM, France
Zsuzsanna Liptak   University of Verona, Italy
Stefano Lonardi   UC Riverside, USA
Veli Mäkinen   University of Helsinki, Finland
Ion Mandoiu   University of Connecticut, USA

Giovanni Manzini	University of Eastern Piedmont, Italy
Paul Medvedev	Pennsylvania State University, USA
Irmtraud Meyer	University of British Columbia, Canada
István Miklós	Renyi Institute, Budapest, Hungary
Satoru Miyano	University of Tokyo
Bernard Moret	EPFL, Lausanne, Switzerland
Vincent Moulton	University of East Anglia, UK
Luay Nakhleh	Rice University, USA
Nadia Pisanti	Università di Pisa, Italy
Mihai Pop	University of Maryland, USA
Teresa Przytycka	NIH, USA
Sven Rahmann	University of Duisburg-Essen, Germany
Marie-France Sagot	Inria Grenoble Rhône-Alpes and Université de Lyon, France
S. Cenk Sahinalp	Simon Fraser University, Canada
David Sankoff	University of Ottawa, Canada
Russell Schwartz	Carnegie Mellon University, USA
Joao Setubal	University of Sao Paulo, Brazil
Peter F. Stadler	University of Leipzig, Germany
Jens Stoye	Bielefeld University, Germany
Krister Swenson	Université de Montréal/McGill University, Canada
Jijun Tang	University of South Carolina, USA
Lusheng Wang	City University of Kong Kong, SAR China
Louxin Zhang	National University of Singapore
Michal Ziv-Ukelson	Ben Gurion University of the Negev, Israel

Additional Reviewers

Claudia Landi	Gunnar W. Klau
Pedro Real Jurado	Annelyse Thévenin
Yuri Pirola	João Paulo Pereira Zanetti
James Cussens	Andrea Farruggia
Seyed Hamid Mirebrahim	Faraz Hach
Mateusz Łacki	Alberto Policriti
Anton Polishko	Sławomir Lasota
Giosue Lo Bosco	Heng Li
Christian Komusiewicz	Daniel Doerr
Piotr Dittwald	Leo van Iersel
Giovanna Rosone	Thu-Hien To
Flavio Mignone	Rayan Chikhi
Marco Cornolti	Hind Alhakami
Fábio Henrique Viduani Martinez	Ermin Hodzic
Yu Lin	Guillaume Holley
Salem Malikic	Ibrahim Numanagic

Inken Wohlers
Rachid Ounit
Blerina Sinaimeri
Vincent Lacroix
Linda Sundermann
Gianluca Della Vedova

Pedro Feijão
Cornelia Caragea
Johanna Christina
Agata Charzyńska

WABI Steering Committee

Bernard Moret   EPFL, Switzerland
Vincent Moulton   University of East Anglia, UK
Jens Stoye   Bielefeld University, Germany
Tandy Warnow   The University of Texas at Austin, USA

ALGO Organizing Committee

Marcin Bieńkowski   Tomasz Jurdziński
Jarosław Byrka (Co-chair)   Krzysztof Loryś
Agnieszka Faleńska   Leszek Pacholski (Co-chair)

Table of Contents

QCluster: Extending Alignment-Free Measures with Quality Values for Reads Clustering

Matteo Comin, Andrea Leoni, and Michele Schimd

Department of Information Engineering, University of Padova, Padova, Italy
comin@dei.unipd.it

Abstract. The data volume generated by Next-Generation Sequencing (NGS) technologies is growing at a pace that is now challenging the storage and data processing capacities of modern computer systems. In this context an important aspect is the reduction of data complexity by collapsing redundant reads in a single cluster to improve the run time, memory requirements, and quality of post-processing steps like assembly and error correction. Several alignment-free measures, based on k-mers counts, have been used to cluster reads.

Quality scores produced by NGS platforms are fundamental for various analysis of NGS data like reads mapping and error detection. Moreover future-generation sequencing platforms will produce long reads but with a large number of erroneous bases (up to 15%). Thus it will be fundamental to exploit quality value information within the alignment-free framework.

In this paper we present a family of alignment-free measures, called D^q-type, that incorporate quality value information and k-mers counts for the comparison of reads data. A set of experiments on simulated and real reads data confirms that the new measures are superior to other classical alignment-free statistics, especially when erroneous reads are considered. These measures are implemented in a software called QCluster (http://www.dei.unipd.it/~ciompin/main/qcluster.html).

Keywords: alignment-free measures, reads quality values, clustering reads.

1 Introduction

The data volume generated by Next-Generation Sequencing (NGS) technologies is growing at a pace that is now challenging the storage and data processing capacities of modern computer systems [1]. Current technologies produce over 500 billion bases of DNA per run, and the forthcoming sequencers promise to increase this throughput. The rapid improvement of sequencing technologies has enabled a number of different sequencing-based applications like genome resequencing, RNA-Seq, ChIP-Seq and many others [2]. Handling and processing such large files is becoming one of the major challenges in most genome research projects.

D. Brown and B. Morgenstern (Eds.): WABI 2014, LNBI 8701, pp. 1–13, 2014.

Alignment-based methods have been used for quite some time to establish similarity between sequences [3]. However there are cases where alignment methods can not be applied or they are not suited. For example the comparison of whole genomes is impossible to conduct with traditional alignment techniques, because of events like rearrangements that can not be captured with an alignment [4–6]. Although fast alignment heuristics exist, another drawback is that alignment methods are usually time consuming, thus they are not suited for large-scale sequence data produced by Next-Generation Sequencing technologies (NGS)[7, 8]. For these reasons a number of alignment-free techniques have been proposed over the years [9].

The use of alignment-free methods for comparing sequences has proved useful in different applications. Some alignment-free measures use the patterns distribution to study evolutionary relationships among different organisms [4, 10, 11]. Several alignment-free methods have been devised for the detection of enhancers in ChIP-Seq data [12–14] and also of entropic profiles [15, 16]. Another application is the classification of protein remotely related, which can be addressed with sophisticate word counting procedures [17, 18]. The assembly-free comparison of genomes based on NGS reads has been investigated only recently [7, 8]. For a comprehensive review of alignment-free measures and applications we refer the reader to [9].

In this study we want to explore the ability of alignment-free measures to cluster reads data. Clustering techniques are widely used in many different applications based on NGS data, from error correction [19] to the discovery of groups of microRNAs [20]. With the increasing throughput of NGS technologies another important aspect is the reduction of data complexity by collapsing redundant reads in a single cluster to improve the run time, memory requirements, and quality of subsequent steps like assembly.

In [21] Solovyov *et. al.* presented one of the first comparison of alignment-free measures when applied to NGS reads clustering. They focused on clustering reads coming from different genes and different species based on k-mer counts. They showed that D-type measures (see section 2), in particular D_2^*, can efficiently detect and cluster reads from the same gene or species (as opposed to [20] where the clustering is focused on errors). In this paper we extend this study by incorporating quality value information into these measures.

Quality scores produced by NGS platforms are fundamental for various analysis of NGS data: mapping reads to a reference genome [22]; error correction [19]; detection of insertion and deletion [23] and many others. Moreover future-generation sequencing technologies will produce long and less biased reads with a large number of erroneous bases [24]. The average number of errors per read will grow up to 15%, thus it will be fundamental to exploit quality value information within the alignment-free framework and the *de novo* assembly where longer and less biased reads could have dramatic impact.

In the following section we briefly review some alignment-free measures. In section 3 we present a new family of statistics, called D^q-type, that take advantage of quality values. The software QCluster is discussed in section 4 and relevant results on simulated and real data are presented in section 5. In section 6 we summarize the findings and we discuss future directions of investigation.

2 Previous Work on Alignment-Free Measures

One of the first papers that introduced an alignment-free method is due to Blaisdell in 1986 [25]. He proposed a statistic called D_2, to study the correlation between two sequences. The initial purpose was to speed up database searches, where alignment-based methods were too slow. The D_2 similarity is the correlation between the number of occurrences of all k-mers appearing in two sequences. Let X and Y be two sequences from an alphabet Σ. The value X_w is the number of times w appears in X, with possible overlaps. Then the D_2 statistic is:

$$D_2 = \sum_{w \in \Sigma^k} X_w Y_w.$$

This is the inner product of the word vectors X_w and Y_w, each one representing the number of occurrences of words of length k, *i.e.* k-mers, in the two sequences. However, it was shown by Lippert *et al.* [26] that the D_2 statistic can be biased by the stochastic noise in each sequence. To address this issue another popular statistic, called D_2^z, was introduced in [13]. This measure was proposed to standardize the D_2 in the following manner:

$$D_2^z = \frac{D_2 - \mathbb{E}(D_2)}{\mathbb{V}(D_2)},$$

where $\mathbb{E}(D_2)$ and $\mathbb{V}(D_2)$ are the expectation and the standard deviation of D_2, respectively. Although the D_2^z similarity improves D_2, it is still dominated by the specific variation of each pattern from the background [27, 28]. To account for different distributions of the k-mers, in [27] and [28] two other new statistics are defined and named D_2^* and D_2^s. Let $\tilde{X}_w = X_w - (n - k + 1) * p_w$ and $\tilde{Y}_w = Y_w - (n - k + 1) * p_w$ where p_w is the probability of w under the null model. Then D_2^* and D_2^s can be defined as follows:

$$D_2^* = \sum_{w \in \Sigma^k} \frac{\tilde{X}_w \tilde{Y}_w}{(n - k + 1) p_w}.$$

and,

$$D_2^s = \sum_{w \in \Sigma^k} \frac{\tilde{X}_w \tilde{Y}_w}{\sqrt{\tilde{X}_w^2 + \tilde{Y}_w^2}}$$

This latter similarity measure responds to the need of normalization of D_2. These set of alignment-free measures are usually called D-type statistics. All these statistics have been studied by Reinert *et al.* [27] and Wan *et al.* [28] for the detection of regulatory sequences. From the word vectors X_w and Y_w several other measures can be computed like L_2, Kullback-Leibler divergence (KL), symmetrized KL [21] etc.

3   Comparison of Reads with Quality Values

3.1   Background on Quality Values

Upon producing base calls for a read x, sequencing machines also assign a *quality score* $Q_x(i)$ to each base in the read. These scores are usually given as *phred-scaled* probability [29] of the i-th base being wrong

$$Q_x(i) = -10 \log_{10} Prob\{\text{the base } i \text{ of read } x \text{ is wrong }\}.$$

For example, if $Q_x(i) = 30$ then there is 1 in 1000 chance that base i of read x is incorrect. If we assume that quality values are produced independently to each other (similarly to [22]), we can calculate the probability of an entire read x being correct as:

$$P_x\{\text{the read } x \text{ is correct}\} = \prod_{j=0}^{n-1} (1 - 10^{-Q_x(j)/10})$$

where n is the length of the read x. In the same way we define the probability of a word w of length k, occuring at position i of read x being correct as:

$$P_{w,i}\{\text{the word } w \text{ at position } i \text{ of read } x \text{ is correct}\} = \prod_{j=0}^{k-1} (1 - 10^{-Q_x(i+j)/10}).$$

In all previous alignment-free statistics the k-mers are counted such that each occurrence contributed as 1 irrespective of its quality. Here we can use the quality of that occurrence instead to account also for erroneous k-mers. The idea is to model sequencing as the process of reading k-mers from the reference and assigning a probability to them. Thus this formula can be used to weight the occurrences of all k-mers used in the previous statistics.

3.2   New D^q-Type Statistics

We extend here D-type statistics [27, 28] to account for quality values. By defining X_w^q as the sum of probabilities of all the occurrences of w in x:

$$X_w^q = \sum_{i \in \{i|\ w \text{ occurs in } x \text{ at position } i\}} P_{w,i}$$

we assign a weight (*i.e.* a probability) to each occurrence of w. Now X_w^q can be used instead of X_w to compute the alignment-free statistics. Note that, by using X_w^q, every occurrence is not counted as 1, but with a value in $[0,1]$ depending of the reliability of the read. We can now define a new alignment-free statistic as :

$$D_2^q = \sum_{w \in \Sigma^k} X_w^q Y_w^q.$$

This is the extension of the D_2 measure, in which occurrences are weighted based on quality scores. Following section 2 we can also define the centralized k-mers counts as follows:

$$\widetilde{X_w^q} = X_w^q - (n-k+1)p_w E(P_w)$$

where $n = |x|$ is the length of x, p_w is the probability of the word w in the i.i.d. model and the expected number of occurrences $(n-k+1)p_w$ is multiplied by $E(P_w)$ which represents the expected probability of k-mer w based on the quality scores.

We can now extend two other popular alignment-free statistics:

$$D_2^{*q} = \sum_{w \in \Sigma^k} \frac{\tilde{X}_w^q \tilde{Y}_w^q}{(n-k+1)p_w E(P_w)}$$

and,

$$D_2^{sq} = \sum_{w \in \Sigma^k} \frac{\tilde{X}_w^q \tilde{Y}_w^q}{\sqrt{\tilde{X}_w^{q^2} + \tilde{Y}_w^{q^2}}}$$

We call these three alignment-free measures D^q-type. Now, $E(P_w)$ depends on w and on the actual sequencing machine, therefore it can be very hard, if not impossible, to calculate precisely. However, if the set $\mathbb{D}$ of all the reads is large enough we can estimate the prior probability using the posterior relative frequency, i.e. the frequency observed on the actual set $\mathbb{D}$, similarly to [22]. We assume that, given the quality values, the error probability on a base is independent from its position within the read and from all other quality values (see [22]). We defined two different approximations, the first one estimates $E(P_w)$ as the average error probability of the k-mer w among all reads $x \in \mathbb{D}$:

$$E(P_w) \approx \frac{\sum_{x \in \mathbb{D}} X_w^q}{\sum_{x \in \mathbb{D}} X_w} \tag{1}$$

while the second defines, for each base j of w, the average quality observed over all occurrences of w in $\mathbb{D}$:

$$\overline{Q_w}[j] = \frac{\sum_{x \in \mathbb{D}} \sum_{i \in \{i| \ w \text{ occurs in } x \text{ at position } i\}} Q_x(i+j)}{\sum_{x \in \mathbb{D}} X_w}$$

and it uses the average quality values to compute the expected word probability.

$$E(P_w) \approx \prod_{j=0}^{k-1} (1 - 10^{-\overline{Q_w}(j)/10}) \tag{2}$$

We called the first approximation *Average Word Probability (AWP)* and the second one *Average Quality Probability (AQP)*. Both these approximations are implemented within the software QCluster and they will tested in section 5.

3.3 Quality Value Redistribution

If we consider the meaning of quality values it is possible to further exploit it to extend and improve the above statistics. Let's say that the base A has quality 70%, it means that there is a 70% probability that the base is correct. However there is also another 30% probability that the base is incorrect. Let's ignore for the moment insertion and deletion errors, if the four bases are equiprobable, this means that with uniform probability 10% the wrong base is a C, or a G or a T. It's therefore possible to redistribute the "missing quality" among other bases. We can perform a more precise operation by redistributing the missing quality among other bases in proportion to their frequency in the read. For example, if the frequencies of the bases in the read are A=20%, C=30%, G=30%, T=20%, the resulting qualities, after the redistribution, will be: A=70%, $C = 30\%*30\%/(30\%+30\%+20\%) = 11\%$, $G = 30\%*30\%/(30\%+30\%+20\%) = 11\%$, $T = 30\% * 20\%/(30\% + 30\% + 20\%) = 7,5\%$.

The same redistribution, with a slight approximation, can be extended to k-mers quality. More in detail, we consider only the case in which only one base is wrong, thus we redistribute the quality of only one base at a time. Given a k-mer, we generate all neighboring words that can be obtained by substitution of the wrong base. The quality of the replaced letter is calculated as in the previous example and the quality of the entire word is again given by the product of the qualities of all the bases in the new k-mers. We increment the corresponding entry of the vector X_w^q with the score obtained for the new k-mer. This process is repeated for all bases of the original k-mer. Thus every time we are evaluating the quality of a word, we are also scoring neighboring k-mers by redistributing the qualities. We didn't consider the case where two or more bases are wrong simultaneously, because the computational cost would be too high and the quality of the resulting word would not appreciably affect the measures.

4   QCluster: Clustering of Reads with D^q-Type Measures

All the described algorithms were implemented in the software QCluster. The program takes in input a fastq format file and performs centroid-based clustering (k-means) of the reads based on the counts and the quality of k-mers. The software performs centroid-based clustering with KL divergence and other distances like L_2 (Euclidean), D_2, D_2^*, symmetrized KL divergence etc. When using the D^q-type measures, one needs to choose the method for the computation of the expected word probability, AWP or AQP, and the quality redistribution.

Since some of the implemented distances (symmetrized KL, D_2^*) do not guarantee to converge, we implemented a stopping criteria. The execution of the algorithm interrupts if the number of iterations without improvements exceeds a certain threshold. In this case, the best solution found is returned. The maximum number of iterations may be set by the user and for our experiments we use the value 5. Several other options like reverse complement and different normalization are available. All implemented measures can be computed in linear time and space, which is desirable for large NGS datasets. The QCluster[1] software has been implemented in C++ and compiled and tested using GNU GCC.

5   Experimental Results

Several tests have been performed in order to estimate the effectiveness of the different distances, on both simulated and real datasets. In particular, we had to ensure that, with the use of the additional information of quality values, the clustering improved compared to that produced by the original algorithms.

For simulations we use the dataset of human mRNA genes downloaded from NCBI[2], also used in [21]. We randomly select 50 sets of 100 sequences each of human mRNA, with the length of each sequence ranged between 500 and 10000 bases. From each sequence, 10000 reads of length 200 were simulated using Mason[3] [30] with different parameters, e.g. percentage of mismatches, read length. We apply QCluster using different distances, to the whole set of reads and then we measure the quality of the clusters produced by evaluating the extent to which the partitioning agrees with the natural splitting of the sequences. In other words, we measured how well reads originating from the same sequence are grouped together. We calculate the recall rate as follows, for each mRNA sequence S we identified the set of reads originated from S. We looked for the cluster C that contains most of the reads of S. The percentage of the S reads that have been grouped in C is the recall value for the sequence S. We repeat the same operation for each sequence and calculate the average value of recall rate over all sequences.

Several clustering were produced by using the following distance types: D_2^*, D_2, L_2, KL, symmetrized KL and compared with D_2^{*q} in all its variants, using the expectation formula (1) AWP or (2) AQP, with and without quality redistribution (q-red). In order to avoid as much as possible biases due to the initial random generation of centroids, each algorithm was executed 5 times with different random seeds and the clustering with the lower distortion was chosen.

Table 1 reports the recall while varying error rates, number of clusters and the parameters k. For all distances the recall rate decreases with the number of clusters, as expected. For traditional distances, if the reads do not contain errors then D_2^* preforms consistently better then the others D_2, L_2, KL. When the sequencing process becomes more noisy, the KL distances appears to be less sensitive to sequencing errors. However if quality information are used, D_2^{*q} outperforms all other methods and the advantage grows with the error rate. This confirms that the use of quality values can improve clustering accuracy. When the number of clusters increases then the advantage of D_2^{*q} becomes more evident. In these experiments the use of AQP for expectation within D_2^{*q} is more stable and better performing compared with formula AWP. The contribution of

---

[1] http://www.dei.unipd.it/~ciompin/main/qcluster.html
[2] ftp://ftp.ncbi.nlm.nih.gov/refseq/H-sapiens/mRNA-Prot/
[3] http://seqan.de/projects/mason.html

Table 1. Recall rates of clustering of mRNA simulated reads (10000 reads of length 200) for different measures, error rates, number of clusters and parameter k

Distance	No Errors	3%	5%	10%	No Errors	3%	5%	10%
	2 clusters				2 clusters			
D_2^*	**0,815**	0,813	0,810	0,801	**0,822**	0,819	0,814	0,794
D_2^{*q} AQP	**0,815**	**0,815**	**0,813**	**0,810**	**0,822**	**0,822**	**0,820**	**0,809**
D_2^{*q} AQP q-red	**0,815**	**0,815**	**0,813**	**0,810**	**0,822**	**0,822**	**0,820**	0,807
D_2^{*q} AWP	0,809	0,806	0,805	0,802	0,809	0,807	0,805	0,802
D_2^{*q} AWP q-red	0,809	0,806	0,805	0,802	0,809	0,807	0,805	0,802
L_2	0,811	0,807	0,806	0,801	0,810	0,806	0,805	0,801
KL	0,812	0,809	0,807	0,802	0,812	0,809	0,807	0,802
Symm, KL	0,812	0,809	0,807	0,802	0,812	0,808	0,806	0,802
D_2	0,811	0,807	0,806	0,801	0,809	0,806	0,805	0,800
	3 clusters				3 clusters			
D_2^*	**0,695**	0,689	0,683	0,662	**0,717**	0,707	0,697	0,668
D_2^{*q} AQP	**0,695**	**0,696**	**0,696**	0,689	**0,717**	0,711	**0,705**	0,679
D_2^{*q} AQP q-red	**0,695**	**0,696**	**0,696**	**0,691**	**0,717**	**0,712**	0,704	**0,681**
D_2^{*q} AWP	0,653	0,646	0,646	0,638	0,668	0,662	0,655	0,646
D_2^{*q} AWP q-red	0,653	0,646	0,645	0,637	0,668	0,662	0,655	0,644
L_2	0,682	0,673	0,671	0,657	0,685	0,677	0,674	0,663
KL	0,694	0,687	0,685	0,672	0,696	0,689	0,687	0,675
Symm, KL	0,693	0,686	0,684	0,669	0,695	0,688	0,685	0,673
D_2	0,675	0,668	0,662	0,654	0,675	0,671	0,665	0,655
	4 clusters				4 clusters			
D_2^*	**0,623**	0,613	0,606	0,574	0,627	0,616	0,591	0,551
D_2^{*q} AQP	0,622	0,621	0,618	0,602	**0,628**	**0,617**	0,602	0,572
D_2^{*q} AQP q-red	0,622	**0,622**	**0,619**	**0,605**	**0,628**	**0,617**	**0,603**	**0,573**
D_2^{*q} AWP	0,580	0,563	0,566	0,535	0,582	0,571	0,572	0,555
D_2^{*q} AWP q-red	0,580	0,560	0,565	0,533	0,582	0,570	0,570	0,555
L_2	0,554	0,551	0,547	0,540	0,568	0,565	0,553	0,543
KL	0,555	0,548	0,545	0,536	0,566	0,558	0,547	0,537
Symm, KL	0,556	0,549	0,546	0,538	0,562	0,554	0,547	0,539
D_2	0,553	0,547	0,547	0,538	0,556	0,549	0,548	0,540
	5 clusters				5 clusters			
D_2^*	0,553	0,539	0,532	0,500	0,560	0,534	0,512	0,462
D_2^{*q} AQP	**0,554**	**0,545**	**0,551**	0,532	0,560	0,544	0,524	**0,489**
D_2^{*q} AQP q-red	0,553	0,544	0,550	**0,533**	**0,561**	**0,545**	**0,531**	0,487
D_2^{*q} AWP	0,483	0,475	0,470	0,463	0,509	0,494	0,485	0,470
D_2^{*q} AWP q-red	0,483	0,475	0,470	0,461	0,509	0,494	0,482	0,470
L_2	0,478	0,472	0,465	0,453	0,500	0,495	0,486	0,465
KL	0,498	0,488	0,484	0,468	0,507	0,501	0,492	0,476
Symm, KL	0,498	0,488	0,484	0,468	0,507	0,500	0,491	0,474
D_2	0,470	0,464	0,457	0,449	0,488	0,482	0,476	0,455
	$k=2$				$k=3$			
	(a)				(b)			

quality redistribution (q-red) is limited, although it seems to have some positive effect with the expectation AQP.

The future generation sequencing technologies will produce long reads with a large number of erroneous bases. To this end we study how read length affects these measures. Since the length of sequences under investigation is limited we keep the read length under 400 bases. In Table 2 we report some experiments for the setup with 4 clusters and $k = 3$, while varying the error rate and read length. If we compare these results with Table 1, where the read length is 200, we can observe a similar behavior. As the error rate increases the improvement with respect to the other measures remains evident, in particular the difference in terms of recall of D_2^{*q} with the expectations AQP grows with the length of reads when compared with KL (up to 9%), and it remains constant when compared with D_2^*. With the current tendency of the future sequencing technologies to produce longer reads this behavior is desirable. These performance are confirmed also for other setups with larger k and higher number of clusters (data not shown).

Table 2. Recall rates for clustering of mRNA simulated reads (10000 reads, $k = 3$, 4 clusters) for different measures, error rates and read length

Distance	No Errors	3%	5%	10%	No Errors	3%	5%	10%
	4 clusters				4 clusters			
D_2^*	**0,680**	0,667	0,658	0,625	**0,713**	0,700	0,697	0,672
D_2^{*q} AQP	**0,680**	**0,672**	**0,673**	**0,650**	**0,713**	**0,712**	0,710	0,693
D_2^{*q} AQP q-red	**0,680**	0,671	**0,673**	**0,650**	**0,713**	0,711	**0,711**	**0,694**
D_2^{*q} AWP	0,616	0,610	0,608	0,601	0,643	0,636	0,632	0,623
D_2^{*q} AWP q-red	0,616	0,610	0,607	0,602	0,643	0,635	0,631	0,622
L_2	0,610	0,600	0,602	0,581	0,638	0,630	0,624	0,614
KL	0,617	0,604	0,601	0,577	0,649	0,632	0,628	0,618
Symm, KL	0,613	0,603	0,599	0,576	0,647	0,632	0,627	0,616
D_2	0,601	0,593	0,588	0,575	0,626	0,618	0,615	0,604
	read length=300				read length=400			
	(a)				(b)			

5.1   Boosting Assembly

Assembly is one of the most challenging computational problems in the field of NGS data. It is a very time consuming process with highly variable outcomes for different datasets [31]. Currently large datasets can only be assembled on high performance computing systems with considerable CPU and memory resources. Clustering has been used as preprocessing, prior to assembly, to improve memory requirements as well as the quality of the assembled contigs [20, 21]. Here we test if the quality of assembly of real read data can be improved with clustering. For the assembly component we use Velvet [32], one of the most popular assembly tool for NGS data. We study the *Zymomonas mobilis* genome and download as input the reads dataset *SRR017901* (454 technology) with 23.5Mbases corresponding to 10× coverage. We apply the clustering algorithms, with $k = 3$, and divide the dataset of reads in two clusters. Then we produce an assembly, as a set of contigs, for each cluster using Velvet and we merged the generated contigs. In order to evaluate the clustering quality, we compare this merged set with the assembly, without clustering, using of the whole set of reads. Commonly used metrics such as number of contigs, $N50$ and percentage of mapped contigs are presented in Table 3. When merging contigs from different clusters, some contig might be very similar or they can cover the same region of the genome, this can artificially increase these values. Thus we compute also a less biased measure that is the percentage of the genome that is covered by the contigs (last column).

Table 3. Comparison of assembly with and without clustering preprocess ($k = 3$, 2 clusters). The assembly with Velvet is evaluated in terms of mapped contigs, N50, number of contigs and genome coverage. The dataset used is SRR017901 (23.5M bases, 10x coverage) that contains reads of *Zymomonas mobilis*.

Distance	Mapped Contigs	N50	Number of Contigs	Genome Coverage
No Clustering	93.55%	112	22823	0,828
D_2^*	93.97%	138	28701	0,914
D_2^{*q} AQP	94.09%	141	29065	**0,921**
D_2^{*q} AQP q-red	94.13%	**141**	**29421**	0,920
D_2^{*q} AWP	**94.36%**	137	28425	0,907
D_2^{*q} AWP q-red	**94.36%**	137	28549	0,908
L_2	94.24%	135	28297	0,904
KL	94.19%	135	28171	0,903
Symm, KL	94.27%	134	27999	0,902
D_2	94.33%	134	28019	0,903

In this set of experiments the introduction of clustering as a preprocessing step increases the number of contigs and the N50. More relevant is the fact that the genome coverage is incremented by 10% with respect to the assembly without clustering. The relative performance between the distance measures is very similar to the case with simulated data. In fact D_2^{*q} with expectation AQP and quality redistribution is again the best performing. More experiments should be conducted in order to prove that assembly can benefit from the clustering preprocessing. However this first preliminary tests show that, at least for some configuration, a 10% improvement on the genome coverage can be obtained.

The time required to performed the above experiments are in general less than a minute on a modern laptop with an Intel i7 and 8Gb of ram. The introduction of quality values typically increases the running time by 4% compared to standard alignment-free methods. The reads dataset *SRR017901* is about 54MB and the memory required to cluster this set is 110MB. Also in the other experiments the memory requirements remain linear in the input size.

6   Conclusions

The comparison of reads with quality values is essentials in many genome projects. The importance of quality values will increase in the near future with the advent of future sequencing technologies, that promise to produce long reads, but with 15% errors. In this paper we presented a family of alignment-free measures, called D^q-type, that incorporate quality value information and k-mers counts for the comparison of reads data. A set of experiments on simulated and real reads data confirms that the new measures are superior to other classical alignment-free statistics, especially when erroneous reads are considered. If quality information are used, D_2^{*q} outperforms all other methods and the advantage grows with the error rate and with the length of reads. This confirms that the use of quality values can improve clustering accuracy.

Preliminary experiments on real reads data show that the quality of assembly can also improve when using clustering as preprocessing. All these measures are implemented in a software called QCluster. As a future work we plan to explore other applications like genome diversity estimation and meta-genome assembly in which the impact of reads clustering might be substantial.

Acknowledgments. M. Comin was partially supported by the Ateneo Project CPDA110239 and by the P.R.I.N. Project 20122F87B2.

References

1. Medini, D., Serruto, D., Parkhill, J., Relman, D., Donati, C., Moxon, R., Falkow, S., Rappuoli, R.: Microbiology in the post-genomic era. Nature Reviews Microbiology 6, 419–430 (2008)
2. Jothi, R., et al.: Genome-wide identification of in vivo protein-DNA binding sites from ChIP-Seq data. Nucleic Acids Res. 36, 5221–5231 (2008)
3. Altschul, S., Gish, W., Miller, W., Myers, E.W., Lipman, D.: Basic local alignment search tool. Journal of Molecular Biology 215(3), 403–410 (1990)
4. Sims, G.E., Jun, S.-R., Wu, G.A., Kim, S.-H.: Alignment-free genome comparison with feature frequency profiles (FFP) and optimal resolutions. PNAS 106(8), 2677–2682 (2009)
5. Comin, M., Verzotto, D.: Whole-genome phylogeny by virtue of unic subwords. In: Proc. 23rd Int. Workshop on Database and Expert Systems Applications (DEXA-BIOKDD 2012), pp. 190–194 (2012)
6. Comin, M., Verzotto, D.: Alignment-free phylogeny of whole genomes using underlying subwords. BMC Algorithms for Molecular Biology 7(34) (2012)
7. Song, K., Ren, J., Zhai, Z., Liu, X., Deng, M., Sun, F.: Alignment-Free Sequence Comparison Based on Next-Generation Sequencing Reads. Journal of Computational Biology 20(2), 64–79 (2013)
8. Comin, M., Schimd, M.: Assembly-free Genome Comparison based on Next-Generation Sequencing Reads and Variable Length Patterns. Accepted at RECOMB-SEQ 2014: 4th Annual RECOMB Satellite Workshop at Massively Parallel Sequencing. Proceedings to appear in BMC Bioinformatics (2014)

9. Vinga, S., Almeida, J.: Alignment-free sequence comparison – a review. Bioinformatics 19(4), 513–523 (2003)
10. Gao, L., Qi, J.: Whole genome molecular phylogeny of large dsDNA viruses using composition vector method. BMC Evolutionary Biology 7(1), 41 (2007)
11. Qi, J., Luo, H., Hao, B.: CVTree: a phylogenetic tree reconstruction tool based on whole genomes. Nucleic Acids Research 32 (Web Server Issue), 45–47 (2004)
12. Goke, J., Schulz, M.H., Lasserre, J., Vingron, M.: Estimation of pairwise sequence similarity of mammalian enhancers with word neighbourhood counts. Bioinformatics 28(5), 656–663 (2012)
13. Kantorovitz, M.R., Robinson, G.E., Sinha, S.: A statistical method for alignment-free comparison of regulatory sequences. Bioinformatics 23(13), 249–255 (2007)
14. Comin, M., Verzotto, D.: Beyond fixed-resolution alignment-free measures for mammalian enhancers sequence comparison. Accepted for presentation at The Twelfth Asia Pacific Bioinformatics Conference. Proceedings to appear in IEEE/ACM Transactions on Computational Biology and Bioinformatics (2014)
15. Comin, M., Antonello, M.: Fast Computation of Entropic Profiles for the Detection of Conservation in Genomes. In: Ngom, A., Formenti, E., Hao, J.-K., Zhao, X.-M., van Laarhoven, T. (eds.) PRIB 2013. LNCS, vol. 7986, pp. 277–288. Springer, Heidelberg (2013)
16. Comin, M., Antonello, M.: Fast Entropic Profiler: An Information Theoretic Approach for the Discovery of Patterns in Genomes. IEEE/ACM Transactions on Computational Biology and Bioinformatics 11(3), 500–509 (2014)
17. Comin, M., Verzotto, D.: Classification of protein sequences by means of irredundant patterns. Proceedings of the 8th Asia-Pacific Bioinformatics Conference (APBC), BMC Bioinformatics 11(Suppl.1), S16 (2010)
18. Comin, M., Verzotto, D.: The Irredundant Class method for remote homology detection of protein sequences. Journal of Computational Biology 18(12), 1819–1829 (2011)
19. Hashimoto, W.S., Morishita, S.: Efficient frequency-based de novo short-read clustering for error trimming in next-generation sequencing. Genome Research 19(7), 1309–1315 (2009)
20. Bao, E., Jiang, T., Kaloshian, I., Girke, T.: SEED: efficient clustering of next-generation sequences. Bioinformatics 27(18), 2502–2509 (2011)
21. Solovyov, A., Lipkin, W.I.: Centroid based clustering of high throughput sequencing reads based on n-mer counts. BMC Bioinformatics 14, 268 (2013)
22. Heng, L., Jue, R., Durbin, R.: Mapping short DNA sequencing reads and calling variants using mapping quality scores. Genome Research 18, 1851–1858 (2008)
23. Albers, C., Lunter, G., MacArthur, D.G., McVean, G., Ouwehand, W.H., Durbin, R.: Dindel: accurate indel calls from short-read data. Genome Research 21(6), 961–973 (2011)
24. Carneiro, M.O., Russ, C., Ross, M.G., Gabriel, S.B., Nusbaum, C., DePristo, M.A.: Pacific biosciences sequencing technology for genotyping and variation discovery in human data. BMC Genomics 13, 375 (2012)
25. Blaisdell, B.E.: A measure of the similarity of sets of sequences not requiring sequence alignment. PNAS USA 83(14), 5155–5159 (1986)
26. Lippert, R.A., Huang, H.Y., Waterman, M.S.: Distributional regimes for the number of k-word matches between two random sequences. Proceedings of the National Academy of Sciences of the United States of America 100(13), 13980–13989 (2002)
27. Reinert, G., Chew, D., Sun, F., Waterman, M.S.: Alignment-free sequence comparison (I): statistics and power. Journal of Computational Biology 16(12), 1615–1634 (2009)

28. Wan, L., Reinert, G., Chew, D., Sun, F., Waterman, M.S.: Alignment-free sequence comparison (II): theoretical power of comparison statistics. Journal of Computational Biology 17(11), 1467–1490 (2010)
29. Ewing, B., Green, P.: Base-calling of automated sequencer traces using phred. II. Error probabilities. Genome Research 8(3), 186–194 (1998)
30. Holtgrewe, M.: Mason–a read simulator for second generation sequencing data. Technical Report FU Berlin (2010)
31. Birney, E.: Assemblies: the good, the bad, the ugly. Nature Methods 8, 59–60 (2011)
32. Zerbino, D.R., Birney, E.: Velvet: algorithms for de novo short read assembly using de Bruijn graphs. Genome Research 18, 821–829 (2008)

Improved Approximation for the Maximum Duo-Preservation String Mapping Problem*

Nicolas Boria, Adam Kurpisz, Samuli Leppänen, and Monaldo Mastrolilli

Dalle Molle Institute for Artificial Intelligence (IDSIA), Manno, Switzerland

Abstract. In this paper we present improved approximation results for the MAX DUO-PRESERVATION STRING MAPPING problem (MPSM) introduced in [Chen et al., Theoretical Computer Science, 2014] that is complementary to the well-studied MIN COMMON STRING PARTITION problem (MCSP). When each letter occurs at most k times in each string the problem is denoted by k-MPSM. First, we prove that k-MPSM is APX-Hard even when $k = 2$. Then, we improve on the previous results by devising two distinct algorithms: the first ensures approximation ratio 8/5 for $k = 2$ and ratio 3 for $k = 3$, while the second guarantees approximation ratio 4 for any bigger value of k. Finally, we address the approximation of CONSTRAINED MAXIMUM INDUCED SUBGRAPH (CMIS, a generalization of MPSM, also introduced in [Chen et al., Theoretical Computer Science, 2014]), and improve the best known 9-approximation for 3-CMIS to a 6-approximation, by using a configuration LP to get a better linear relaxation. We also prove that such a linear program has an integrality gap of k, which suggests that no constant approximation (i.e. independent of k) can be achieved through rounding techniques.

Keywords: Polynomial approximation, Max Duo-Preserving String Mapping Problem, Min Common String Partition Problem, Linear Programming, Configuration LP.

1 Introduction

String comparison is a central problem in stringology with a wide range of applications, including data compression, and bio-informatics. There are various ways to measure the similarity of two strings: one may use the Hamming distance which counts the number of positions at which the corresponding symbols are different, the Jaro-Winkler distance, the overlap coefficient, etc. However in computer science, the most common measure is the so called *edit distance* that measures the minimum number of edit operations that must be performed to transform the first string into the second. In biology, this number may provide some measure of the kinship between different species based on the similarities of their DNA. In data compression, it may help to store efficiently a set of similar

---

* Research supported by the Swiss National Science Foundation project 200020_144491/1 "Approximation Algorithms for Machine Scheduling Through Theory and Experiments", and by the Sciex-Project 12.311

D. Brown and B. Morgenstern (Eds.): WABI 2014, LNBI 8701, pp. 14–25, 2014.

yet different data (e.g. different versions of the same object) by storing only one "base" element of the set, and then storing the series of edit operations that result in the other versions of the base element.

The concept of edit distance changes definition based on the set of edit operations that are allowed. When the only edit operation that is allowed is to shift a block of characters, the edit distance can be measured by solving the MIN COMMON STRING PARTITION problem.

The MIN COMMON STRING PARTITION (MCSP) is a fundamental problem in the field of string comparison [7,13], and can be applied more specifically to genome rearrangement issues, as shown in [7]. Consider two strings A and B, both of length n, such that B is a permutation of A. Also, let $\mathcal{P}_A$ denote a *partition* of A, that is, a set of substrings whose concatenation results in A. The MCSP Problem introduced in [13] and [19] asks for a partition $\mathcal{P}_A$ of A and $\mathcal{P}_B$ of B of minimum cardinality such that $\mathcal{P}_A$ is a permutation of $\mathcal{P}_B$. The $k-$MCSP denotes the restricted version of the problem where each letters has at most k occurrences. This problem is NP-Hard and even APX-Hard, also when the number of occurrences of each letter is at most 2 (note that the problem is trivial when this number is at most 1) [13]. Since then, the problem has been intensively studied, especially in terms of polynomial approximation [7,8,9,13,15,16], but also parametric computation [4,17,10,14]. The best approximations known so far are an $O(\log n \log^* n)$-approximation for the general version of the problem [9], and an $O(k)$-approximation for $k-$MCSP [16]. On the other hand, the problem was proved to be Fixed Parameter Tractable (FPT), first with respect to both k and the cardinality ϕ of an optimal partition [4,10,14], and more recently, with respect to ϕ only [17].

In [6], the maximization version of the problem is introduced and denoted by MAX DUO-PRESERVATION STRING MAPPING (MPSM). Reminding that a *duo* denotes a couple of consecutive letters it is clear that when a solution $(\mathcal{P}_A, \mathcal{P}_B)$ for MIN COMMON STRING PARTITION partitions A and B into ϕ substrings, this solution can be translated as a mapping π from A to B that preserves exactly $n - \phi$ duos. Hence, given two strings A and B, the MPSM problem asks for a mapping π from A to B that preserves a maximum number of duos (a formal definition is given in Subsection 3.1). An example is provided in Figure 1.

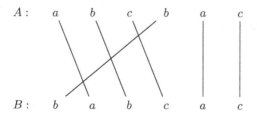

Fig. 1. A mapping π that preserves 3 duos

Considering that MCSP is NP-Hard [13], its maximization counterpart MPSM is also NP-Hard. However, these two problems might have different behaviors in terms of approximation, inapproximability, and parameterized complexity. MAX INDEPENDENT SET and MIN VERTEX COVER provide a good example of how symmetrical problems might have different characteristics: on the one hand, MAX INDEPENDENT SET is inapproximable within ratio $n^{\varepsilon-1}$ for a given $\varepsilon \in (0,1)$ unless $\mathbf{P} = \mathbf{NP}$ [18], and is $W[1]$-Hard [11]; and on the other hand MIN VERTEX COVER is easily 2-approximable in polynomial time by taking all endpoints of a maximal matching [12], and is FPT [5].

The authors of [6] provide some approximation results for MPSM in the following way: a graph problem called CONSTRAINED MAXIMUM INDUCED SUBGRAPH (CMIS) is defined and proved to be a generalization of MPSM. Using a solution to the linear relaxation of CMIS, it is proved that a randomized rounding provides a k^2 expected approximation ratio for k-CMIS (and thus for k-MPSM), and a 2 expected approximation ratio for 2-CMIS (and thus for 2-MPSM).

In what follows, we start by proving briefly that k-MPSM is APX-Hard, even when $k = 2$ (Section 2). Then, we present some improved approximation results for MPSM (Section 3), namely a general approximation algorithm that guarantees approximation ratio 4 regardless of the value of k (Subsection 3.2), and an algorithm that improves on this ratio for small values of k (Subsection 3.3). Finally, we improve on the approximation of 3-CMIS, by using a configuration LP to get a better relaxed solution (Section 4), and analyze the integrality gap of this relaxed solution.

2   Hardness of Approximation

We will show that MPSM is APX–hard, which essentially rules out any polynomial time approximation schemes unless P = NP. The result follows with slight modifications from the known approximation hardness result for MCSP. Indeed, in [13] it is shown that any instance of MAX INDEPENDENT SET in a cubic graph (3–MIS) can be reduced to an instance of 2–MCSP (proof of Theorem 2.1 in [13]). We observe that the construction used in their reduction also works as a reduction from 3–MIS to 2–MPSM. In particular, given a cubic graph with n vertices and independence number α, the corresponding reduction to 2–MPSM has an optimum value of $m = 4n + \alpha$.

Given a ρ–approximation to 2–MPSM, we will hence always find an independent set of size at least $\rho m - 4n$. It is shown in [3] that it is NP–hard to approximate 3–MIS within $\frac{139}{140} + \epsilon$ for any $\epsilon > 0$. Therefore, unless P = NP, for every $\epsilon > 0$ there is an instance I of 3–MIS such that:

$$\frac{\mathrm{APP}_I}{\mathrm{OPT}_I} \leqslant \frac{139}{140} + \epsilon$$

where APP_I is the solution produced by any polynomial time approximation algorithm and OPT_I the optimum value of I. Substituting here we get:

$$\frac{\rho m - 4n}{m - 4n} \leqslant \frac{139}{140} + \epsilon$$

Solving for ρ yields:

$$\rho \leqslant \frac{139}{140} + \frac{4n}{m}\left(\frac{1}{140} - \epsilon\right) + \epsilon \leqslant \frac{139}{140} + \frac{16}{17 \cdot 140} + \frac{1}{17}\epsilon$$

where the last inequality follows from noting that for any cubic graph the maximum independent set α is always at least of size $\frac{1}{4}n$.

3 Approximation Algorithms for MAX DUO-PRESERVATION STRING MAPPING

In this section we present two different approximation algorithms. First, a simple algorithm that provides a 4-approximation ratio for the general version of the problem, and then an algorithm that improves on this ratio for small values of k.

3.1 Preliminaries

For $i = 1, ..., n$, we denote by a_i the ith character of string A, and by b_i the ith character in B. We also denote by $D^A = (D_1^A, ..., D_{n-1}^A)$ and $D^B = (D_1^B, ..., D_{n-1}^B)$ the set of duos of A and B respectively. For $i = 1, ..., n - 1$, D_i^A corresponds to the duo (a_i, a_{i+1}), and D_i^B corresponds to the duo (b_i, b_{i+1}).

A mapping π from A to B is said to be *proper* if it is bijective, and if, $\forall i = 1, ..., n$, $a_i = b_{\pi(i)}$. In other words, each letter of the alphabet in A must be mapped to the same letter in B for the mapping to be proper. A couple of duos $\left(D_i^A, D_j^B\right)$ is said to be *preservable* if $a_i = b_j$ and $a_{i+1} = b_{j+1}$. Given a mapping π, a preservable couple of duos $\left(D_i^A, D_j^B\right)$ is said to be *preserved by* π if $\pi(i) = j$ and $\pi(i + 1) = j + 1$. Finally, two preservable couples of duos $\left(D_i^A, D_j^B\right)$ and $\left(D_h^A, D_l^B\right)$ will be called *conflicting* if there is no proper mapping that preserves both of them. These conflicts can be of two types, w.l.o.g., we suppose that $i \leqslant h$ (resp. $j \leqslant l$):

- Type 1: $i = h$ (resp. $j = l$) and $j \neq l$ (resp. $i \neq h$) (see Figure 2(a))
- Type 2: $i = h - 1$ (resp. $j = l - 1$) and $j \neq l - 1$ (resp. $i \neq h - 1$) (see Figure 2(b))

Let us now define formally the problem at hand:

Definition 1. MAX DUO-PRESERVATION STRING MAPPING *(MPSM):*

- **Instance:** two strings A and B such that B is a permutation of A.
- **Solution:** a proper mapping π from A to B.
- **Objective:** maximizing the number of duos preserved by π, denoted by $f(\pi)$.

Let us finally introduce the concept of *duo-mapping*. A duo-mapping σ is a mapping, which - unlike a mapping π that maps each character in A to a character in B - maps a *subset* of duos of D^A to a *subset* of duos of D^B. Having

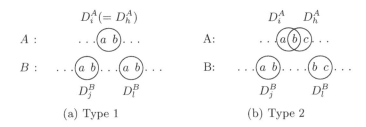

(a) Type 1   (b) Type 2

Fig. 2. Different types of conflicting pairs of duos

$\sigma(i) = j$ means that the duo D_i^A is mapped to the duo D_j^B. Again, a duo-mapping σ is said to be proper if it is bijective, and if $D_i^A = D_{\sigma(i)}^B$ for all duos mapped through σ. Note that a proper duo-mapping might map some conflicting couple of duos. Revisit the example of Figure 2(b): having $\sigma(i) = j$ and $\sigma(h) = l$ defines a proper duo-mapping that maps conflicting couple of duos. Notice however that a proper duo-mapping might generate conflicts of Type 2 only. We finally define the concept of *unconflicting* duo-mapping, which is a proper duo-mapping that does not map any pair of conflicting duos.

Remark 1. An unconflicting duo-mapping σ on some subset of duos of size $f(\sigma)$ immediatly derives a proper mapping π on the whole set of characters with $f(\pi) \geqslant f(\sigma)$: it suffices to map characters mapped by σ in the same way that σ does, and map arbitrarily the remaining characters.

3.2   A 4-Approximation Algorithm for MPSM

Proposition 1. *There exists a 4-approximation algorithm for MPSM that runs in $O(n^{3/2})$ time.*

Proof. Consider the two strings $A = a_1 a_2 ... a_n$ and $B = b_1 b_2 ... b_n$ that one wishes to map while preserving a maximal number of duos, and let D^A and D^B denote their respective sets of duos. Also, denote by π^* an optimal mapping that preserves a maximum number $f(\pi^*)$ of duos.

Build a bipartite graph G in the following way: vertices on the left and the right represent duos of D^A and D^B, respectively. Whenever one duo on the right and one on the left are preservable (same two letters in the same order), we add an edge between the two corresponding vertices. Figure 3 provides an example of this construction.

At this point, notice that there exists a one-to-one correspondence between matchings in G and proper duo-mappings between D^A and D^B. In other words there exists a matching in G with $f(\pi^*)$ edges. Indeed, the set of duos preserved by any solution (and *a fortiori* by the optimal one) can be represented as a matching in G. Hence, denoting by M^* a maximum matching in G, it holds that :

$$f(\pi^*) \leqslant |M^*| \tag{1}$$

Unfortunately, a matching M^* in G does not immediately translate into a proper mapping that preserves $|M^*|$ duos. However, it does correspond to a proper duo-mapping that maps $|M^*|$ duos, which, as noticed earlier, might generate conflicts of Type 2 only.

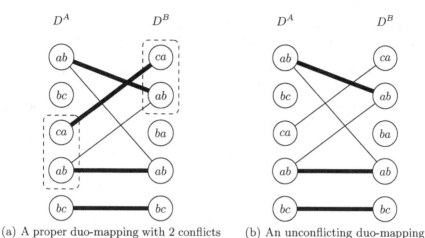

(a) A proper duo-mapping with 2 conflicts   (b) An unconflicting duo-mapping

Fig. 3. The graph G where $A = abcabc$ and $B = cababc$

In G, a conflict of Type 2 corresponds to two consecutive vertices on one side matched to two non-consecutive vertices on the other side. Hence, to generate an unconflicting duo-mapping σ using a matching M^*, it suffices to partition the matching M^* in 4 sub-matchings in the following way : Let $M(even, odd)$ denote the submatching of M^* containing all edges whose left endpoint have even indices, and right endpoint have odd indices; and define $M(odd, even)$, $M(even, even)$, and $M(odd, odd)$ in the same way. Denote by $\hat{M}$ the matching with biggest cardinality among these 4. Obviously, remembering that the four submatchings define a partition of M^*, it holds that $|\hat{M}| \geqslant |M^*|/4$. Considering that $\hat{M}$ does not contain any pair of edges with consecutive endpoints, the corresponding duo-mapping σ has no conflict. Following Remark 1, σ derives a proper mapping π on the characters such that:

$$f(\pi) \geqslant f(\sigma) = |\hat{M}| \geqslant \frac{|M^*|}{4} \overset{(1)}{\geqslant} \frac{f(\pi^*)}{4}$$

A 4-approximate solution can thus be computed by creating the graph G from strings A and B, computing an maximum matching M^* on it, partitioning M^* four ways by indices parity and return the biggest partition $\hat{M}$. Then map the matched duos following the edges $\hat{M}$, and map all the other characters arbitrarily. The complexity of the whole procedure is given by the complexity of computing an optimal matching in G, which is $O(n^{3/2})$.   □

It is likely that the simple edge removal procedure that nullifies all conflicts of Type 2 can be replaced by a more involved heuristic method in order to solve efficiently real life problems.

3.3 An 8/5-Approximation for 2-MPSM

In the following, we make use of a reduction from MSPM to MAX INDEPENDENT SET (MIS) already pointed out in [13]. Given two strings A and B, consider the graph H built in the following way: H has a vertex v_{ij} for each preservable couple of duos $((D_i^A), (D_j^B))$, and H has an edge (v_{ij}, v_{hl}) for each conflicting pair of preservable couple of duos $((D_i^A), (D_j^B))$ $((D_h^A), (D_l^B))$. It is easy to see that there is a 1 to 1 correspondence between independent sets in H and unconflicting duo-mappings between A and B.

Notice that, for a given k, a couple of duos $((D_i^A), (D_j^B))$ can belong to at most $6(k-1)$ conflicting pairs: on the one hand, there can be at most $2(k-1)$ conflicts of Type 1 (one for each other occurrence of the duo D_i^A in D^A and D^B), and on the other hand at most $4(k-1)$ conflicts of Type 2 (one for each possible conflicting occurrence of D_{j-1}^B or D_{j+1}^B in D^A, and one for each possible conflicting occurrence of D_{j-1}^A or D_{j+1}^A in D^B). This bound is tight.

Hence, for a given instance of k-MPSM, the corresponding instance of MIS is a graph with maximum degree $\Delta \leqslant 6(k-1)$. Using the approximation algorithm of [2] and [1] for independent set (which guarantees approximation ratio $(\Delta+3)/5$), this leads to obtaining approximation ratio arbitrarily close to $(6k-3)/5$ for k-MPSM, which already improves on the best known 2-approximation when $k=2$, and also on the 4-approximation of Proposition 1 when $k=3$.

We now prove the following result in order to further improve on the approximation:

Lemma 1. *In a graph H corresponding to an instance of 2-MPSM, there exists an optimal solution for MIS that does not pick any vertex of degree 6.*

Proof. Consider a vertex v_{ij} of degree 6 in such a graph H. This vertex corresponds to a preservable couple of duos that conflicts with 6 other preservable couples. There exists only one possible configuration in the strings A and B that can create this situation, which is illustrated in Figure 4(a).

In return, this configuration always corresponds to the gadget illustrated in Figure 4(b), where vertices v_{ij}, v_{hj}, v_{il}, and v_{hl} have no connection with the rest of the graph.

Now, consider any maximal independent set S that picks some vertex v_{ij} of degree 6 in H. The existence of this degree-6 vertex induces that graph H contains the gadget of Figure 4(a). S is maximal, so it necessarily contains vertex v_{hl} as well. Let $S' = S \setminus (\{v_{ij}\}, \{v_{hl}\}) \cup (\{v_{il}\}, \{v_{hj}\})$. Reminding that v_{il} and v_{hj} have no neighbor outside of the gadget, it is clear that S' also defines an independent set.

Hence, in a maximal (and *a fortiori* optimal) independent set, any pair of degree-6 vertices (in such graphs, degree-6 vertices always appear in pair) can

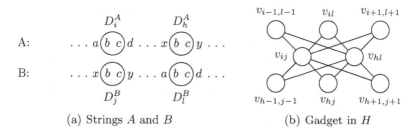

A:

B:

(a) Strings A and B (b) Gadget in H

Fig. 4. A degree 6-vertex in graph H

be replaced by a pair of degree 2 vertices, which concludes the proof of Lemma 1. ☐

Let H' be the subgraph of H induced by all vertices apart from vertices of degree 6. Lemma 1 tells us that an optimal independent set on H' has the same cardinality than an optimal independent set in H. However H' has maximum degree 5 and not 6, which yields a better approximation when using the algorithm described in [2] and [1]:

Proposition 2. 2-MPSM is approximable within ratio arbitrarily close to 8/5 in polynomial time.

Notice that the reduction from k-MPSM to $6(k-1)$-MIS also yields the following simple parameterized algorithm:

Corollary 1. k-MPSM can be solved in $O^*((6(k-1)+1)^\psi)$, where ψ denotes the value of an optimal solution.

Consider an optimal independent set S in H, if some vertex v of H has no neighbour in S then v necessarily belongs to S. Thus, in order to build an optimal solution S, one can go through the decision tree that, for each vertex v that has no neighbour in the current solution, consists of deciding which vertex among v and its set of neighbours will be included in the solution. Any solution S will take one of these $6(k-1)+1$ vertices. Each node of the decision tree has at most $6(k-1)+1$ branches, and the tree has obviously depth ψ, considering that one vertex is added to S at each level.

4   Some Results on 3-CONSTRAINED MAXIMUM INDUCED SUBGRAPH

In this section we consider the CONSTRAINED MAXIMUM INDUCED SUBGRAPH problem (CMIS) which is a generalization of MAX DUO-PRESERVATION STRING MAPPING (MPSM). In [6] the CMIS served as the main tool to analyze its special case, namely the MPSP problem. The problem is expressed as a natural linear

program denoted by NLP, which is used to obtain a randomized k^2 approximation algorithm. In this section we provide a 6-approximation algorithm for the
3-CMIS which improves on the previous 9-approximation algorithm. We do this
by introducing a *configuration-LP* denoted by CLP. Moreover we show that
both NLP and CLP have an integrality gap of at least k which implies that it
is unlikely to construct a better than k-approximation algorithm based on these
linear programs.

We start with a formal definition of the problem.

Definition 2. CONSTRAINED MAXIMUM INDUCED SUBGRAPH *(CMIS)*:

- *Instance:* an *m*-partite graph $G(V, E)$ with parts: $G_1, \ldots, G_m$. Each part G_i
  has n_i^2 vertices organized in an $n_i \times n_i$ grid.
- *Solution:* a subset of vertices such that within each grid in each column and
  each row at most one vertex is chosen.
- *Objective:* maximizing the number of edges in the induced subgraph.

In the constrained k-CMIS problem each grid consists of at most $k \times k$ vertices.

Let v_p^{ij} be the vertex placed in position (i, j) in the pth grid. Consider the
linear program NLP as proposed in [6]. Let x_p^{ij} be the boolean variable which
takes value 1 if the corresponding vertex v_p^{ij} is chosen, and 0 otherwise. Let
$x_{p_{ij}q_{kl}}$ be the edge-corresponding boolean variable such that it takes the value 1
if both the endpoint vertices v_p^{ij} and v_q^{kl} are selected and 0 otherwise. The task
is to choose a subset of vertices, such that within each block, in each column and
each row at most one vertex is chosen. The objective is to maximize the number
of edges in the induced subgraph. The LP formulation is the following:

$$
\begin{aligned}
NLP: \\
Max \quad & \sum_{(v_p^{ij} v_q^{kl}) \in E} x_{p_{ij}q_{kl}} \\
s.t. \quad & x_{p_{ij}q_{kl}} \leqslant x_p^{ij} \qquad \text{for } i, j, k, l = [n_p], \quad p, q = [m], \\
& \sum_{i=1}^{n_p} x_p^{ij} = 1 \qquad \text{for } j = [n_p], \quad p = [m], \\
& \sum_{j=1}^{n_p} x_p^{ij} = 1 \qquad \text{for } i = [n_p], \quad p = [m], \\
& 0 \leqslant x_{p_{ij}q_{kl}} \leqslant 1 \quad \text{for } i, j, k, l = [n_p], \quad p, q = [m], \\
& 0 \leqslant x_p^{ij} \leqslant 1 \qquad \text{for } i, j = [n_p], \quad p = [m].
\end{aligned}
\tag{2}
$$

Note that when the size of each grid is constant, the CLP is of polynomial size.
The first constraint ensures that the value of the edge-corresponding variable
is not greater than the value of the vertex-corresponding variable of any of its
endpoints. The second and the third constraints ensure that within each grid at
most one vertex is taken in each column, each row, respectively.

Notice that within each grid there are $k!$ possible ways of taking a feasible
subset of vertices. We call a *configuration*, a feasible subset of vertices for a given
grid. Let us denote by $\mathcal{C}_p$ the set of all possible configurations for a grid p. Now,

consider that we have boolean variable x_{C_p} for each possible configuration. The variable x_{C_p} takes value 1 if all the vertices contained in C_p are chosen and 0 otherwise. The induced linear program is called *Configuration-LP*, (CLP). The CLP formulation for the $CMIS$ problem is the following:

$$
\begin{aligned}
&CLP(K): \\
&Max \qquad \sum_{(v_p^{ij} v_q^{kl}) \in E} x_{p_{ij} q_{kl}} \\
&s.t. \qquad x_{p_{ij} q_{kl}} \leqslant x_p^{ij} \qquad \text{for } i,j,k,l = [n_p], \quad p,q = [m], \\
&\qquad\quad x_p^{ij} = \sum_{v_p^{ij} \in C_p \in \mathcal{C}_p} x_{C_p} \text{ for } i,j = [n_p], \quad p, = [m], \qquad (3) \\
&\qquad\quad \sum_{C_p \in \mathcal{C}_p} x_{C_p} = 1 \qquad \text{for } p = [m], \\
&\qquad\quad 0 \leqslant x_{p_{ij} q_{kl}} \leqslant 1 \qquad \text{for } i,j,k,l = [n_p], \quad p,q = [m], \\
&\qquad\quad 0 \leqslant x_{C_p} \leqslant 1 \qquad \text{for } C_p \in \mathcal{C}_p, \quad p = [m],
\end{aligned}
$$

The first constraint is the same as in NLP. The second one ensures that the value of the vertex-corresponding variable is equal to the summation of the values of the configuration-corresponding variables containing considered vertex. The third constraint ensures that within each grid exactly one configuration can be taken. Notice that the vertex variables are redundant and serve just as an additional description. In particular the first and the second constraints could be merged into one constraint without vertex variables.

One can easily see that the CLP is at least as strong as the NLP formulation: a feasible solution to CLP always translates to a feasible solution to NLP.

Proposition 3. *There exists a randomized 6-approximation algorithm for the* 3-CONSTRAINED MAXIMUM INDUCED SUBGRAPH *problem.*

Proof. Consider a randomized algorithm that, in each grid G_p, takes the vertices from configuration C with a probability $\dfrac{\sqrt{x_C}}{\sum_{C_p \in \mathcal{C}_p} \sqrt{x_{C_p}}}$.

Consider any vertex, w.l.o.g. $v_p^{1,1}$. Each vertex is contained in two configurations, w.l.o.g. let $v_p^{1,1}$ be contained in C_p^1 and C_p^2. The probability that $v_p^{1,1}$ is chosen is:

$$
\Pr\left(v_p^{1,1} \text{is taken}\right) = \frac{\sqrt{x_{C_p^1}} + \sqrt{x_{C_p^2}}}{\sum_{C_p \in \mathcal{C}_p} \sqrt{x_{C_p}}}
$$

Optimizing the expression $\sqrt{x_{C_p^1}} + \sqrt{x_{C_p^2}}$ under the condition $x_{C_p^1} + x_{C_p^2} = x_p^{1,1}$, we have that the minimum is when either $x_{C_p^1} = 0$ or $x_{C_p^2} = 0$ which implies $\sqrt{x_{C_p^1}} + \sqrt{x_{C_p^2}} = \sqrt{x_p^{1,1}}$. Thus:

$$
\Pr\left(v_p^{1,1} \text{is taken}\right) \geqslant \frac{\sqrt{x_p^{1,1}}}{\sum_{C_p \in \mathcal{C}_p} \sqrt{x_{C_p}}}
$$

Using a standard arithmetic inequality we can get that:

$$\frac{\sum_{C_p \in \mathcal{C}_p} \sqrt{x_{C_p}}}{6} \leqslant \sqrt{\frac{\sum_{C_p \in \mathcal{C}_p} x_{C_p}}{6}} = \sqrt{\frac{1}{6}}$$

which implies that:

$$\Pr\left(v_p^{1,1} \text{ is taken}\right) \geqslant \frac{\sqrt{x_p^{1,1}}}{\sqrt{6}}$$

Now let us consider any edge and the corresponding variable, $x_{p_{ij}q_{kl}}$. The probability that the edge is taken can be lower bounded by:

$$\Pr\left(x_{p_{ij}q_{kl}} \text{ is taken}\right) = \Pr\left(v_p^{ij} \text{ is taken}\right) \cdot \Pr\left(v_q^{kl} \text{ is taken}\right) \geqslant \frac{\sqrt{x_p^{ij}}}{\sqrt{6}} \cdot \frac{\sqrt{x_q^{kl}}}{\sqrt{6}} \geqslant$$

$$\frac{1}{6} \min\{x_p^{ij}, x_q^{kl}\} \geqslant \frac{1}{6} x_{p_{ij}q_{kl}}$$

Since our algorithm takes in expectation every edge with probability $\frac{1}{6}$ of the fractional value assigned to the corresponding edge-variable by the CLP it is a randomized 6-approximation algorithm.   □

4.1   Integrality Gap of NLP and CLP

We now show that the linear relaxation NLP has an integrality gap of at least k. Consider the following instance of k-CMIS. Let the input graph $G(V, E)$ consists of two grids, G_1, G_2. Both grids consist of k^2 vertices. Every vertex from one grid is connected to all the vertices in the second grid and vice versa. Thus the number of edges is equal to k^4. By putting all the LP variables to $\frac{1}{k}$ one can easily notice that this solution is feasible and the objective value for this solution is k^3. On the other hand any feasible integral solution for this instance must return at most k vertices from each grid, each of which is connected to at most k vertices from the other grid. Thus the integral optimum is at most k^2. This produces the intergality gap of k. Moreover by putting the configuration-corresponding variables in $CLP(K)$ to $\frac{1}{k!}$ we can construct a feasible solution to $CLP(K)$ with the same integrality gap of k.

References

1. Berman, P., Fujito, T.: On Approximation Properties of the Independent Set Problem for Low Degree Graphs. Theory of Computing Systems 32(2), 115–132 (1999)
2. Berman, P., Fürer, M.: Approximating Maximum Independent Set in Bounded Degree Graphs. In: Sleator, D.D. (ed.) SODA, pp. 365–371. ACM/SIAM (1994)
3. Berman, P., Karpinski, M.: On Some Tighter Inapproximability Results (Extended Abstract). In: Wiedermann, J., Van Emde Boas, P., Nielsen, M. (eds.) ICALP 1999. LNCS, vol. 1644, pp. 200–209. Springer, Heidelberg (1999)

4. Bulteau, L., Fertin, G., Komusiewicz, C., Rusu, I.: A Fixed-Parameter Algorithm for Minimum Common String Partition with Few Duplications. In: Darling, A., Stoye, J. (eds.) WABI 2013. LNCS, vol. 8126, pp. 244–258. Springer, Heidelberg (2013)

5. Chen, J., Kanj, I.A., Jia, W.: Vertex Cover: Further Observations and Further Improvements. In: Widmayer, P., Neyer, G., Eidenbenz, S. (eds.) WG 1999. LNCS, vol. 1665, pp. 313–324. Springer, Heidelberg (1999)

6. Chen, W., Chen, Z., Samatova, N.F., Peng, L., Wang, J., Tang, M.: Solving the maximum duo-preservation string mapping problem with linear programming. Theoretical Computer Science 530, 1–11 (2014)

7. Chen, X., Zheng, J., Fu, Z., Nan, P., Zhong, Y., Lonardi, S., Jiang, T.: Assignment of Orthologous Genes via Genome Rearrangement. Transactions on Computational Biology and Bioinformatics 2(4), 302–315 (2005)

8. Chrobak, M., Kolman, P., Sgall, J.: The Greedy Algorithm for the Minimum Common String Partition Problem. In: Jansen, K., Khanna, S., Rolim, J.D.P., Ron, D. (eds.) RANDOM 2004 and APPROX 2004. LNCS, vol. 3122, pp. 84–95. Springer, Heidelberg (2004)

9. Cormode, G., Muthukrishnan, S.: The string edit distance matching problem with moves. ACM Transactions on Algorithms 3(1) (2007)

10. Damaschke, P.: Minimum Common String Partition Parameterized. In: Crandall, K.A., Lagergren, J. (eds.) WABI 2008. LNCS (LNBI), vol. 5251, pp. 87–98. Springer, Heidelberg (2008)

11. Downey, R.G., Fellows, M.R.: Parameterized Complexity, p. 530. Springer (1999)

12. Garey, M.R., Johnson, D.S.: Computers and Intractability: A Guide to the Theory of NP-Completeness. W.H. Freeman and Co., San Francisco (1979)

13. Goldstein, A., Kolman, P., Zheng, J.: Minimum Common String Partition Problem: Hardness and Approximations. In: Fleischer, R., Trippen, G. (eds.) ISAAC 2004. LNCS, vol. 3341, pp. 484–495. Springer, Heidelberg (2004)

14. Jiang, H., Zhu, B., Zhu, D., Zhu, H.: Minimum common string partition revisited. Journal of Combinatorial Optimization 23(4), 519–527 (2012)

15. Kolman, P., Walen, T.: Approximating reversal distance for strings with bounded number of duplicates. Discrete Applied Mathematics 155(3), 327–336 (2007)

16. Kolman, P., Walen, T.: Reversal Distance for Strings with Duplicates: Linear Time Approximation using Hitting Set. Electronic Journal of Combinatorics 14(1) (2007)

17. Bulteau, L., Komusiewicz, C.: Minimum common string partition parameterized by partition size is fixed-parameter tractable. In: SODA, pp. 102–121 (2014)

18. Lund, C., Yannakakis, M.: The Approximation of Maximum Subgraph Problems. In: Lingas, A., Karlsson, R.G., Carlsson, S. (eds.) ICALP 1993. LNCS, vol. 700, pp. 40–51. Springer, Heidelberg (1993)

19. Swenson, K.M., Marron, M., Earnest-DeYoung, J.V., Moret, B.M.E.: Approximating the true evolutionary distance between two genomes. ACM Journal of Experimental Algorithmics 12 (2008)

A Faster 1.375-Approximation Algorithm for Sorting by Transpositions

Luís Felipe I. Cunha[1], Luis Antonio B. Kowada[2],
Rodrigo de A. Hausen[3], and Celina M.H. de Figueiredo[1]

[1] Universidade Federal do Rio de Janeiro, Brasil
{lfignacio,celina}@cos.ufrj.br
[2] Universidade Federal Fluminense, Brasil
luis@vm.uff.br
[3] Universidade Federal do ABC, Brasil
hausen@compscinet.org

Abstract. Sorting by Transpositions is an NP-hard problem for which several polynomial time approximation algorithms have been developed. Hartman and Shamir (2006) developed a 1.5-approximation algorithm, whose running time was improved to $O(n \log n)$ by Feng and Zhu (2007) with a data structure they defined, the permutation tree. Elias and Hartman (2006) developed a 1.375-approximation algorithm that runs in $O(n^2)$ time. In this paper, we propose the first correct adaptation of this algorithm to run in $O(n \log n)$ time.

Keywords: comparative genomics, genome rearrangement, sorting by transpositions, approximation algorithms.

1 Introduction

By comparing the orders of common genes between two organisms, one may estimate the series of mutations that occurred in the underlying evolutionary process. In a simplified genome rearrangement model, each mutation is a transposition, and the sole chromosome of each organism is modeled by a permutation, which means that there are no duplicated or deleted genes. A *transposition* is a rearrangement of the gene order within a chromosome, in which two contiguous blocks are swapped. The transposition distance is the minimum number of transpositions required to transform one chromosome into another. Bulteau *et al.* [3] proved that the problem of determining the transposition distance between two permutations – or Sorting by Transpositions (SBT) – is NP-hard.

Several approaches to handle the SBT problem have been considered. Our focus is to explore approximation algorithms for estimating the transposition distance between permutations, providing better practical results or lowering time complexities.

Bafna and Pevzner [2] designed a 1.5-approximation $O(n^2)$ algorithm, based on the cycle structure of the *breakpoint graph*. Hartman and Shamir [10], by

D. Brown and B. Morgenstern (Eds.): WABI 2014, LNBI 8701, pp. 26–37, 2014.
© Springer-Verlag Berlin Heidelberg 2014

considering *simple permutations*, proposed an easier 1.5-approximation algorithm and, by exploiting a balanced tree data structure, decreased the running time to $O(n^{\frac{3}{2}}\sqrt{\log n})$. Feng and Zhu [7] developed the balanced *permutation tree* data structure, further decreasing the complexity of Hartman and Shamir's 1.5-approximation algorithm to $O(n \log n)$.

Elias and Hartman [6] obtained, by a thorough computational case analysis of cycles of the breakpoint graph, a 1.375-approximation $O(n^2)$ algorithm. Firoz *et al.* [8] tried to lower the running time of this algorithm to $O(n \log n)$ via a simple application of permutation trees, but we later found counter-examples [5] that disprove the correctness of Firoz *et al.*'s strategy.

In this paper, we propose a new algorithm that uses the strategy of Elias and Hartman towards *bad full configurations*, implemented using permutation trees and achieving both a 1.375 approximation ratio and $O(n \log n)$ time complexity. Section 2 contains basic definitions, Section 3 presents a strategy to find in linear time a sequence of two transpositions in which both are 2-moves, if it exists, and Section 4 describes our 1.375-approximation algorithm for SBT.

2 Background

For our purposes, a gene is represented by a unique integer and a chromosome with n genes is a *permutation* $\pi = [\boldsymbol{\pi_0}\, \pi_1\, \pi_2 \ldots \pi_n\, \boldsymbol{\pi_{n+1}}]$, where $\boldsymbol{\pi_0} = 0$, $\boldsymbol{\pi_{n+1}} = n+1$ and each π_i is a unique integer in the range $1, \ldots, n$. The *transposition* $t(i,j,k)$, where $1 \leq i < j < k \leq n+1$ *over* π, is the permutation $\pi \cdot t(i,j,k)$ where the product interchanges the two contiguous blocks $\pi_i\, \pi_{i+1} \ldots \pi_{j-1}$ and $\pi_j\, \pi_{j+1} \ldots \pi_{k-1}$. A sequence of q transpositions *sorts* a permutation π if $\pi\, t_1\, t_2 \cdots t_q = \iota$, where every t_i is a transposition and ι is the identity permutation $[0\, 1\, 2 \ldots n\, \boldsymbol{n+1}]$. The *transposition distance* of π, denoted $d(\pi)$, is the length of a minimum sequence of transpositions that sorts π.

Given a permutation π, the *breakpoint graph* of π is $G(\pi) = (V, R \cup D)$; the set of vertices is $V = \{0, -1, +1, -2, +2, \ldots, -n, +n, -(n+1)\}$, and the edges are partitioned into two sets, the directed *reality edges* $R = \{ \overrightarrow{i} = (+\pi_i, -\pi_{i+1}) \mid i = 0, \ldots, n\}$ and the undirected *desire edges* $D = \{(+i, -(i+1)) \mid i = 0, \ldots, n\}$. Fig. 1 shows $G([0\,10\,9\,8\,7\,1\,6\,11\,5\,4\,3\,2\,12])$, the horizontal lines represent the edges in R and the arcs represent the edges in D.

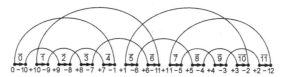

Fig. 1. $G([0\ 10\ 9\ 8\ 7\ 1\ 6\ 11\ 5\ 4\ 3\ 2\ 12])$. The cycles $C_2 = \langle 1\,3\,6 \rangle$ and $C_3 = \langle 5\,8\,10 \rangle$ intersect, but C_2 and C_3 are not interleaving; the cycles $C_1 = \langle 0\,2\,4 \rangle$ and $C_2 = \langle 1\,3\,6 \rangle$ are interleaving, and so are $C_3 = \langle 5\,8\,10 \rangle$ and $C_4 = \langle 7\,9\,11 \rangle$.

Every vertex in $G(\pi)$ has degree 2, so $G(\pi)$ can be partitioned into disjoint cycles. We shall use the terms *a cycle in* π and *a cycle in* $G(\pi)$ interchangeably to denote the latter. A cycle in π has length ℓ (or it is an ℓ-cycle), if it has exactly ℓ reality edges. A permutation π is a *simple permutation* if every cycle in π has length at most 3.

Non-trivial bounds on the transposition distance were obtained by using the breakpoint graph [2], after applying a transposition t, the number of cycles of odd length in $G(\pi)$, denoted $c_{odd}(\pi)$, is changed such that $c_{odd}(\pi t) = c_{odd}(\pi) + x$, where $x \in \{-2, 0, 2\}$ and t is said to be an x-move for π. Since $c_{odd}(\iota) = n+1$, we have the lower bound $d(\pi) \geq \left\lceil \frac{(n+1) - c_{odd}(\pi)}{2} \right\rceil$, where the equality holds if, and only if, π can be sorted with only 2-moves.

Hannenhalli and Pevzner [9] proved that every permutation π can be transformed into a simple one $\hat{\pi}$, by inserting new elements on appropriate positions of π, preserving the lower bound for the distance, $\left\lceil \frac{(n+1) - c_{odd}(\pi)}{2} \right\rceil = \left\lceil \frac{(m+1) - c_{odd}(\hat{\pi})}{2} \right\rceil$ where m is such that $\hat{\pi} = [0\hat{\pi}_1 \ldots \hat{\pi}_m m+1]$. Additionally, a sequence that sorts $\hat{\pi}$ can be transformed into a sequence that sorts π, which implies that $d(\pi) \leq d(\hat{\pi})$. This method is commonly used in the literature, as in Hartman and Shamir's [10] and Elias and Hartman's [6] approximation algorithms.

A transposition $t(i, j, k)$ *affects* a cycle C if it contains one of the following reality edges: $\overrightarrow{i+1}$, or $\overrightarrow{j+1}$, or $\overrightarrow{k+1}$. A cycle is *oriented* if there is a 2-move that affects it (name given by the relative order of such a triplet of reality edges), otherwise it is *unoriented*. If there exists a 2-move that may be applied to π, then π is *oriented*, otherwise π is *unoriented*.

A sequence of q transpositions in which exactly r transpositions are 2-moves is a (q, r)-*sequence*. A $\frac{q}{r}$-sequence is a (x, y)-sequence such that $x \leq q$ and $\frac{x}{y} \leq \frac{q}{r}$.

A cycle in π is determined by its reality edges, in the order that they appear, starting from the leftmost edge. The notation $C = \langle x_1 x_2 \ldots x_\ell \rangle$, where $\overrightarrow{x_1}$, $\overrightarrow{x_2}$, $\ldots$, $\overrightarrow{x_\ell}$ are reality edges, and $x_1 = \min\{x_1, x_2, \ldots, x_\ell\}$, characterizes an ℓ-cycle.

Let $\overrightarrow{x}, \overrightarrow{y}, \overrightarrow{z}$, where $x < y < z$, be reality edges in a cycle C, and $\overrightarrow{a}, \overrightarrow{b}, \overrightarrow{c}$, where $a < b < c$ be reality edges in a different cycle C'. The pair of reality edges $\overrightarrow{x}, \overrightarrow{y}$ *intersects* the pair $\overrightarrow{a}, \overrightarrow{b}$ if these four edges occur in an alternating order in the breakpoint graph, i.e. $x < a < y < b$ or $a < x < b < y$. Similarly, two triplets of reality edges $\overrightarrow{x}, \overrightarrow{y}, \overrightarrow{z}$ and $\overrightarrow{a}, \overrightarrow{b}, \overrightarrow{c}$ are *interleaving* if these six edges occur in an alternating order, i.e. $x < a < y < b < z < c$ or $a < x < b < y < c < z$. Two cycles C and C' *intersect* if there is a pair of reality edges in C that intersects with a pair of reality edges in C', and two 3-cycles are *interleaving* if their respective triplets of reality edges are interleaving. See Fig. 1.

A *configuration* of π is a subset of the cycles in $G(\pi)$. A configuration $\mathcal{C}$ is *connected* if, for any two cycles C_1 and C_k in $\mathcal{C}$, there are cycles $C_1, \ldots, C_{k-1} \in \mathcal{C}$ such that, for each $i \in \{1, 2, \ldots, k-1\}$, the cycles C_i and C_{i+1} are either intersecting or interleaving. If the configuration $\mathcal{C}$ is connected and maximal, then $\mathcal{C}$ is a *component*. Every permutation admits a unique decomposition into disjoint components. For instance, in Fig. 1, the configuration $\{C_1, C_2, C_3, C_4\}$ is a component, but the configuration $\{C_1, C_2, C_3\}$ is connected but not a component.

Let C be a 3-cycle in a configuration $\mathcal{C}$. An *open gate* is a pair of reality edges of C that does not intersect any other pair of reality edges in $\mathcal{C}$. If a configuration

$\mathcal{C}$ has only 3-cycles and no open gates, then $\mathcal{C}$ is a *full configuration*. Some full configurations, such as the one in Fig. 2(a), do not correspond to the breakpoint graph of any permutation [6].

A configuration $\mathcal{C}$ that has k edges is in the *cromulent form*[1] if every edge from $\overrightarrow{0}$ to $\overrightarrow{k-1}$ is in $\mathcal{C}$. Given a configuration $\mathcal{C}$ having k edges, a *cromulent relabeling* (Fig. 2b) of $\mathcal{C}$ is a configuration $\mathcal{C}'$ such that $\mathcal{C}'$ is in the cromulent form and there is a function σ satisfying that, for every pair of edges $\overrightarrow{i}$, $\overrightarrow{j}$ in $\mathcal{C}$ such that $i < j$, we have that $\overrightarrow{\sigma(i)}, \overrightarrow{\sigma(j)}$ are in $\mathcal{C}'$ and $\sigma(i) < \sigma(j)$.

Given an integer x, a *circular shift* of a configuration $\mathcal{C}$, which is in the cromulent form and has k edges, is a configuration denoted $\mathcal{C} + x$ such that every edge $\overrightarrow{i}$ in $\mathcal{C}$ corresponds to $\overrightarrow{i + x} \pmod{k}$ in $\mathcal{C} + x$. Two configurations $\mathcal{C}$ and $\mathcal{K}$ are *equivalent* if there is an integer x such that $\mathcal{C}' + x = \mathcal{K}'$, where $\mathcal{C}'$ and $\mathcal{K}'$ are their respective cromulent relabelings.

(a) (b)

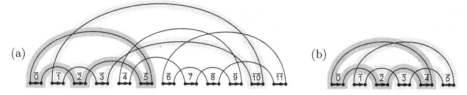

Fig. 2. (a) Full configuration $\{C_1, C_2, C_3, C_4\} = \{\langle 0\,2\,5\rangle, \langle 1\,3\,10\rangle, \langle 4\,7\,9\rangle, \langle 6\,8\,11\rangle\}$. (b) The cromulent relabeling of $\{C_1, C_2\}$ is $\{\langle 0\,2\,4\rangle, \langle 1\,3\,5\rangle\}$.

Elias and Hartman's algorithm Elias and Hartman [6] performed a systematic enumeration of all components having nine or less cycles, in which all cycles have length 3. Starting from single 3-cycles, components were obtained by applying a series of *sufficient extensions*, as described next. An *extension* of a configuration $\mathcal{C}$ is a connected configuration $\mathcal{C} \cup \{C\}$, where $C \notin \mathcal{C}$. A *sufficient extension* is an extension that either: 1) closes an open gate; or 2) extends a full configuration such that the extension has at most one open gate. A configuration obtained by a series of sufficient extensions is named *sufficient configuration*, which has an (x, y)-, or $\frac{x}{y}$-, sequence if it is possible to apply such a sequence to its cycles.

Lemma 1. *[6] Every unoriented sufficient configuration of nine cycles has an $\frac{11}{8}$-sequence.*

Components with less than nine cycles are called *small components*. Elias and Hartman showed that there are just five kinds of small components that do not have an $\frac{11}{8}$-sequence; these components are called *bad small components*. Small components that have an $\frac{11}{8}$-sequence are *good small components*.

Lemma 2. *[6] The bad small components are:* $A = \{\langle 0\,2\,4\rangle, \langle 1\,3\,5\rangle\}$; $B = \{\langle 0\,2\,10\rangle, \langle 1\,3\,5\rangle, \langle 4\,6\,8\rangle, \langle 7\,9\,11\rangle\}$; $C = \{\langle 0\,5\,7\rangle, \langle 1\,9\,11\rangle, \langle 2\,4\,6\rangle, \langle 3\,8\,10\rangle\}$; $D = \{\langle 0\,2\,4\rangle, \langle 1\,12\,14\rangle, \langle 3\,5\,7\rangle, \langle 6\,8\,10\rangle, \langle 9\,11\,13\rangle\}$; *and* $E = \{\langle 0\,2\,16\rangle, \langle 1\,3\,5\rangle, \langle 4\,6\,8\rangle, \langle 7\,9\,11\rangle, \langle 10\,12\,14\rangle, \langle 13\,15\,17\rangle\}$.

---

[1] *cromulent*: neologism coined by David X. Cohen, meaning "normal" or "acceptable."

If a permutation has bad small components, it is still possible to find $\frac{11}{8}$-sequences, as Lemma 3 states.

Lemma 3. *[6] Let π be a permutation with at least eight cycles and containing only bad small components. Then π has an $(11, 8)$-sequence.*

Corollary 1. *[6] If every cycle in $G(\pi)$ is a 3-cycle, and there are at least eight cycles, then π has an $\frac{11}{8}$-sequence.*

Lemmas 1 and 3, and Corollary 1 form the theoretical basis for Elias and Hartman's $\frac{11}{8} = 1.375$-approximation algorithm for SBT, shown in Algorithm 1.

Algorithm 1. Elias and Hartman's Sort(π)

1 Transform permutation π into a simple permutation $\hat{\pi}$.
2 Check if there is a $(2, 2)$-sequence. If so, apply it.
3 While $G(\hat{\pi})$ contains a 2-cycle, apply a 2-move.
4 $\hat{\pi}$ consists of 3-cycles. Mark all 3-cycles in $G(\hat{\pi})$.
5 **while** $G(\hat{\pi})$ *contains a marked 3-cycle C* **do**
6   **if** C *is oriented* **then**
7   Apply a 2-move to it.
8   **else**
9   Try to sufficiently extend C eight times (to obtain a configuration with at most 9 cycles).
10   **if** *sufficient configuration with 9 cycles has been achieved* **then**
11   Apply an $\frac{11}{8}$-sequence.
12   **else** It is a small component
13   **if** *it is a good component* **then**
14   Apply an $\frac{11}{8}$-sequence.
15   **else**
16   Unmark all cycles of the component.

17 (Now $G(\hat{\pi})$ has only bad small components.)
18 **while** $G(\hat{\pi})$ *contains at least eight cycles* **do**
19   Apply an $(11, 8)$-sequence
20 While $G(\hat{\pi})$ contains a 3-cycle, apply a $(3, 2)$-sequence.
21 Mimic the sorting of π using the sorting of $\hat{\pi}$.

Feng and Zhu's permutation tree Feng and Zhu [7] introduced the *permutation tree*, a binary balanced tree that represents a permutation, and provided four algorithms: to *build* a permutation tree in $O(n)$ time, to *join* two permutation trees into one in $O(h)$ time, where h is the height difference between the trees, to *split* a permutation tree into two in $O(\log n)$ time, and to *query* a permutation tree and find reality edges that intersect a given pair $\overrightarrow{i}, \overrightarrow{j}$ in $O(\log n)$ time.

Operations *split* and *join* allow applying a transposition to a permutation π and updating the tree in time $O(\log n)$. Lemma 4 provides a way to determine, in logarithmic time, which transposition should be applied to a permutation, and serves as the basis for the *query* procedure. This method was applied [7] to Hartman and Shamir's 1.5-approximation algorithm [10], to find a $(3,2)$-sequence that affects a pair of intersecting or interleaving cycles.

Lemma 4. *[2] Let $\overrightarrow{i}$ and $\overrightarrow{j}$ be two reality edges in an unoriented cycle C, $i < j$. Let $\pi_k = \max_{i < m \leq j} \pi_m$, $\pi_\ell = \pi_k + 1$, then $\overrightarrow{k}$ and $\overleftarrow{\ell - 1}$ belong to the same cycle, and the pair $\overrightarrow{k}, \overleftarrow{\ell - 1}$ intersects the pair $\overrightarrow{i}, \overrightarrow{j}$.*

Firoz's et al. use of the permutation tree Firoz *et al.* [8] suggested the use of the permutation tree to reduce the running time of Elias and Hartman's [6] algorithm. In [5], we showed that this strategy fails to extend some full configurations.

Firoz *et al.* [8] stated that extensions can be done in $O(\log n)$ time. To do that, they categorized sufficient extensions of a configuration A into *type 1 extensions* – those that add a cycle that closes open gates – and *type 2 extensions* – those that extend a full configuration by adding a cycle C such that $A \cup \{C\}$ has at most one open gate.

A type 1 extension can be performed in logarithmic time by running *query* for an open gate. In a type 2 extension, since there are no open gates, Firoz *et al.* claimed that it is sufficient to perform queries on all pairs of reality edges belonging to the same cycle in a configuration that is being extended. But, as shown in [5], there is an infinite family of configurations for which this strategy fails; some instances are subsets of two cycles of $[0\,10\,9\,8\,7\,1\,6\,11\,5\,4\,3\,2\,12]$ (Fig. 1). Consider the configuration $A = \{C_1\}$; try to sufficiently extend A (step 9 in Algorithm 1) using the steps proposed by Firoz *et al.*:

1. Configuration A has three open gates. Executing the *query* for an open gate results in a pair of edges belonging the cycle C_2. Therefore, we add this cycle to the configuration A, which becomes $A = \{C_1, C_2\}$.
2. Configuration A has no more open gates. Executing the *query* for every pair of edges in the same cycle of A, we observe that the *query* will return a pair that is already in A. So far, Firoz *et al.*'s method has failed to extend A.

3   Finding a $(2,2)$-Sequence in Linear Time

Elias and Hartman [6] proved that, given a simple permutation, a $(2,2)$-sequence can be found in $O(n^2)$ time. Firoz *et al.* [8] described a strategy for finding and applying a $(2,2)$-sequence in $O(n \log n)$ time using permutation trees and the result in Lemma 5; see below. But, according to their strategy, it is still necessary to search for an oriented cycle in $O(n)$ time and, after applying the first 2-move, checking for the existence of an oriented cycle, again in $O(n)$ time. However, these steps must be performed $O(n)$ times in the worst case, which implies that Firoz *et al.*'s strategy also takes $O(n^2)$ time.

Algorithm 2. Search $(2,2)$-sequence from K_1

---

1 **for** $i = \min K_1 + 1, \ldots, \operatorname{mid} K_1 - 1$ **do**
2   **if** $\overrightarrow{i}$ *belongs to an oriented cycle* K_j **then**
3   **if** $\operatorname{mid} K_j < \operatorname{mid} K_1$ *or* $\max K_j < \max K_1$ **then**
4   **return** $(2,2)$-sequence that affects K_1 and K_j.

5   **if** $\overrightarrow{i}$ *belongs to an unoriented cycle* L_j **then**
6   **if** $mid K_1 < mid L_j < max K_1 < max L_j$ **then**
7   **return** $(2,2)$-sequence that affects K_1 and L_j.

8 **for** $i = \operatorname{mid} K_1 + 1, \ldots, \max K_1 - 1$ **do**
9   **if** $\overrightarrow{i}$ *belongs to an oriented cycle* K_j **then**
10   **if** $\operatorname{mid} K_1 < \min K_j$ **then**
11   **return** $(2,2)$-sequence that affects K_1 and K_j.

12 **for** $i = \max K_1 + 1, \ldots, n - 1$ **do**
13   **if** $\overrightarrow{i}$ *belongs to an oriented cycle* K_j **then**
14   **if** $\max K_1 \leq \min K_j$ **then**
15   **return** $(2,2)$-sequence affecting K_1 and K_j.

Algorithm 4 summarizes our approach towards finding and applying a $(2,2)$-sequence in $O(n)$ time.

Lemma 5. *[2,4,6] Given a breakpoint graph of a simple permutation, there exists a $(2,2)$-sequence if any of the following conditions is met:*

1. *there are either four 2-cycles, or two intersecting 2-cycles, or two non intersecting 2-cycles, and the resulting graph contains an oriented cycle after the first transposition is applied;*
2. *there are two non interleaving oriented 3-cycles;*
3. *there is an oriented cycle interleaving an unoriented cycle.*

Our strategy to find a $(2,2)$-sequence in linear time starts with checking whether a breakpoint graph satisfies Lemma 5, as described in detail in Algorithm 2. It differs from previous approaches [6,8] in that the leftmost oriented cycle, dubbed K_1, is fixed when verifying conditions 2 and 3, avoiding comparisons between every pair of cycles.

Given a simple permutation π, it is trivial to enumerate all of its cycles in linear time. The size of each cycle, and whether it is oriented, are both determined in constant time.

Christie [4] proved that every permutation has an even number (possibly zero) of even cycles; he also showed that, given a simple permutation, when the number of even cycles is not zero, there exists a $(2,2)$-sequence that affects those cycles if, and only if, there are either four 2-cycles, or there are two intersecting even cycles. Therefore, in these cases, a $(2,2)$-sequence can be applied in $O(\log n)$

using permutation trees. If there is only a pair of non-intersecting 2-cycles, it remains to check if there is a 3-cycle intersecting both even cycles: i) if the 3-cycle is oriented, then first we apply the 2-move over the 3-cycle, and the second 2-move is over the 2-cycles; ii) if the 3-cycle is unoriented, then first we apply the 2-move over the 2-cycles, and the second 2-move is over the 3-cycle, which turns oriented after the first transposition. There is also a $(2,2)$-sequence if there is an oriented cycle intersecting at most one even cycle.

However, if no even cycle satisfies the previous conditions, but there is an oriented cycle, the 3-cycles must be scanned for the existence of a $(2,2)$-sequence, as required conditions 2 and 3 in Lemma 5.

To check, in linear time, for the existence of a pair of cycles satisfying either condition 2 or 2 in Lemma 5, consider the oriented cycles of the breakpoint graph, in the order $K_1 = \langle a_1\,b_1\,c_1 \rangle, K_2 = \langle a_2\,b_2\,c_2 \rangle, \ldots$ such that $a_1 < a_2 < \ldots$, and the unoriented cycles in the order $L_1 = \langle x_1\,y_1\,z_1 \rangle, L_2 = \langle x_2\,y_2\,z_2 \rangle, \ldots$ such that $x_1 < x_2 < \ldots$. Given any 3-cycle $C = \langle a\,b\,c \rangle$, let $\min C = a$, $\mathrm{mid}\,C = \min\{b, c\}$ and $\max C = \max\{b, c\}$. The main idea is:

1. Check for the existence of an oriented cycle K_j non-interleaving K_1 or an unoriented cycle L_j interleaving K_1. Algorithm 2 solves that: between $\min K_1$ and $\mathrm{mid}\,K_1$, between $\mathrm{mid}\,K_1$ and $\max K_1$, and to the right of $\max K_1$, search for an oriented cycle K_i non-interleaving K_1 or an unoriented cycle L_i interleaving K_1.

2. If every oriented cycle interleaves K_1 and no unoriented cycle interleaves K_1, then check for the existence of two oriented cycles K_i, K_j that are intersecting but not interleaving. Notice that if there is a pair of non-interleaving oriented cycles, then the cycles intersect each other, otherwise one of the cycles would be non-interleaving K_1, and Algorithm 2 would have this case already covered (see Fig. 3). Algorithm 3 describes how to verify the existence of two intersecting oriented cycles that are also interleaving K_1.

Fig. 3. Oriented cycles represented by their reality edges. All oriented cycles interleave K_1, but K_i and K_j non-interleave each other.

4   Sufficient Extensions Using *Query*

At the end of Section 2, we discussed Firoz's *et al.* use of the permutation tree, and as proven in [5], their strategy does not account for configurations with less than nine cycles that are not components, since successive invocations of the *query* procedure may result in a full configuration with less than nine cycles that is not a small component. Our proposed strategy generalizes the definitions related to small components by defining a *small configuration*, a configuration with less than nine cycles.

Algorithm 3. Finding intersecting oriented cycles interleaving K_1.

---

1 s_1 = sequence of edges belonging to oriented cycles from left to right between min K_1 and mid K_1.

2 s_2 = sequence of edges belonging to oriented cycles from left to right between mid K_1 and max K_1.

3 **if** s_1 *and* s_2 *are different* **then**

4 $\quad\lfloor$ There is a pair of intersecting oriented cycles, exists a $(2,2)$-sequence.

5 **else**

6 $\quad\lfloor$ All oriented cycles are mutually interleaving.

Algorithm 4. Find and Apply (2,2)-sequence

---

1 **if** *there are four 2-cycles* **then**

2 $\quad\lfloor$ Apply $(2,2)$-sequence.

3 **else if** *there is a pair of intersecting 2-cycles* **then**

4 $\quad\lfloor$ Apply $(2,2)$-sequence.

5 **else if** *there is a 3-cycle intersecting a pair of 2-cycles* **then**

6 $\quad\lfloor$ Apply $(2,2)$-sequence.

7 **else if** *there is a pair of 2-cycles* **and** *an oriented 3-cycle intersecting at most one of them* **then**

8 $\quad\lfloor$ Apply $(2,2)$-sequence.

9 **else if** *Search $(2,2)$-sequence from K_1 returns a sequence* **then**

10 $\quad\lfloor$ Apply $(2,2)$-sequence.

11 **else if** *Finding intersecting oriented cycles interleaving K_1* **then**

12 $\quad\lfloor$ Apply $(2,2)$-sequence.

13 **else**

14 $\quad\lfloor$ There are no $(2,2)$-sequences to apply.

A small configuration is said to be *full* if it has no open gates. Small configurations are also classified as *good* if they have an $\frac{11}{8}$-sequence, or as *bad* otherwise.

Algorithm 1 applies an $\frac{11}{8}$-sequence to every sufficient unoriented configuration of nine cycles, and also to every good small component. After that, the permutation contains just bad small components, and Lemma 3 states the existence of an $(11,8)$-sequence in every combination of bad small components with at least 8 cycles.

By doing extensions using the *query* procedure, we can deal with bad small full configurations, which may or may not be bad small components. The possible bad small full configurations are the bad small components A, B, C, D and E, from Lemma 2, and one more full configuration

$$F = \{\langle 0\,7\,9 \rangle, \langle 1\,3\,6 \rangle, \langle 2\,4\,11 \rangle, \langle 5\,8\,10 \rangle\},$$

which is the only bad small full configuration that is not a component [6].

Our strategy (Algorithm 5) is similar to Elias and Hartman's (Algorithm 1): we apply an $\frac{11}{8}$-sequence to every sufficient unoriented configuration of nine cycles, and additionally to every good small full configuration; the main difference is that, whenever a combination of bad small full configuration is found, a decision to apply an $\frac{11}{8}$-sequence is made according to Lemma 6.

Lemma 6. *Every combination of F with one or more copies of either B, C, D or E has an $\frac{11}{8}$-sequence.*

Proof. Consider all breakpoint graphs of F and its circular shifts combined with B, C, D, E, and their circular shifts. A combination of a pair of small full configurations is obtained by starting from one small full configuration and inserting a new one in different positions in the breakpoint graph. Altogether, there are 324 such graphs. A computerized case analysis, in [1], enumerates every possible breakpoint graph and provides an $\frac{11}{8}$-sequence for each of them. □

Notice that Lemma 6 considers neither combinations of F with F, nor combinations of F with A. We have found that almost every combination of F with F has an $\frac{11}{8}$-sequence. Let $F_i F^j$ be the configuration obtained by inserting the circular shift $F + j$ between the edges $\overrightarrow{i}$ and $\overrightarrow{i+1}$ of F.

Lemma 7. *There exists an $\frac{11}{8}$-sequence for $F_i F^j$, if:*

- $i \in \{0, 4\}$ and $j \in \{0, 1, 2, 3, 4, 5\}$;
- $i \in \{1, 2, 3\}$ and $j \in \{1, 2, 3, 4, 5\}$; or
- $i = 5$ and $j \in \{1, 5\}$.

Proof. The $\frac{11}{8}$-sequences for the cases enumerated above were also found through a computerized case analysis [1]. Note that $F_i F^j$ is equivalent to $F_{i+6} F^j$ for $i = \{0, 1, \ldots, 5\}$, which simplifies our analysis. □

The combinations of F with F for which our branch-and-bound case analysis cannot find an $\frac{11}{8}$-sequence are: $F_1 F^0$, $F_2 F^0$, $F_3 F^0$, $F_5 F^0$, $F_5 F^2$, $F_5 F^3$ and $F_5 F^4$.

All combinations of one copy of F and one of A have less than eight cycles. It only remains to analyse the combinations of F and two copies of A, denoted F–A–A. The *good F–A–A combinations* are the F–A–A combinations for which an $\frac{11}{8}$-sequence exists. Out of 57 combinations of F–A–A, only 31 are good. The explicit list of combinations is in [1].

Combinations of F and A, B, C, D, E that have an $\frac{11}{8}$-sequence are called *well-behaved combinations*: the ones in Lemmas 6, 7 and the good $F-A-A$ combinations. The remaining combinations having F are called *naughty*.

For extensions that yield a bad small configuration, Algorithm 5 adds their cycles to a set $\mathcal{S}$ (line 18). Later, if a well-behaved combination is found among the cycles in $\mathcal{S}$, an $\frac{11}{8}$-sequence is applied (line 21) and the set is emptied. The set $\mathcal{S}$ may just contain naughty combinations and in the next iteration (line 6) another bad small configuration may be obtained and added to $\mathcal{S}$. We have shown [1] that every combination of three copies of F is well-behaved, even if each pair of F is naughty; the same can also be said of every combination of F and three copies of A such that each triple $F-A-A$ is naughty. Therefore, at most 12 cycles are in $\mathcal{S}$, since there are in the worst case three copies of F; or one copy of F and three copies of A. In all these cases we apply $\frac{11}{8}$-sequences as proved in [1].

New Algorithm. The previous results allow us to devise Algorithm 5, that basically obtains configurations using the *query* procedure, and applies $\frac{11}{8}$-sequences to configurations of size at most 9. It differs from Algorithm 1 not only in the use of permutation trees, but also because we continuously deal with bad small full configurations instead of only at the end.

Algorithm 5. New algorithm based on Elias and Hartman's algorithm

---

1 Transform permutation π into a simple permutation $\hat{\pi}$.
2 Find and Apply (2,2)-sequence (Algorithm 4).
3 While $G(\hat{\pi})$ contains a 2-cycle, apply a 2-move.
4 $\hat{\pi}$ consists of 3-cycles. Mark all 3-cycles in $G(\hat{\pi})$.
5 Let $\mathcal{S}$ be an empty set.
6 **while** $G(\hat{\pi})$ *contains at least eight* 3-*cycles* **do**
7   Start a configuration $\mathcal{C}$ with a marked 3-cycle.
8   **if** *the cycle in* $\mathcal{C}$ *is oriented* **then**
9   Apply a 2-move to it.
10   **else**
11   Try to sufficiently extend $\mathcal{C}$ eight times.
12   **if** $\mathcal{C}$ *is a sufficient configuration with 9 cycles* **then**
13   Apply an $\frac{11}{8}$-sequence.
14   **else** $\mathcal{C}$ is a small full configuration
15   **if** $\mathcal{C}$ *is a good small configuration* **then**
16   Apply an $\frac{11}{8}$-sequence.
17   **else** $\mathcal{C}$ is a bad small configuration.
18   Add every cycle in $\mathcal{C}$ to $\mathcal{S}$.
19   Unmark all cycles in $\mathcal{C}$.
20   **if** $\mathcal{S}$ *contains a well-behaved combination* **then**
21   Apply an $\frac{11}{8}$-sequence.
22   Mark the remaining 3-cycles in $\mathcal{S}$.
23   Remove all cycles from $\mathcal{S}$.
24 While $G(\hat{\pi})$ contains a 3-cycle, apply a $(4,3)$-sequence or a $(3,2)$-sequence.
25 Mimic the sorting of π using the sorting of $\hat{\pi}$.

Theorem 1. *Algorithm 5 runs in* $O(n \log n)$ *time.*

Proof. Steps 1 through 5 can be implemented to run in linear time (proofs in [6] and in Sect. 3). Step 11 runs in $O(\log n)$ time using permutation trees. The comparisons in Steps 12, 14, 15, 17 and 20 are done in constant time using lookup tables of size bound by a constant. Updating the set $\mathcal{S}$ also requires constant time, since it has at most 12 cycles. Every sequence of transpositions of size bounded by a constant can be applied in time $O(\log n)$ due to the use of permutation trees. The time complexity of the loop between Steps 6 to 23 is $O(n \log n)$, since the number of 3-cycles is linear in n, and the number cycles decreases, in the worst case, once in three iterations. In Step 24, the search for a $(4, 3)$ or a $(3, 2)$-sequence is done in constant time, since the number of cycles is bounded by a constant. Steps 24 and 25 also run in time $O(n \log n)$.   □

5   Conclusion

The goal of this paper is to lower the time complexity of Elias and Hartman's [6] 1.375-approximation algorithm down to $O(n \log n)$. Our new approach provides, so far, both the lowest fixed approximation ratio and time complexity of any non-trivial algorithm for sorting by transpositions.

We have previously shown that a simple application of permutation trees [7], as claimed in [8], does not suffice to correctly improve the running time of Elias and Hartman's algorithm. In order to lower the time complexity, it is necessary to add more configurations [1] to the original analysis in [6], and also to perform some changes in the sorting procedure, as shown in Algorithm 5.

References

1. http://compscinet.org/research/sbt1375 (2014)
2. Bafna, V., Pevzner, P.A.: Sorting by transpositions. SIAM J. Discrete Math. 11(2), 224–240 (1998)
3. Bulteau, L., Fertin, G., Rusu, I.: Sorting by transpositions is difficult. SIAM J. Discrete Math. 26(3), 1148–1180 (2012)
4. Christie, D.A.: Genome Rearrangement Problems. Ph.D. thesis, University of Glasgow, UK (1999)
5. Cunha, L.F.I., Kowada, L.A.B., de A. Hausen, R., de Figueiredo, C.M.H.: On the 1.375-approximation algorithm for sorting by transpositions in $O(n \log n)$ time. In: Setubal, J.C., Almeida, N.F. (eds.) BSB 2013. LNCS, vol. 8213, pp. 126–135. Springer, Heidelberg (2013)
6. Elias, I., Hartman, T.: A 1.375-approximation algorithm for sorting by transpositions. IEEE/ACM Trans. Comput. Biol. Bioinformatics 3(4), 369–379 (2006)
7. Feng, J., Zhu, D.: Faster algorithms for sorting by transpositions and sorting by block interchanges. ACM Trans. Algorithms 3(3), 1549–6325 (2007)
8. Firoz, J.S., Hasan, M., Khan, A.Z., Rahman, M.S.: The 1.375 approximation algorithm for sorting by transpositions can run in $O(n \log n)$ time. J. Comput. Biol. 18(8), 1007–1011 (2011)
9. Hannenhalli, S., Pevzner, P.A.: Transforming cabbage into turnip: Polynomial algorithm for sorting signed permutations by reversals. J. ACM 46(1), 1–27 (1999)
10. Hartman, T., Shamir, R.: A simpler and faster 1.5-approximation algorithm for sorting by transpositions. Inf. Comput. 204(2), 275–290 (2006)

A Generalized Cost Model for DCJ-Indel Sorting

Phillip E.C. Compeau*

Department of Mathematics, UC San Diego, 9500 Gilman Drive #0112,
San Diego, CA 92093, United States
pcompeau@math.ucsd.edu

Abstract. The double-cut-and-join operation (DCJ) is a fundamental graph operation that is used to model a variety of genome rearrangements. However, DCJs are only useful when comparing genomes with equal (or nearly equal) gene content. One obvious extension of the DCJ framework supplements DCJs with insertions and deletions of chromosomes and chromosomal intervals, which implies a model in which DCJs receive unit cost, whereas insertions and deletions receive a nonnegative cost of ω. This paper proposes a unified model finding a minimum-cost transformation of one genome (with circular chromosomes) into another genome for any value of ω. In the process, it resolves the open case $\omega > 1$.

1 Introduction

Large scale chromosomal mutations were observed indirectly via the study of linkage maps near the beginning of the 20th Century, and these genome rearrangements were first directly observed by Dobzhansky and Sturtevant in 1938 (see [8]). Yet only in the past quarter century has the combinatorial study of genome rearrangements taken off, as researchers have attempted to create and adapt discrete genomic models along with distance functions modeling the evolutionary distance between two genomes. See [9] for an overview of the combinatorial methods used to compare genomes.

Recent research has moved toward multichromosomal genomic models as well as distance functions that allow for mutations involving more than one chromosome. Perhaps the most commonly used such model represents an ordered collection of disjoint chromosomal intervals along a chromosome as either a path or cycle, depending on whether the chromosome is linear or circular. For genomes with equal gene content, the double cut and join operation (DCJ), introduced in [11], incorporates a wide class of operations into a simple graph operation. It has led to a large number of subsequent results over the last decade, beginning with a linear-time algorithm for the problem of DCJ sorting (see [3]), in which we attempt to transform one genome into another using a minimum number of DCJs.

For genomes with unequal gene content, the incorporation of insertions and deletions of chromosomes and chromosomal intervals (collectively called "indels") into the DCJ framework was discussed in [12] and solved in [5]. The latter

---

* The author would like to thank Pavel Pevzner and the reviewers for very insightful comments.

D. Brown and B. Morgenstern (Eds.): WABI 2014, LNBI 8701, pp. 38–51, 2014.

authors provided a linear-time algorithm for the problem of DCJ-indel sorting, which aims to find a minimum collection of DCJs and indels required to transform one genome into another. An alternate linear-time algorithm for DCJ-indel sorting can be found in [7].

We can generalize DCJ-indel sorting to the problem of finding a minimum cost transformation of one genome into another by DCJs and indels in which DCJs receive unit cost and indels receive a cost of ω, where ω is a nonnegative constant. The case $\omega \leq 1$ was resolved by the authors in [10]. The current paper aims to find a unifying framework that will solve the problem of DCJ-indel sorting for all $\omega \geq 0$ when the input genomes have only circular chromosomes. In the process, it resolves the open case $\omega > 1$ for genomes with circular chromosomes.

In Section 2, we will discuss the theoretical foundation required to address DCJ-indel sorting. In Section 3, we show that in any minimum-cost transformation with DCJs and indels (for any value of ω), each indel can be encoded and thus amortized as a DCJ, which helps to explain why DCJ-indel sorting has the same computational complexity as DCJ sorting. In Section 4, we will address the problem of DCJ-indel sorting genomes with circular chromosomes for all $\omega \geq 0$ for a particular subclass of genome pairs. In Section 5, we generalize this result to all pairs of genomes with circular chromosomes.

2 Preliminaries

A **genome** Π is a graph containing an even number of labeled nodes and comprising the edge-disjoint union of two perfect matchings: the **genes**[1] of Π, denoted $g(\Pi)$; and the **adjacencies** of Π, denoted $a(\Pi)$. Consequently, each node of Π has degree 2, and the connected components of Π form cycles that alternate between genes and adjacencies; these cycles are called **chromosomes**. This genomic model, in which chromosomes are circular, offers a reasonable and commonly used approximation of genomes having linear chromosomes.

A **double cut and join operation (DCJ)** on Π, introduced in [11], forms a new genome by replacing two adjacencies of Π with two new adjacencies on the same four nodes. Despite being simply defined, the DCJ incorporates the **reversal** of a chromosomal segment, the **fusion** of two chromosomes into one chromosome, and the **fission** of one chromosome into two chromosomes (Fig. 1).[2] For genomes Π and Γ with the same genes, the **DCJ distance**, denoted $d(\Pi, \Gamma)$, is the minimum number of DCJs needed to transform Π into Γ.

The **breakpoint graph** of Π and Γ, denoted $B(\Pi, \Gamma)$ (introduced in [2]), is the edge-disjoint union of $a(\Pi)$ and $a(\Gamma)$ (Fig. 2). The line graph of the breakpoint graph is the *adjacency graph*, which was introduced in [3] and is also commonly used in genome rearrangement studies. Note that the connected components of $B(\Pi, \Gamma)$ form cycles (of length at least 2) that alternate between adjacencies of Π and Γ, and so we will let $c(\Pi, \Gamma)$ denote the number of cycles in

---

[1] In practice, gene edges typically represent synteny blocks containing a large number of contiguous genes.

[2] When the DCJ is applied to circularized linear chromosomes, it encompasses a larger variety of operations. See [5] for details.

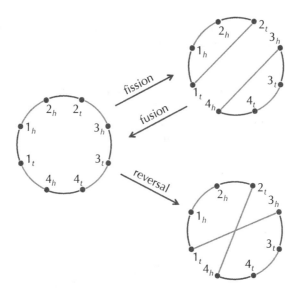

Fig. 1. DCJs replace two adjacencies of a genome and incorporate three operations on circular chromosomes: reversals, fissions, and fusions. Genes are shown in black, and adjacencies are shown in red.

$B(\Pi, \Gamma)$. A consequence of the definition of a DCJ as a rearrangement involving only two edges is that if Π' is obtained from Π by a single DCJ, then $|c(\Pi', \Gamma) - c(\Pi, \Gamma)| \leq 1$. The authors in [11] provided a greedy algorithm for sorting Π into Γ that reduces the number of cycles in the breakpoint graph by 1 at each step, which implies that the DCJ distance is given by

$$d(\Pi, \Gamma) = |g(\Pi)| - c(\Pi, \Gamma) \ . \tag{1}$$

The DCJ distance offers a useful metric for measuring the evolutionary distance between two genomes having the same genes, but we strive toward a genomic model that incorporates insertions and deletions as well. A **deletion** in Π is defined as the removal of either an entire chromosome or chromosomal interval of Π, i.e., if adjacencies $\{v, w\}$ and $\{x, y\}$ are contained in the order (v, w, x, y) on some chromosome of Π, then a deletion replaces the path connecting v to y with the single adjacency $\{v, y\}$. An **insertion** is simply the inverse operation of a deletion. The term **indels** refers collectively to insertions and deletions.

To consider genomes with unequal gene content, we will henceforth assume that any pair of genomes Π and Γ satisfy $g(\Pi) \cup g(\Gamma) = \mathcal{G}$, where $\mathcal{G}$ is a perfect matching on a collection of nodes $\mathcal{V}$. A **transformation** of Π into Γ is a sequence of DCJs and indels such that any deleted node must belong to $\mathcal{V} - V(\Gamma)$ and any inserted node must belong to $\mathcal{V} - V(\Pi)$.[3]

---

[3] This assumption follows the lead of the authors in [5]. It prevents, among other things, a trivial transformation of genome Π into genome Γ of similar gene content in which we simply delete all the chromosomes of Π and replace them with the chromosomes of Γ.

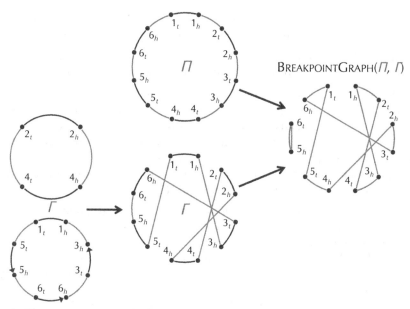

Fig. 2. The construction of the breakpoint graph of genomes Π and Γ having the same genes. First, the nodes of Γ are rearranged so that they have the same position in Π. Then, the adjacency graph is formed as the disjoint union of adjacencies of Π (red) and Γ (blue).

The **cost** of a transformation $\mathbb{T}$ is equal to the weighted sum of the number of DCJs in $\mathbb{T}$ plus ω times the number of indels in $\mathbb{T}$, where ω is some nonnegative constant determined in advance. The **DCJ-indel distance** between Π and Γ, denoted $d_\omega(\Pi, \Gamma)$, is the minimum cost of any transformation of Π into Γ. Note that since a transformation of Π into Γ can be inverted to yield a transformation of Γ into Π, the DCJ-indel distance is symmetric by definition. Yet unlike the DCJ distance, the DCJ-indel distance does not form a metric, as the triangle inequality does not hold; see [4] for a discussion in the case that $\omega = 1$.

Although we would like to compute DCJ-indel distance, here we are interested in the more difficult problem of **DCJ-indel sorting**, or producing a minimum cost transformation of Π into Γ. The case $\omega = 1$ was resolved by the authors in [5]; this result was extended to cover all values $0 \leq \omega \leq 1$ in [10]. This work aims to use the ideas presented in [7] as a stepping stone for a generalized presentation that will solve the problem of DCJ-indel sorting for all $\omega \geq 0$, thus resolving the open case that $\omega > 1$.

3   Encoding Indels as DCJs

A chromosome of Π (Γ) sharing no genes with Γ (Π) is called a **singleton**. We use the notation $\mathrm{sing}_\Gamma(\Pi)$ to denote the number of singletons of Π with respect to Γ and the notation $\mathrm{sing}(\Pi, \Gamma)$ to denote the sum $\mathrm{sing}_\Gamma(\Pi) + \mathrm{sing}_\Pi(\Gamma)$. We will deal with singletons later; for now, we will show that in the absence of singletons, the insertion or deletion of a chromosomal interval is the only type of indel that we need to consider for the problem of DCJ-indel sorting.

Theorem 1. *If* $\text{sing}(\Pi, \Gamma) = 0$, *then any minimum-cost transformation of* Π *into* Γ *cannot include the insertion or deletion of entire chromosomes.*

Proof. We proceed by contradiction. Say that we have a minimum-cost transformation of Π into Γ in which (without loss of generality) we are deleting an entire chromosome C. Because Π has no singletons, C must have been produced as the result of the deletion of a chromosomal interval or as the result of a DCJ. C cannot have been produced from the deletion of an interval, since we could have simply deleted the chromosome that C came from. Thus, assuming C was produced as the result of a DCJ, there are now three possibilities:

1. The DCJ could be a reversal. In this case, we could have simply deleted the chromosome to which the reversal was appiled, yielding a transformation of strictly smaller cost.
2. The DCJ could be a fission of a chromosome C' that produced C along with another chromosome. In this case, the genes of C appeared as a contiguous interval of C', which we could have simply deleted at lesser total cost.
3. The DCJ could be the fusion of two chromosomes, C_1 and C_2. This case is somewhat more difficult to deal with and is handled by Lemma 2.

Lemma 2. *If* $\text{sing}_\Gamma(\Pi) = 0$, *then any minimum-cost transformation of* Π *into* Γ *cannot include the deletion of a chromosome that was produced by a fusion.*

Proof. Suppose for the sake of contradiction that a minimum-cost transformation $\mathbb{T}$ of Π into Γ involves k fusions of $k + 1$ chromosomes $C_1, C_2, \ldots, C_{k+1}$ to form a chromosome C, which is then deleted. Without loss of generality, we may assume that this collection of fusions is "maximal", i.e., none of the C_i is produced as the result of a fusion.

Because Π has no singletons, each C_i must have been produced as a result of a DCJ. Similar reasoning to that used in the main proof of Theorem 1 shows that this DCJ cannot be a reversal, and by the assumption of maximality, it cannot be the result of a fusion. Thus, each C_i is produced by a fission applied to some chromosome C_i' to produce C_i in addition to some other chromosome C_i^*.

Now, let Π' be the genome in $\mathbb{T}$ occurring immediately before these $2k + 2$ operations. Assume that the $k + 1$ fissions applied to the C_i' replace adjacencies $\{v_i, w_i\}$ and $\{x_i, y_i\}$ with $\{v_i, y_i\}$ and $\{w_i, x_i\}$.[4] Furthermore, assume that the ensuing k fusions are as follows:

$$\{v_1, y_1\}, \{v_2, y_2\} \rightarrow \{y_1, v_2\}, \{v_1, y_2\}$$
$$\{y_1, v_2\}, \{v_3, y_3\} \rightarrow \{y_1, v_3\}, \{v_2, y_3\}$$
$$\{y_1, v_3\}, \{v_4, y_4\} \rightarrow \{y_1, v_4\}, \{v_3, y_4\}$$
$$\vdots$$
$$\{y_1, v_k\}, \{v_{k+1}, y_{k+1}\} \rightarrow \{y_1, v_{k+1}\}, \{v_k, y_{k+1}\}$$

---

[4] It can be verified that these $4k + 4$ nodes must be distinct by the assumption that $\mathbb{T}$ has minimum cost.

The genome resulting from these $2k + 1$ operations, which we call $\Pi_{\mathbb{T}}''$, is identical to Π' except that for each i ($1 \leq i \leq k + 1$), it has replaced the adjacencies $\{v_i, w_i\}$ and $\{x_i, y_i\}$ in C_i' with the adjacencies $\{w_i, x_i\} \in C_i^*$ and $\{v_i, y_{i+1 \bmod (k+1)}\} \in C$. In $\mathbb{T}$, we then delete C from $\Pi_{\mathbb{T}}''$.

Now consider the transformation $\mathbb{U}$ that is identical to $\mathbb{T}$ except that when we reach Π', $\mathbb{U}$ first applies the following k DCJs:

$$\{v_1, w_1\}, \{x_2, y_2\} \rightarrow \{v_1, y_2\}, \{w_1, x_2\}$$
$$\{v_2, w_2\}, \{x_3, y_3\} \rightarrow \{v_2, y_3\}, \{w_2, x_3\}$$
$$\{v_3, w_3\}, \{x_4, y_4\} \rightarrow \{v_3, y_4\}, \{w_3, x_4\}$$
$$\vdots$$
$$\{v_k, w_k\}, \{x_{k+1}, y_{k+1}\} \rightarrow \{v_k, y_{k+1}\}, \{w_k, x_{k+1}\}$$

$\mathbb{U}$ then applies k subsequent DCJs as follows:

$$\{x_1, y_1\}, \{w_1, x_2\} \rightarrow \{y_1, x_2\}, \{w_1, x_1\}$$
$$\{y_1, x_2\}, \{w_2, x_3\} \rightarrow \{y_1, x_3\}, \{w_2, x_2\}$$
$$\{y_1, x_3\}, \{w_3, x_4\} \rightarrow \{y_1, x_4\}, \{w_3, x_3\}$$
$$\vdots$$
$$\{y_1, x_k\}, \{w_k, x_{k+1}\} \rightarrow \{y_1, x_{k+1}\}, \{w_k, x_k\}$$

The resulting genome, which we call $\Pi_{\mathbb{U}}''$, has the exact same adjacencies as $\Pi_{\mathbb{T}}''$ except that it contains the adjacencies $\{y_1, x_{k+1}\}$ and $\{v_{k+1}, w_{k+1}\}$ instead of $\{v_{k+1}, y_1\}$ and $\{w_{k+1}, x_{k+1}\}$. Because two genomes on the same genes are equivalent if and only if they share the same adjacencies, a single DCJ on $\{y_1, x_{k+1}\}$ and $\{v_{k+1}, w_{k+1}\}$ would change $\Pi_{\mathbb{U}}''$ into $\Pi_{\mathbb{T}}''$. Furthermore, in $\Pi_{\mathbb{T}}''$, $\{v_{k+1}, y_1\}$ belongs to C and $\{w_{k+1}, x_{k+1}\}$ belongs to C_{k+1}^*, so that this DCJ in question must be a fission producing C and C_{k+1}^*. In $\mathbb{U}$, rather than applying this fission, we simply delete the chromosomal interval containing the genes of C. As a result, $\mathbb{U}$ is identical to $\mathbb{T}$ except that it replaces $2k + 1$ DCJs and a deletion by $2k$ DCJs and a deletion. Hence, $\mathbb{U}$ has strictly smaller cost than $\mathbb{T}$, which provides the desired contradiction. □

Following Theorem 1, we recall the observation in [1] that we can view the deletion of a chromosomal interval replacing adjacencies $\{v, w\}$ and $\{x, y\}$ with the single adjacency as a fission replacing $\{v, w\}$ and $\{x, y\}$ by the two adjacencies $\{w, x\}$ and $\{v, y\}$, thus forming a circular chromosome containing $\{v, y\}$ that is scheduled for later removal. By viewing this operation as a DCJ, we establish a bijective correspondence between the deletions of a minimum cost transformation of Π into Γ (having no singletons) and a collection of chromosomes sharing no genes with Π. (Insertions are handled symmetrically.)

Therefore, define a **completion** of genomes Π and Γ as a pair of genomes (Π', Γ') such that Π is a subgraph of Π', Γ is a subgraph of Γ', and $g(\Pi') = g(\Gamma') = \mathcal{G}$. Each of $\Pi' - \Pi$ and $\Gamma' - \Gamma$ is formed of alternating cycles called **new chromosomes**; in other words, the chromosomes of Π' comprise the chromosomes of Π in addition to some new chromosomes that are disjoint from Π.

By our bijective correspondence, we use $\text{ind}(\Pi', \Gamma')$ to denote the total number of new chromosomes of Π' and Γ'. We will amortize the cost of a deletion by charging unit cost for the DCJ that produces a new chromosome, followed by $1 - \omega$ for the removal of this chromosome, yielding our next result. For simplicity, we henceforth set $N = |\mathcal{V}|/2$, the number of genes of Π and Γ.

Theorem 3. *If* $\text{sing}(\Pi, \Gamma) = 0$, *then*

$$d_\omega(\Pi, \Gamma) = \min_{(\Pi', \Gamma')} \{d(\Pi', \Gamma') + (\omega - 1) \cdot \text{ind}(\Pi', \Gamma')\} \tag{2}$$

$$= N - \max_{(\Pi', \Gamma')} \{c(\Pi', \Gamma') + (1 - \omega) \cdot \text{ind}(\Pi', \Gamma')\} \tag{3}$$

where the optimization is taken over all completions of Π *and* Γ. $\qquad\square$

A completion (Π^*, Γ^*) is called **optimal** if it achieves the maximum in (3). We plan to use Theorem 3 to construct an optimal completion for genomes lacking singletons. Once we have formed an optimal completion (Π^*, Γ^*), we can simply invoke the $O(N)$-time sorting algorithm described in [3] to transform Π^* into Γ^* via a minimum collection of DCJs.

4   DCJ-Indel Sorting Genomes without Singletons

Define a node $v \in \mathcal{V}$ to be **Π-open** (**Γ-open**) if $v \notin \Pi$ ($v \notin \Gamma$). When forming adjacencies of Π^* (Γ^*), we connect pairs of Π-open (Γ-open) nodes. Given genomes Π and Γ with unequal gene content, we can still define the breakpoint graph $B(\Pi, \Gamma)$ as the edge-disjoint union of $a(\Pi)$ and $a(\Gamma)$; however, because the adjacencies of Π and Γ are not necessarily perfect matchings on $\mathcal{V}$, $B(\Pi, \Gamma)$ may contain paths (of positive length) in addition to cycles.

We can view the problem of constructing an optimal completion (Π^*, Γ^*) as adding edges to $B(\Pi, \Gamma)$ to form $B(\Pi^*, \Gamma^*)$. Our hope is to construct these edges via direct analysis of $B(\Pi, \Gamma)$. First, note that cycles of $B(\Pi, \Gamma)$ must embed as cycles of $B(\Pi^*, \Gamma^*)$, whereas odd-length paths of $B(\Pi, \Gamma)$ end in either two Π-open nodes or two Γ-open nodes, and even-length paths of $B(\Pi, \Gamma)$ end in a Π-open node and a Γ-open node. The paths of $B(\Pi, \Gamma)$ must be **linked** in some way by edges in $a(\Pi^*) - a(\Pi)$ or $a(\Gamma^*) - a(\Gamma)$ to form cycles alternating between edges of $a(\Pi^*)$ and $a(\Gamma^*)$. Our basic intuition is to do so in such a way as to create as many cycles as possible, at least when ω is small; this intuition is confirmed by the following two results.

Proposition 4. *If* $0 < \omega < 2$ *and* $\text{sing}(\Pi, \Gamma) = 0$, *then for any optimal completion* (Π^*, Γ^*) *of* Π *and* Γ, *every path of length* $2k - 1$ *in* $B(\Pi, \Gamma)$ *($k \geq 1$) embeds into a cycle of length* $2k$ *in* $B(\Pi^*, \Gamma^*)$.

Proof. Let P be a path of length $2k - 1$ in $B(\Pi, \Gamma)$. Without loss of generality, assume that P has Π-open nodes v and w as endpoints. Suppose that for some completion (Π', Γ'), P does not embed into a cycle of length $2k$ in $B(\Pi', \Gamma')$ (i.e.,

$\{v, w\}$ is not an adjacency of Π'); in this case, we must have distinct adjacencies $\{v, x\}$ and $\{w, y\}$ in Π' belonging to the same cycle of $B(\Pi, \Gamma)$.

Consider the completion Π'' that is formed from Π' by replacing $\{v, x\}$ and $\{w, y\}$ with $\{v, w\}$ and $\{x, y\}$. It is clear that $c(\Pi'', \Gamma') = c(\Pi', \Gamma') + 1$. Furthermore, since we have changed only two edges of Π' to produce Π'', we have increased or decreased the number of new chromosomes of Π'' by at most 1, and so $|\text{ind}(\Pi'', \Gamma') - \text{ind}(\Pi', \Gamma')| < 1$. Thus, it follows from (3) that (Π', Γ') cannot be optimal. □

As a result of Proposition 4, when $0 < \omega < 2$, any cycle of $B(\Pi^*, \Gamma^*)$ that is not induced from a cycle or odd-length path of $B(\Pi, \Gamma)$ must be a **k-bracelet**, which contains k even-length paths of $B(\Pi, \Gamma)$, where k is even. We use the term **bracelet links** to refer to adjacencies of a bracelet belonging to new chromosomes; each k-bracelet in $B(\Pi^*, \Gamma^*)$ contains $k/2$ bracelet links from $\Pi^* - \Pi$ and $k/2$ bracelet links from $\Gamma^* - \Gamma$. According to (3), we need to make $c(\Pi^*, \Gamma^*)$ large, which means that when indels are inexpensive, we should have bracelets containing as few bracelet links as possible.

Proposition 5. *If $0 < \omega < 2$ and $\text{sing}(\Pi, \Gamma) = 0$, then for any optimal completion (Π^*, Γ^*) of Π and Γ, all of the even-length paths of $B(\Pi, \Gamma)$ embed into 2-bracelets of $B(\Pi^*, \Gamma^*)$.*

Proof. Suppose that a completion (Π', Γ') of Π and Γ contains a k-bracelet for $k \geq 4$. This bracelet must contain two bracelet adjacencies $\{v, w\}$ and $\{x, y\}$ belonging to $a(\Pi')$, where these four nodes are contained in the order (v, w, x, y) in the bracelet. Consider the genome Π'' that is obtained from Π' by replacing $\{v, w\}$ and $\{x, y\}$ with $\{v, y\}$ and $\{w, x\}$. As in the proof of Proposition 4, $c(\Pi'', \Gamma') = c(\Pi', \Gamma') + 1$ and $|\text{ind}(\Pi'', \Gamma') - \text{ind}(\Pi', \Gamma')| < 1$, so that (Π', Γ') cannot be optimal. □

The conditions provided by the previous two propositions are very strong. To resolve the case that $0 < \omega < 2$, note that if we must link the endpoints of any odd-length path in $B(\Pi, \Gamma)$ to construct an optimal completion, then we may first create some new chromosomes before dealing with the case of even-length paths. Let $k_\Gamma(\Pi)$ be the number of new chromosomes formed by linking the endpoints of odd-length paths of $B(\Pi, \Gamma)$ that end with Π-open nodes, and set $k(\Pi, \Gamma) = k_\Gamma(\Pi) + k_\Pi(\Gamma)$. After linking the endpoints of odd-length paths, if we can link pairs of even-length paths (assuming any exist) into 2-bracelets so that one *additional* new chromosome is created in each of Π^* and Γ^*, then we will have constructed an optimal completion. This construction is guaranteed by the following proposition.

Proposition 6. *If $0 < \omega < 2$ and $\text{sing}(\Pi, \Gamma) = 0$, then any optimal completion (Π^*, Γ^*) of Π and Γ has the property that one new chromosome of Π^* (Γ^*) contains all of the bracelet adjacencies of Π^* (Γ^*).*

Proof. By Proposition 4, we may assume that we have started forming an optimal completion (Π^*, Γ^*) by linking the endpoints of any odd-length paths in

$B(\Pi, \Gamma)$ to each other. Given any even-length path P in $B(\Pi, \Gamma)$, there is exactly one other even-length path P_1 that would form a new chromosome in Γ^* if linked with P, and exactly one other even-length path P_2 in $B(\Pi, \Gamma)$ that would form a new chromosome in Π^* if linked with P (P_1 and P_2 may be the same). As long as there are more than two other even-length paths to choose from, we can simply link P to any path other than P_1 or P_2. We then iterate this process until two even-length paths remain, which we link to complete the construction of Π^* and Γ^*; each of these genomes has one new chromosome containing all of that genome's bracelet adjacencies. □

It is easy to see that the conditions in the preceding three propositions are sufficient (but not necessary) when constructing an optimal completion for the boundary cases $\omega = 0$ and $\omega = 2$. We are now ready to state our first major result with respect to DCJ-indel sorting.

Algorithm 7. *When $0 \leq \omega \leq 2$ and $\mathrm{sing}(\Pi, \Gamma) = 0$, the following algorithm solves the problem of DCJ-indel sorting Π into Γ in $O(N)$ time.*

1. *Link the endpoints of any odd-length path in $B(\Pi, \Gamma)$, which may create some new chromosomes in Π^* and Γ^*.*
2. *Arbitrarily select an even-length path P of $B(\Pi, \Gamma)$ (if one exists).*
   (a) *If there is more than one additional even-length path in $B(\Pi, \Gamma)$, link P to an even-length path that produces no new chromosomes in Π^* or Γ^*.*
   (b) *Otherwise, link the two remaining even-length paths in $B(\Pi, \Gamma)$ to form a new chromosome in each of Π^* and Γ^*.*
3. *Iterate Step 2 until no even-length paths of $B(\Pi, \Gamma)$ remain. The resulting completion is (Π^*, Γ^*).*
4. *Apply the $O(N)$-time algorithm for DCJ sorting from [11] to transform Π^* into Γ^*.*

Let $p_{\mathrm{odd}}(\Pi, \Gamma)$ and $p_{\mathrm{even}}(\Pi, \Gamma)$ equal the number of odd- and even-length paths in $B(\Pi, \Gamma)$, respectively. The optimal completion (Π^*, Γ^*) constructed by Algorithm 7 has the following properties:

$$c(\Pi^*, \Gamma^*) = c(\Pi, \Gamma) + p_{\mathrm{odd}}(\Pi, \Gamma) + \frac{p_{\mathrm{even}}(\Pi, \Gamma)}{2} \tag{4}$$

$$\mathrm{ind}(\Pi^*, \Gamma^*) = k(\Pi, \Gamma) + \min\{2, p_{\mathrm{even}}(\Pi, \Gamma)\} \tag{5}$$

These formulas, when combined with Theorem 3, yield a formula for the DCJ-indel distance as a function of Π, Γ, and ω alone.

Corollary 8. *If $0 \leq \omega \leq 2$ and $\mathrm{sing}(\Pi, \Gamma) = 0$, the DCJ-indel distance between Π and Γ is given by the following equation:*

$$d_\omega(\Pi, \Gamma) = N - \left[\left(c(\Pi, \Gamma) + p_{\mathrm{odd}}(\Pi, \Gamma) + \frac{p_{\mathrm{even}}(\Pi, \Gamma)}{2} \right) + (1 - \omega) \cdot \right.$$
$$\left. \left(k(\Pi, \Gamma) + \min\{2, p_{\mathrm{even}}(\Pi, \Gamma)\} \right) \right] \tag{6}$$

□

We now turn our attention to the case $\omega > 2$. Intuitively, as ω grows, we should witness fewer indels. Let $\delta_\Gamma(\Pi)$ be equal to 1 if $g(\Pi) - g(\Gamma)$ is nonempty and 0 otherwise; then, set $\delta(\Pi, \Gamma) = \delta_\Gamma(\Pi) + \delta_\Pi(\Gamma)$. Note that $\delta(\Pi, \Gamma)$ is a lower bound on the number of indels in any transformation of Π into Γ. The following result shows that in the absence of singletons, this bound is achieved by every minimum-cost transformation when $\omega > 2$.

Theorem 9. *If $\omega > 2$ and $\mathrm{sing}(\Pi, \Gamma) = 0$, then any minimum-cost transformation of Π into Γ has at most one insertion and at most one deletion. As a result,*

$$d_\omega(\Pi, \Gamma) = N - \max_{\mathrm{ind}(\Pi', \Gamma') = \delta(\Pi, \Gamma)} \{c(\Pi', \Gamma') + (1 - \omega) \cdot \delta(\Pi, \Gamma)\} \ . \qquad (7)$$

Proof. Suppose for the sake of contradiction that $\mathbb{T}$ is a minimum-cost transformation of Π into Γ and that (without loss of generality) $\mathbb{T}$ contains two deletions of chromosomal intervals P_1 and P_2, costing 2ω. Say that one of these deletions replaces adjacencies $\{v, w\}$ and $\{x, y\}$ with $\{v, y\}$ (deleting the interval connecting w to x) and the other deletion replaces adjacencies $\{a, b\}$ and $\{c, d\}$ with $\{a, d\}$ (deleting the interval connecting b to c).

Consider a second transformation $\mathbb{U}$ that is otherwise identical to $\mathbb{T}$, except that it replaces the deletions of P_1 and P_2 with three operations. First, a DCJ replaces $\{v, w\}$ and $\{a, b\}$ with $\{v, a\}$ and $\{w, b\}$; the new adjacency $\{w, b\}$ joins P_1 and P_2 into a single chromosomal interval P. Second, a deletion removes P and replaces adjacencies $\{c, d\}$ and $\{x, y\}$ with the single adjacency $\{d, y\}$. Third, another DCJ replaces $\{v, a\}$ and $\{d, y\}$ with the adjacencies $\{v, y\}$ and $\{a, d\}$, yielding the same genome as the first scenario at a cost of $2 + \omega$. Because $\mathbb{U}$ is otherwise the same as $\mathbb{T}$, $\mathbb{U}$ will have strictly lower cost precisely when $\omega > 2$, in which case $\mathbb{T}$ cannot have minimum cost. □

One can verify that the condition in Theorem 9 is sufficient but not necessary to guarantee a minimum-cost transformation when $\omega = 2$. Furthermore, a consequence of Theorem 9 is that the optimal completion is independent of the value of ω. In other words, if a completion achieves the maximum in (7), then this completion is automatically optimal for *all* values of $\omega \geq 2$.

Fortunately, Algorithm 7 already describes the construction of a completion (Π', Γ') that is optimal when $\omega = 2$. Of course, we cannot guarantee that this completion has the desired property that $\mathrm{ind}(\Pi', \Gamma') = \delta(\Pi, \Gamma)$. However, if $\mathrm{ind}(\Pi', \Gamma') > \delta(\Pi, \Gamma)$, then we can apply $\mathrm{ind}(\Pi', \Gamma') - \delta(\Pi, \Gamma)$ total fusions to Π' and Γ' in order to obtain a different completion (Π^*, Γ^*). Each of these fusions reduces the number of new chromosomes by 1 and (by (3)) must also decrease the number of cycles in the breakpoint graph by 1, since (Π', Γ') is optimal for $\omega = 2$. As a result, $c(\Pi^*, \Gamma^*) - \mathrm{ind}(\Pi^*, \Gamma^*) = c(\Pi', \Gamma') - \mathrm{ind}(\Pi', \Gamma')$. Thus, (Π^*, Γ^*) is optimal for $\omega = 2$, and since $\mathrm{ind}(\Pi^*, \Gamma^*) = \delta(\Pi, \Gamma)$, we know that (Π^*, Γ^*) must be optimal for any $\omega > 2$ as already noted. This discussion immediately implies the following algorithm.

Algorithm 10. *If $\omega \geq 2$ and $\text{sing}(\Pi, \Gamma) = 0$, then the following algorithm solves the problem of DCJ-indel sorting Π into Γ in $O(N)$ time.*

1. *Follow the first three steps of Algorithm 7 to construct a completion (Π', Γ') that is optimal for $\omega = 2$.*
2. *Apply a total of $\text{ind}(\Pi', \Gamma') - \delta(\Pi, \Gamma)$ fusions to Π' and Γ' in order to produce a completion (Π^*, Γ^*) having $\text{ind}(\Pi^*, \Gamma^*) = \delta(\Pi, \Gamma)$.*
3. *Apply the $O(N)$-time algorithm for DCJ sorting from [11] to transform Π^* into Γ^*. Any DCJ involving a new chromosome can be viewed as an indel.*

The optimal completion (Π^*, Γ^*) returned by Algorithm 10 has the property that

$$c(\Pi^*, \Gamma^*) = c(\Pi, \Gamma) + p_{\text{odd}}(\Pi, \Gamma) + \frac{p_{\text{even}}(\Pi, \Gamma)}{2} - \left[\text{ind}(\Pi', \Gamma') - \delta(\Pi, \Gamma) \right] , \quad (8)$$

where (Π', Γ') is the optimal completion for $\omega = 2$ returned by Algorithm 7. Combining this equation with (5) and (7) yields a closed formula for the DCJ-indel distance when $\omega > 2$ in the absence of singletons.

Corollary 11. *If $\omega \geq 2$ and $\text{sing}(\Pi, \Gamma) = 0$, then the DCJ-indel distance between Π and Γ is given by the following equation:*

$$
\begin{aligned}
d_\omega(\Pi, \Gamma) = N - \Bigg[\bigg(c(\Pi, \Gamma) + p_{\text{odd}}(\Pi, \Gamma) + \frac{p_{\text{even}}(\Pi, \Gamma)}{2} - k(\Pi, \Gamma) - \\
\min\left\{2, p_{\text{even}}(\Pi, \Gamma)\right\} \bigg) + (2 - \omega) \cdot \delta(\Pi, \Gamma) \Bigg]
\end{aligned}
\quad (9)
$$

$\square$

5   Incorporating Singletons into DCJ-Indel Sorting

We have thus far avoided genome pairs with singletons because Theorem 1, which underlies the main results in the preceding section, only applied in the absence of singletons. Yet fortunately, genomes with singletons will be relatively easy to incorporate into a single DCJ-indel sorting algorithm. As we might guess, different values of ω produce different results.

Theorem 12. *If $\Pi^\emptyset$ and $\Gamma^\emptyset$ are produced from genomes Π and Γ by removing all singletons, then*

$$
\begin{aligned}
d_\omega(\Pi, \Gamma) = d_\omega(\Pi^\emptyset, \Gamma^\emptyset) + \min\left\{1, \omega\right\} \cdot \text{sing}(\Pi, \Gamma) + \max\left\{0, \omega - 1\right\} \cdot \\
\Big[(1 - \delta_{\Gamma^\emptyset}(\Pi^\emptyset)) \cdot \min\left\{1, \text{sing}_\Gamma(\Pi)\right\} + \\
(1 - \delta_{\Pi^\emptyset}(\Gamma^\emptyset)) \cdot \min\left\{1, \text{sing}_\Pi(\Gamma)\right\} \Big]
\end{aligned}
\quad (10)
$$

Proof. Any transformation of $\Pi^\emptyset$ into $\Gamma^\emptyset$ can be supplemented by the deletion of each singleton of Π and the insertion of each singleton of Γ to yield a collection of DCJs and indels transforming Π into Γ. As a result, for any value of ω,

$$d_\omega(\Pi, \Gamma) \leq d_\omega(\Pi^\emptyset, \Gamma^\emptyset) + \omega \cdot \text{sing}(\Pi, \Gamma) \ . \tag{11}$$

Next, we will view an arbitrary transformation $\mathbb{T}$ of Π into Γ as a sequence $(\Pi_0, \Pi_1, \ldots, \Pi_n)$ $(n \geq 1)$, where $\Pi_0 = \Pi$, $\Pi_n = \Gamma$, and Π_{i+1} is obtained from Π_i as the result of a single DCJ or indel. Consider a sequence $(\Pi_0^\emptyset, \Pi_1^\emptyset, \ldots, \Pi_n^\emptyset)$, where $\Pi_i^\emptyset$ is constructed from Π_i by removing the subgraph of Π_i induced by the nodes of the singletons of Π and Γ under the stipulation that whenever we remove a path P connecting v to w, we replace adjacencies $\{v, x\}$ and $\{w, y\}$ in Π_i with $\{x, y\}$ in $\Pi_i^\emptyset$. Certainly, $\Pi_0^\emptyset = \Pi^\emptyset$ and $\Pi_n^\emptyset = \Gamma^\emptyset$. Furthermore, for every i in range, if $\Pi_{i+1}^\emptyset$ is not the result of a DCJ or indel applied to $\Pi_i^\emptyset$, then $\Pi_{i+1}^\emptyset = \Pi_i^\emptyset$. Thus, $(\Pi_0^\emptyset, \Pi_1^\emptyset, \ldots, \Pi_n^\emptyset)$ can be viewed as encoding a transformation of $\Pi^\emptyset$ into Γ using *at most* n DCJs and indels. One can verify that $\Pi_{i+1}^\emptyset = \Pi_i^\emptyset$ precisely when Π_{i+1} is produced from Π_i either by a DCJ that involves an adjacency belonging to a singleton or by an indel containing genes that all belong to singletons. At least $\text{sing}(\Pi, \Gamma)$ such operations must always occur in $\mathbb{T}$; hence,

$$d_\omega(\Pi, \Gamma) \geq d_\omega(\Pi^\emptyset, \Gamma^\emptyset) + \min\{1, \omega\} \cdot \text{sing}(\Pi, \Gamma) \ . \tag{12}$$

In the case that $\omega \leq 1$, the bounds in (11) and (12) immediately yield (10).

Assume, then, that $\omega > 1$. If $\delta_{\Gamma^\emptyset}(\Pi^\emptyset) = 0$, then $g(\Pi^\emptyset) \subseteq g(\Gamma^\emptyset)$, meaning that every deleted gene of Π must belong to a singleton of Π. In this case, the total cost of removing any singletons of Π is trivially minimized by $\text{sing}_\Gamma(\Pi) - 1$ fusions consolidating the singletons of Π into a single chromosome, followed by the deletion of this chromosome. Symmetric reasoning applies to the singletons of Γ if $\delta_{\Pi^\emptyset}(\Gamma^\emptyset) = 0$.

On the other hand, assume that $\omega > 1$ and that $\delta_{\Gamma^\emptyset}(\Pi^\emptyset) = 1$, so that $g(\Pi^\emptyset) - g(\Gamma^\emptyset)$ is nonempty. In this case, if Π has any singletons, then we can create a minimum-cost transformation by applying $\text{sing}_\Gamma(\Pi) - 1$ fusions consolidating the singletons of Π into a single chromosome, followed by another fusion that consolidates these chromosomes into a chromosomal interval of Π that is about to be deleted. Symmetric reasoning applies to the singletons of Γ if $\delta_{\Pi^\emptyset}(\Gamma^\emptyset) = 1$.

Regardless of the particular values of $\delta_{\Gamma^\emptyset}(\Pi^\emptyset)$ and $\delta_{\Pi^\emptyset}(\Gamma^\emptyset)$, we will obtain the formula in (10). □

This proof immediately provides us with an algorithm incorporating the case of genomes with singletons into the existing DCJ-indel sorting framework.

Algorithm 13. *The following algorithm solves the general problem of DCJ-indel sorting genomes Π and Γ for any indel cost $\omega \geq 0$ in $O(N)$ time.*

1. *Case 1: $\omega \leq 1$.*
   (a) *Delete any singletons of Π, then insert any singletons of Γ.*

    (b) *Apply Algorithm 7 to transform the resulting genome into Γ.*

2. *Case 2: $\omega > 1$.*

    (a) *If Π has any singletons, apply $\mathrm{sing}_\Gamma(\Pi) - 1$ fusions to consolidate the singletons of Π into a single chromosome C_Π.*

       i. *If $g(\Pi^\emptyset) \subseteq g(\Gamma^\emptyset)$, delete C_Π.*

       ii. *Otherwise, save C_Π for later.*

    (b) *If Γ has any singletons, apply $\mathrm{sing}_\Pi(\Gamma) - 1$ fusions to consolidate the singletons of Γ into a single chromosome C_Γ.*

       i. *If $g(\Gamma^\emptyset) \subseteq g(\Pi^\emptyset)$, delete C_Γ.*

       ii. *Otherwise, save C_Γ for later.*

    (c) *Apply a sorting algorithm as needed to construct an optimal completion (Π^*, Γ^*) for $\Pi^\emptyset$ and $\Gamma^\emptyset$.*

       i. *If $1 < \omega \leq 2$, apply the first three steps of Algorithm 7.*

       ii. *If $\omega > 2$, apply the first two steps of Algorithm 10.*

    (d) *If $g(\Pi^\emptyset) - g(\Gamma^\emptyset)$ is nonempty, apply a fusion incorporating C_Π into a new chromosome of Π^*. If $g(\Gamma^\emptyset) - g(\Pi^\emptyset)$ is nonempty, apply a fusion incorporating C_Γ into a new chromosome of Γ^*.*

    (e) *Apply the final step of Algorithm 7 or Algorithm 10, depending on the value of ω.*

6 Conclusion

With the problem of DCJ-indel sorting genomes with circular chromosomes unified under a general model, we see three obvious future applications of this work.

First, an extension of these results for genomes with linear chromosomes would prevent us from having to first circularize linear chromosomes when comparing eukaryotic genomes. This work promises to be extremely tedious (if it is indeed possible) without offering dramatic new insights.

Second, we would like to implement the linear-time method for DCJ-indel sorting described in Algorithm 13 and publish the code publicly. Evolutionary study analysis on real data would hopefully determine appropriate choices of ω.

Third, we are currently attempting to extend these results to fully characterize the space of all solutions to DCJ-indel sorting, which would generalize the result in [6] to arbitrary values of ω.

References

1. Arndt, W., Tang, J.: Emulating insertion and deletion events in genome rearrangement analysis. In: 2011 IEEE International Conference on Bioinformatics and Biomedicine, pp. 105–108 (2011)
2. Bafna, V., Pevzner, P.A.: Genome rearrangements and sorting by reversals. SIAM J. Comput. 25(2), 272–289 (1996)
3. Bergeron, A., Mixtacki, J., Stoye, J.: A unifying view of genome rearrangements. In: Bücher, P., Moret, B.M.E. (eds.) WABI 2006. LNCS (LNBI), vol. 4175, pp. 163–173. Springer, Heidelberg (2006)

4. Braga, M., Machado, R., Ribeiro, L., Stoye, J.: On the weight of indels in genomic distances. BMC Bioinformatics 12(suppl. 9), S13 (2011)
5. Braga, M.D.V., Willing, E., Stoye, J.: Genomic distance with DCJ and indels. In: Moulton, V., Singh, M. (eds.) WABI 2010. LNCS, vol. 6293, pp. 90–101. Springer, Heidelberg (2010)
6. Compeau, P.: DCJ-indel sorting revisited. Algorithms for Molecular Biology 8(1), 6 (2013)
7. Compeau, P.E.C.: A simplified view of DCJ-indel distance. In: Raphael, B., Tang, J. (eds.) WABI 2012. LNCS, vol. 7534, pp. 365–377. Springer, Heidelberg (2012)
8. Dobzhansky, T., Sturtevant, A.H.: Inversions in the chromosomes of drosophila pseudoobscura. Genetics 23(1), 28–64 (1938)
9. Fertin, G., Labarre, A., Rusu, I., Tannier, E., Vialette, S.: Combinatorics of Genome Rearrangements. MIT Press (2009)
10. da Silva, P.H., Braga, M.D.V., Machado, R., Dantas, S.: DCJ-indel distance with distinct operation costs. In: Raphael, B., Tang, J. (eds.) WABI 2012. LNCS, vol. 7534, pp. 378–390. Springer, Heidelberg (2012)
11. Yancopoulos, S., Attie, O., Friedberg, R.: Efficient sorting of genomic permutations by translocation, inversion and block interchange. Bioinformatics 21(16), 3340–3346 (2005)
12. Yancopoulos, S., Friedberg, R.: DCJ path formulation for genome transformations which include insertions, deletions, and duplications. Journal of Computational Biology 16(10), 1311–1338 (2009)

Efficient Local Alignment Discovery
amongst Noisy Long Reads

Gene Myers[*]

MPI for Molecular Cell Biology and Genetics, 01307 Dresden, Germany
myers@mpi-cbg.de

Abstract. Long read sequencers portend the possibility of producing reference quality genomes not only because the reads are long, but also because sequencing errors and read sampling are almost perfectly random. However, the error rates are as high as 15%, necessitating an efficient algorithm for finding local alignments between reads at a 30% difference rate, a level that current algorithm designs cannot handle or handle inefficiently. In this paper we present a very efficient yet highly sensitive, threaded *filter*, based on a novel *sort and merge* paradigm, that proposes *seed points* between pairs of reads that are likely to have a significant local alignment passing through them. We also present a *linear* expected-time heuristic based on the classic $O(nd)$ difference algorithm [1] that finds a local alignment passing through a seed point that is exceedingly sensitive, failing but once every billion base pairs. These two results have been combined into a software program we call DALIGN that realizes the fastest program to date for finding overlaps and local alignments in very noisy long read DNA sequencing data sets and is thus a prelude to *de novo* long read assembly.

1   Introduction and Summary

The PacBio RS II sequencer is the first operational "long read" DNA sequencer [2]. While its error rate is relatively high ($\epsilon = 12\text{-}15\%$ error), it has two incredibly powerful offsetting properties, namely, that (a) the set of reads produced is a nearly Poisson sampling of the underlying genome, and (b) the location of errors within reads is truly randomly distributed. Property (a), by the Poisson theory of Lander and Waterman [3], implies that for any minimum target coverage level k, there exists a level of sequencing coverage c that guarantees that every region of the underlying genome is covered k times. Property (b), from the early work of Churchill and Waterman [4], implies that the accuracy of the consensus sequence of k such sequences is $O(\epsilon^k)$ which goes to 0 as k increases. Therefore, provided the reads are long enough that repetitive genome elements do not confound assembling them, then in principle a (near) perfect *de novo* reconstruction of a genome at any level of accuracy is possible given *enough* coverage c.

These properties of the reads are in stark contrast to those of existing technologies where neither property is true. All previous technologies make reproducible

---

[*] Supported by the Klaus Tschira Stiftung, Heidelberg, Germany

D. Brown and B. Morgenstern (Eds.): WABI 2014, LNBI 8701, pp. 52–67, 2014.

sequencing errors. A typical rate of occurrence for these errors is about 10^{-4} implying at best a Q40 reconstruction is possible, whereas in principle any desired reconstruction accuracy is possible with the long reads, e.g., a Q60 reconstruction has been demonstrated for *E. coli* [5]. All earlier technologies also exhibit clear sampling biases, typically due to a biased amplification or selection step, implying that many regions of a target genome are not sequenced. For example, some PCR-based instruments often fail to sequence GC rich stretches. So because their error and sampling are unbiased, the new long read technologies are poised to enable a dramatic shift in the state of the art of *de novo* DNA sequencing.

The questions then are (a) what level of coverage c is required for great assembly, i.e. how cost-effectively can one get near the theoretical ideal above, and (b) how does one build an assembler that works with such high error rates and long reads? The second question is important because most current assemblers do not work on such data as they assume much lower error rates and much shorter reads, e.g. error rates less than 2% and read lengths of 100-250bp. Moreover, the algorithms within these assemblers are specifically tuned for these operating points and some approaches, such as the de-Bruijn graph [6] would catastrophically fail at rates over 10%.

Finding overlaps is typically the first step in an overlap-layout-consensus (OLC) assembler design [7] and is the efficiency bottleneck for such assemblers. In this paper, we develop an efficient algorithm and software for finding all significant local alignments between reads in the presence of the high error rates of the long reads. Finding local alignments is more general then finding overlaps, and we do so because it allows us to find repeats, chimers, undetected vector sequence and other artifacts that must be detected in order to achieve near perfect assemblies. To this authors knowledge, the only previous algorithm and software that can effectively accommodate the level of error in question is BLASR [8] which was original designed as a tool to map long reads to a reference genome, but can also be used for the assembly problem. Empirically our program, DALIGN, is more sensitive while being typically 20 to 40 times faster depending on the data set.

We make use of the same basic filtration concept as BLASR, but realize it with a series of highly optimized threaded radix sorts (as opposed to a BWT index [9]). While we did not make a direct comparison here, we believe the cache coherence and thread ability of the simpler sorting approach is more time efficient then using a more sophisticated but cache incoherent data structure such as a Suffix Array or BWT index. But the real challenge is improving the speed of finding local alignments at a 30-40% difference rate about a seed hit from the filter, as this step consumes the majority of the time, e.g. 85% or more in the case of DALIGN. To find overlaps about a seed hit, we use a novel method of adaptively computing furthest reaching waves of the classic $O(nd)$ algorithm [1] augmented with information that describes the match structure of the last p columns of the alignment leading to a given furthest reaching point. Each wave on average contains a small number of points, e.g. 8, so that in effect an alignment is detected in time linear in the number of columns in the alignment.

In practice DALIGN achieved the following CPU and wall-clock times on 3 publicly available PacBio data sets [10]. The 48X *E coli* data set can be compared against itself in less than 5.4 wall clock minutes on a Macbook Pro with 16Gb of memory and a 4-core i7-processor. The 89X Arabadopsis data set can be processed in 71 CPU hours or 10 wall-clock minutes on our modest 480 core HPC cluster, where each node is a pair of 6-core Intel Xeon E5-2640's at 2.5GHz with 128Gb (albeit only 50-60Gb is used per node). Finally on the 54X human genome data set, 15,600 CPU hours or 32-33 wall-clock hours are needed. Thus our algorithm and software enables the assembly of gigabase genomes in a "reasonable" amount of compute time (e.g., compared to the 404,000 CPU hours reported for BLASR).

2 Preliminaries: Edit Graphs, Alignments, Paths, and F.R.-Waves

DALIGN takes as input a block $\mathcal{A}$ of M long reads $A^1, A^2, \ldots A^M$ and another block $\mathcal{B}$ of N long reads $B^1, B^2, \ldots B^N$ over alphabet $\Sigma = 4$, and seeks read subset pairs $P = (a, i, g) \times (b, j, h)$ such that $len(P) = ((g - i) + (h - j))/2 \geq \tau$ and the optimal alignment between $A^a[i+1, g]$ and $B^b[j+1, h]$ has no more than $2\epsilon \cdot len(P)$ differences where a difference can be either an insertion, a deletion, or a substitution. Both τ and ϵ are user settable parameters, where we call τ the minimum alignment length and ϵ the average error rate. We further will speak of $1 - 2\epsilon$ as the *correlation* or *percent identity* of the alignment. It will also be convenient throughout to introduce $\Sigma_{\mathcal{A}} = \sum_{a=1}^M |A^a|$, the total number of base pairs in $\mathcal{A}$, and $\max_{\mathcal{A}} = \max_a |A^a|$ the length of the longest read in $\mathcal{A}$.

Most readers will recall that an *edit graph* for read $A = a_1 a_2 \ldots a_m$ versus $B = b_1 b_2 \ldots b_n$ is a graph with an $(m+1) \times (n+1)$ array of vertices $(i, j) \in [0, M] \times [0, N]$ and the following edges:

(a)   *deletion* edges $(i - 1, j) \to (i, j)$ with label $\begin{bmatrix} a_i \\ - \end{bmatrix}$ if $i > 0$.

(b)   *insertion* edges $(i, j - 1) \to (i, j)$ with label $\begin{bmatrix} - \\ b_j \end{bmatrix}$ if $j > 0$.

(c)   *diagonal* edges $(i - 1, j - 1) \to (i, j)$ with label $\begin{bmatrix} a_i \\ b_j \end{bmatrix}$ if $i, j > 0$

A simple exercise in induction reveals that the sequence of labels on a path from (i, j) to (g, h) in the edit graph spells out an alignment between $A[i + 1, g]$ and $B[j + 1, h]$. Let a *match* edge be a diagonal edge for which $a_i = b_j$ and otherwise call the diagonal edge a *substitution* edge. Then if match edges have weight 0 and all other edges have weight 1, it follows that the weight of a path is the number of differences in the alignment it models. So our goal in edit graph terms is to find read subset pairs P such that $len(P) \geq \tau$ and the lowest scoring path between (i, j) and (g, h) in the edit graph of A^a versus B^b has cost no more than $2\epsilon \cdot len(P)$.

In 1986 we presented a simple $O(nd)$ algorithm [1] for comparing two sequences that centered on the idea of computing progressive "waves" of *furthest reaching* (f.r.) points. Starting from a point $\rho = (i, j)$ in *diagonal* $\kappa = i - j$ of the edit graph of two sequences, the goal is to find the longest possible paths starting at ρ, first with 0-differences, then with 1-differences, 2-differences, and so on. Note carefully that after d differences, the possible paths can end in diagonals $\kappa \pm d$. In each of these $2d+1$ diagonals we want to know the furthest point on the diagonal that can be reached from ρ with exactly d differences which we denote by $F_\rho(d, k)$. We call these points collectively the d-wave emanating from ρ and formally $W_\rho(d) = \{F_\rho(d, \kappa - d), \ldots F_\rho(d, \kappa + d)\}$. We will more briefly refer to $F_\rho(d, k)$ as the f.r. d-point on k where ρ will be implicitly understood from context. In the 1986 paper we proved that:

$$F(d, k) = Slide(k, max\{F(d-1, k-1)+(1, 0), F(d-1, k)+(1, 1), F(d-1, k+1)+(0, 1)\} \tag{1}$$

where $Slide(k, (i, j)) = (i, j) + max\{\Delta : a_{i+1}a_{i+2} \ldots a_{i+\Delta} = b_{j+1}b_{j+2} \ldots b_{j+\Delta}\}$. In words, the f.r. d-point on k can be computed by first finding the furthest of (a) the f.r. $(d - 1)$-point on $k - 1$ followed by an insertion, or (b) the f.r. $(d - 1)$-point on k followed by a substitution, or (c) the f.r. $(d - 1)$-point on $k + 1$ followed by a deletion, and thereafter progressing as far as possible along match edges (a "slide"). Formally a point (i, j) is furthest if its anti-diagonal, $i + j$, is greatest. Next, it follows easily that the best alignment between A and B is the smallest d such that $(m, n) \in W_{(0,0)}(d)$ where m and n are the length of A and B, respectively. So the $O(nd)$ algorithm simply computes d-waves from $(0, 0)$ in order of d until the goal point (m, n) is reached in the dth wave. It can further be shown that the expected complexity is actually $O(n + d^2)$ under the assumption that A and B are non-repetiitve sequences. In what follows we will be computing waves adaptively and in both the forward direction, as just described, and in the reverse direction, which is conceptually simply a matter of reversing the direction of the edges in the edit graph.

3   Rapid Seed Detection: Concept

Given blocks $\mathcal{A}$ and $\mathcal{B}$ of long, noisy reads, we seek to find local alignments between reads that are sufficiently long (parameter τ) and sufficiently stringent (parameter ϵ). For our application ϵ is much larger than typically contemplated in prior work, 10-15%, but the reads are very long, 10Kbp, so τ is large, 1 or 2Kbp. Here we build a filter that eliminates read pairs that cannot possibly contain a local alignment of length τ or more, by counting the number of conserved k-mers between the reads. A careful and detailed analysis of the statistics of conserved k-mers in the operating range of ϵ and τ required by long read data, has previously been given in the paper about the **BLASR** program [8]. So here we just illustrate the idea by giving a rough estimate assuming all k-mer matches are independent events. Under this simplifying assumption, it follows that a given k-mer is conserved with probability $\pi = (1 - 2\epsilon)^k$ and the number of conserved k-mers in an alignment of τ base pairs is roughly a Bernouilli distribution with rate π and thus an average of $\tau \cdot \pi$ conserved k-mers are expected between two reads that share a local alignment of length τ. As an example, if $k = 14$, $\epsilon = 15\%$, and $\tau = 1500$, then $\pi \approx .7^{14} = .0067$, we expect to have 10.0 14-mers conserved on average, and only .046% of the sought read pairs have 1 or fewer hits between them and only .26% have 2 or fewer hits. Thus a filter with expected sensitivity 99.74% examines only read pairs that have 3 or more conserved 14-mers. BLASR and DALIGN effectively use this strategy where one controls sensitivity and specificity by selecting k and the number of k-mers that must be found. Beyond this point our methods are completely different.

First, we improve specificity by (a) computing the number of conserved k-mers in bands of diagonals of width 2^s between two reads (as opposed to the entire reads) where a typical choice for s is 6, and (b) thresholding a hit on the number of bases, h, in conserved k-mers (as opposed to the number of k-mers). Condition (a) increases specificity as it limits the set of k-mers to be counted at a potentially slight loss of sensitivity because an alignment can have an insertion or deletion bias and so can drift across bands rather than staying in a single band. To understand condition (b) note that 3 consecutive matching k-mers involve a total of $k + 2$ matching bases, whereas 3 disjoint matching k-mers involve a total of $3k$ matching bases. Under our simplifying assumption the first situation happens with probability $\pi^{1+2/k}$ and the second with probability π^3, i.e. one is much more specific than the other. By counting the number of bases involved in k-mer hits we ensure that all filters hits have roughly the same statistical frequency.

There are many ways to find matching k-mers over an alphabet Σ, specifically of size 4 in this work, most involving indices such as Suffix Arrays [11] or BWT indices [9]. We have found in practice that a much simpler series of highly optimized sorts can similarly deliver the number of bases in k-mers in a given diagonal band between two reads. Given blocks $\mathcal{A}$ and $\mathcal{B}$ we proceed as follows:

1. Build the list $List_A = \{(kmer(A^a, i), a, i)\}_{a,i}$ of all k-mers of the $\mathcal{A}$ block and their positions, where $kmer(R, i)$ is the k-mer, $R[i - k + 1, i]$.
2. Similarly build the list $List_B = \{(kmer(B^b, j), b, j)\}_{b,j}$.
3. Sort both lists in order of their k-mers.
4. In a merge sweep of the two k-mer sorted lists build $List_M = \{(a, b, i, j) : kmer(A^a, i) = kmer(B^b, j)\}$ of read and position pairs that have the same k-mer.
5. Sort $List_M$ lexicographically on a, b, and i where a is most significant and i least.

To keep the following analysis simple, let us assume that the sizes of the two blocks are both roughly the same, say N. Steps 1 and 2 are easily seen to take $O(N)$ time and space. The sorts of steps 3 and 5 are in theory $O(L \log L)$ where L is the list size. The only remaining complexity question is how large is $List_M$. First note that there is a contribution (i) from k-mers that are purely random chance, and (ii) from conserved k-mers that are due to the reads actually being correlated. The first term is N^2/Σ^k as we expect to see a given k-mer N/Σ^k

times in each block. For case (ii), suppose that the data set is a c-fold covering of an underlying genome, and, in the worst case, the $\mathcal{A}$ and $\mathcal{B}$ blocks are the same block and contain all the data. The genome is then of size N/c and each position of the genome is covered by c reads by construction. Because c/π k-mers are on average conserved amongst the c reads covering a given position, there are thus $N/c \cdot (c/\pi)^2 = (Nc/\pi^2)$ matching k-mer pairs by non-random correlations. In most projects c is typically 50-100 whereas π is typically $1/100$ (e.g. $k = 14$ and $\epsilon = 15\%$) implying somewhat counter-intuitively that the non-random contribution is dominated by the random contributions! Thus $List_M$ is $O(N^2/\Sigma^k)$ in size and so in expectation the time for the entire procedure is dominated by Step 5 which takes $O(N^2 log N/\Sigma^k)$. Finally, suppose the total amount of data is M and we divide it into blocks of size Σ^k all of which are compared against each other. Then the time for each block comparison is $O(k\Sigma^k)$ using $O(\Sigma^k)$ space, that is linear time and space in the block size. Finally, there are M/Σ^k blocks implying the total time for comparing all blocks is $O(kM \cdot (M/\Sigma^k))$. So our filter, like all others, still has a quadratic component in terms of the number of occurrences of a given k-mer in a data set. With linear times indices such as BWT's the time can theoretically be improved by a factor of k. However, in practice the k arises from a radix sort that actually makes only $k/4$ passes and is so highly optimized, threaded, and cache coherent that we believe it likely outperforms a BWT approach by a considerable margin. At the current time all we can say is that DALIGN which *includes alignment finding* is 20-40 times faster than BLASR which uses a BWT (see Table 6).

For the sorted list $List_M$, note that all entries involving a given read pair (a, b) are in a single contiguous segment of the list after the sort in Step 5. Given parameters h and s, for each pair in such a segment, we place each entry (a, b, i, j) in both *diagonal bands* $d = \lfloor (i-j)/2^s \rfloor$ and $d+1$, and then determine the number of bases in the A-read covered by k-mers in each pair of bands diagonal band, i.e. $Count(a, b, d) = | \cup \{w(A^a, a, i) : (a, b, i, j) \in List_M$ and $\lfloor (i - j)/2^s \rfloor = d$ or $d + 1\}|$. Doing so is easy in linear time in the number of relevant entries as they are sorted on i. If $Count(a, b, d) \geq h$ then we have a hit and we call our local alignment finding algorithm to be described, with each position (i, j) in the bucket d *unless* the position i is already within the range of a local alignment found with an index pair searched before it. This completes the description of our filtration strategy and we now turn to its efficient realization.

4   Rapid Seed Detection: Algorithm Engineering

Todays processors have multiple cores and typically a 3-level cache hierarchy implying memory fetch times vary by up to 100 for L1 cache hits versus a miss at all three cache levels. We therefore seek a realization of the algorithm above that is parallel over T threads and is cache coherent. Doing so is easy for steps 1, 2, and 4 and we optimize the encoding of the lists by squeezing their elements into 64-bit integers. So the key problem addressed in the remainder of this section is how to realize a threadable, memory coherent sort of an array $src[0..N-1]$ of N 64-bit integers in steps 3 and 5.

We chose a *radix sort* [12] where each number is considered as a vector of $P = \lceil hbits/B \rceil$, B-bit digits, $(x_P, x_{P-1}, \ldots, x_1)$ and B is a free parameter to be optimally chosen empirically later. A radix sort sorts the numbers by *stably* sorting the array on the first B-bit digit x_1, then on the second x_2, and so on to x_P in P sorting passes. Each B-bit sort is achieved with a *bucket sort* [12] with 2^B buckets. Often this basic sort is realized with a linked list, but a much better strategy sequentially moves the integers in *src* into pre-computed segments, $trg[bucket[b] .. bucket[b + 1] - 1]$ of an auxiliary array $trg[0..N - 1]$ where, for the pth pass, $bucket[b] = \{i : src[i]_p < b\}$ for each $b \in [0, 2^B - 1]$. In code, the pth bucket sort is:

```
for i = 0 to N-1 do
  { b = src[i]_p
    trg[bucket[b]] = src[i]
    bucket[b] += 1
  }
```

Asymptotically the algorithm takes $O(P(N + 2^B))$ time but B and P are fixed small numbers so the algorithm is effectively $O(N)$.

While papers on threaded sorts are abundant [13], we never the less present our pragmatic implementation of a threaded radix sort, because it uses half the number of passes over the array that other methods use, and accomplishing this is non-trivial as follows. In order to exploit the parallelism of T threads, we let each thread sort a contiguous segment of size $part = \lceil N/T \rceil$ of the array *src* into the appropriate locations of *trg*. This requires that each thread $t \in [0, T-1]$ has its own bucket array $bucket[t]$ where now $bucket[t][b] = \{i : src[i] < b$ or $src[i] = b$ and $i/part < t\}$. In order to reduce *the number of sweeps over the arrays by half*, we produce the bucket array for the *next* pass while performing the current pass. But this is a bit complex because each thread must count the number of B-bit numbers in the next pass that will be handled by not only itself but every other thread separately! That is, if the number at index i will be at index j and bucket b in the next pass then the count in the current pass must be recorded not for the thread $i/part$ currently sorting the number, but for the thread $j/part$ that will sort the number in the next pass. To do so requires that we actually count the number of such events in $next[j/part][i/part][b]$ where now $next$ is a $T \times T \times 2^B$ array. It remains to note that when $src[i]$ is about to be moved in the pth pass, then $j = bucket[src[i]_p]$ and $b = src[i]_{p+1}$. The complete algorithm is presented below in C-style pseudo-code where unbound variables are assumed to vary over the range of the variable. It is easily seen to take $O(N/T + T^2)$ time assuming B and P are fixed.

```
int64 MASK = 2^B-1

sort_thread(int t, int bit, int N, int64 *src, int64 *trg, int *bucket, int *next)
{ for i = t*N to (t+1)*N-1 do
    { c = src[i]
      b = c >> bit
      x = bucket[b & MASK] += 1
      trg[x] = c
```

```
        next[x/N][(b >> B) & MASK] += 1
      }
}

int64 *radix_sort(int T, int N, int hbit, int64 src[0..N-1], int64 trg[0..N-1])
{ int bucket[0..T-1][0..2^B-1], next[0..T-1][0..T-1][0..2^B-1]
  part = (N-1)/T + 1
  for l = 0 to hbit-1 in steps of B do
    { if (l != lbit)
        bucket[t,b] = Sum_t next[u,t,b]
      else
        bucket[t,b] = | { i : i/part == t and src[i] & MASK == b  } |
      bucket[t,b] = Sum_u,(c<b) bucket[u,c] + Sum_(u<t) bucket[u,b]
      next[u,t,b] = 0
      in parallel: sort_thread(t,l,part,src,trg,bucket[t],next[t]])
      (src,trg) = (trg,src)
    }
  return src
}
```

We conclude by emphasizing why this approach to sorting is a particularly efficient realization of a very large array sort. Each bucket sort involves two small arrays *bucket* and *next* that will typically fit in the fastest L1 cache. Each bucket sort makes a single *sweep* through *src* while making 2^B sweeps through the bucket segments of *trg*. Thus $2^B + 1$ cache-coherent sweeps occur during each bucket sort pass. Each sweep can be prefetched as long as their number does not exceed the interleaving of the cache architecture. So the smaller B is the better the caching and prefetching behavior will be, but this is counter balanced by the increasing number of passes $hbit/B$ that are required. We found that on most processors, e.g. an Intel i7, the minimum total time for our radix sort occurs with $B = 8$ which conveniently is the number of bits in a byte.

The number of threads T to employ is a complex question despite the fact that there is no communication or synchronization required between threads in our algorithm and the non-parallel overhead is only $O(T^2)$. The reason is that every thread does not have its own set of caches. They generally have an independent L1 cache, but then share the L2 and L3 caches. This means that the actual number of sweeps taking place is $T(2^B + 1)$ which at the level of the L2 cache begins to induce interleaving interference. Nonetheless, speed up is still very good. For example, on a 4-core Intel i7, the speedup achieved was 3.6 !

5   Rapid Local Alignment Discovery from a Seed

We now turn to finding local alignments of length τ or more and correlation $1 - 2\epsilon$ or better given a seed-hit (i, j) between two reads A and B reported by the filter above. The basic idea is to compute f.r. waves in both the forward and reverse direction from the seed point $\rho = (i, j)$ (see Section 2). The problem of course is that the d-wave from ρ spans $2d + 1$ diagonals, that is, waves become wider and wider as one progresses away from ρ in each direction. We know that only one point in each wave will actually be in the local alignment ultimately reported, but we only know these points after all the relevant waves have been

computed. We use several strategies to trim the span of a wave by removing f.r.
points that are extremely unlikely to be in the desired local alignment.

A key idea is that a desired local alignment should not over any reasonable
segment have an exceedingly low correlation. To this end imagine keeping a bit
vector $B(d, k)$ that actually models the last, say $C = 60$ columns, of the best
path/alignment from ρ to a given f.r. point $F(d, k)$ in the d-wave. That is a
0 will denote a mismatch in a column of the alignment and a 1 will denote a
match. This is actually relatively easy to do: left-shift in a 0 when taking an
indel or substitution edge and then left-shift in a 1 with each matching edge
of a snake. One can further keep track of exactly how many matches $M(d, k)$
there are in the alignment by observing the bit that gets shifted out when a new
bit is shifted in. The pseudo-code below computes $W_\rho(d + 1)[low - 1, hgh + 1]$
from $W_\rho(d)[low, hgh]$ assuming that $[low, hgh] \subseteq [\kappa - d, \kappa + d]$ is the interval
of $W_\rho(d)$ that we have decided to retain (to be described below). Note that the
code computes the information for each wave in place within the arrays W, B,
and M where W simply records the B-coordinate, j, of each f.r. point (i, j) as we
know the diagonal k of the point, and hence that $i = j + k$.

```
MASKC  = 1 << (C-1)
W[low-2] = W[hgh+2] = W[hgh+1] = y = yp = -1
for k = low-1 to hgh+1 do
  { (ym,y,yp) = (y,yp+1,W[d+1]+1)
      if (ym = min(ym,y,yp))
          (y,m,b) = (ym,M[k-1],B[k-1])
      else if (yp = min(ym,y,yp)
          (y,m,b) = (yp,M[k+1],B[k+1])
      else
          (y,m,b) = (y,M[k],B[k])
      if (b & MASKC != 0)
        m -= 1
      b <<= 1
      while (B[y] == A[y+k])
        { y += 1
          if (b & MASKC == 0)
            m += 1
          b = (b << 1) | 1
        }
      (W[k],M[k],B[k]) = (y,m,b)
  }
```

A very simple principle for *trimming* a wave is to remove f.r. points for which
the last C columns of the alignment have less than say $\mathcal{M}$ matches, we call this
the *regional alignment quality*. For example, if $\epsilon = .15$ then one almost certainly
does not want a local alignment that contains a C column segment for which
$M[k] < .55C = 33$ if $C = 60$. A second trimming principle is to keep only f.r.
points which are within $\mathcal{L}$ anti-diagonals of the maximal anti-diagonal reached
by its wave. Intuitively, the f.r. point (i, j) on diagonal $k^\star$ *on the desired path* is
on a greater anti-diagonal $i + j$ than those of the points on either side of it in the

same wave, and as one progresses away from diagonal k^*, the anti-diagonal values of the wave recede rapidly, giving the wave the appearance of an arrowhead. The higher the correlation rate of the alignment, the sharper the arrow head becomes and the points far enough behind the tip of the arrow are almost certainly not points on an optimal local alignment. So for each portion of a wave computed from the previous trimmed wave, we trim away f.r. points from $[low-1, hgh+1]$ that either have $\mathtt{M}[j] < \mathcal{M}$ or $(2\mathtt{W}[k^*]+k^*) - (2\mathtt{W}[j]+j) > \mathcal{L}$. In the experimental section we show that $\mathcal{L} = 30$ is a universally good value for trimming.

While not a formal proof per se, the following argument explains why in the empirical results section we see that the average wave size $hgh-low$ is a constant for any fixed value of ϵ, and hence why the alignment finding algorithm is linear expected time in the alignment length. Imagine the extension of an f.r. point that is actually on the path of an alignment with correlation $1 - 2\epsilon$ or better. For the next wave, this point jumps forward one difference and then "slides" on average $\alpha = (1 - \epsilon)^2/(1 - (1 - \epsilon)^2)$ matching diagonals. Contrast this to an f.r. point off the alignment path which jumps one difference and then only slides $\beta = 1/(\Sigma-1)$ diagonals, assuming every base is equally likely. On average then, an entry d diagonals away from the alignment path, has involved d jumps from f.r. points off the path, and hence is $d(\alpha - \beta)$ behind the f.r. point on the alignment path in the same wave. Thus the average width of a wave trimmed with lag cutoff $\mathcal{L}$ would be less than $2\mathcal{L}/(\alpha - \beta)$. This last step of the argument is incorrect as the statistics of average random path length under the difference model is more complex then assuming all random steps are the same, but there is a definite expected value of path length with d-differences, and therefore the basis of the argument holds, albeit with a different value for β. Since α increases as ϵ decreases, it further explains why the wave becomes more pointy and narrower as ϵ goes to zero.

The computation of successive waves eventually ends because either (a) the boundary of the edit graph of A and B is reached, or (b) all the f.r. points fail the regional alignment quality criterion in which case one can assume that the two reads no longer correlate with each other. In case (b), one should not report the best point in the last wave, as the trimming criterion is overly permissive (e.g. the last 5 columns could all be mismatches!) Because we seek alignments that have an average correlation rate of $1 - 2\epsilon$, we choose to end the path at a *polished point* with greatest anti-diagonal for which the last $E \leq C$ columns are such that *every suffix* of the last E columns have a correlation of $1 - 2\epsilon$ or better. We call such alignments *suffix positive* (at rate ϵ) for reasons that will become obvious momentarily. We must then keep track of the polished f.r. point with greatest anti-diagonal as the waves are computed, which in turns means that we must test the alignment bit-vector of the leading f.r. point(s) for the suffix positive property in each wave.

One can in $O(1)$ time determine if an alignment bit-vector e is suffix positive by building a 2^E-element table $SP[e]$ as follows. Let $Score(\epsilon) = 0$ and recursively let $Score(1b) = Score(b) + \alpha$ and $Score(0b) = Score(b) - \beta$ where $\alpha = 2\epsilon$ and $\beta = 1 - 2\epsilon$. Note that if bit-vector b has m matches and d differences, then

$Score(b) = \alpha m - \beta d$ and if this is non-negative then it implies that $m/(m + d) \geq 1 - 2\epsilon$, i.e. b's alignment has correlation $1 - 2\epsilon$ or better. Let $SP[e] = min\{Score(b) : b$ is a suffix of $e\}$. Clearly $SP[e] \geq 0$ if and only if e is suffix positive (at rate ϵ). By computing $Score$ over the trie of all length E bit vectors and recording the minimum along each path of the trie, the table SP can be built in linear time.

However if E is large, say 30 (as we generally prefer to set it), then the table gets too big. If so, then pick a size D (say 15) for which the SP-table size is reasonable and consider an E-bit vector e to consist of $X = E/D$, D-bit segments $e_X \cdot e_{X-1} \cdot \ldots \cdot e_1$. Precompute the table SP, but for only D bits, and a table SC for bit-vectors of the same size where $SC[b] = Score(b)$. Given these two 2^D tables one can then determine if the longer bit-vector e is suffix positive in $O(X)$ time by calculating whether $Polish(X)$ is true or not with the following recurrences:

$$
\begin{aligned}
Score(x) &= \begin{cases} Score(x-1) + SC[e_x] & \text{if } x \geq 1 \\ 0 & \text{if } x = 0 \end{cases} \\
Polish(x) &= \begin{cases} Polish(x-1) \text{ and } Score(x-1) + SP[e_x] \geq 0 & \text{if } x \geq 1 \\ true & \text{if } x = 0 \end{cases}
\end{aligned}
\tag{2}
$$

In summary, we compute waves of f.r. points keeping only those that are locally part of a good alignment and not too far behind the leading f.r. point. The waves stop either when a boundary is reached, in which case the boundary point is taken as the end of the alignment, or all possible points are eliminated, in which case the furthest polished f.r. point is taken as the end of the alignment (in the given direction). The search takes place both in the forward direction and the reverse direction from a seed tip ρ. The intervals of A and B at which the forward and reverse searches end is reported as a local alignment if the alignment has length τ or more.

Clearly the algorithm is heuristic: (a) it could fail to find an alignment by virtue of eliminating incorrectly an f.r. point on the alignment, and (b) it could over report alignments whose correlation is less than $1 - 2\epsilon$ as local segments of worse quality are permitted depending on the setting of $\mathcal{M}$. We will examine the sensitivity and specificity of the algorithm in the Empirical Performance section, but for the moment indicate that with reasonable choices of $\mathcal{M}$ and $\mathcal{L}$ the algorithm fails less, than once in a billion base pairs, i.e. (a) almost never happens. It is our belief that this heuristic variation of the $O(nd)$ algorithm is superior to any other filter verification approach for local alignments in the case of identity matching over DNA while simultaneously being extremely sensitive. Intuitively this is because the heuristic explores many fewer vertices of the edit graph than dynamic programming based approaches because in expectation the span $hgh - low$ of trimmed waves is a small constant, that is, an alignment is found in linear expected time with near certainty.

6   Empirical Performance

All trials reported in this section were run on a Macbook Pro with a 2.7GHz Intel Core i7 and the code was compiled with gcc version 4.2.1 with the -O4 level of optimization set.

For a given setting of ϵ, we ran trials to determine the sensitivity of the local alignment algorithm in terms of the trimming parameters $\mathcal{M}$ and $\mathcal{L}$. Each trial consisted of generating a 1Mbp random DNA sequence (with every base equally likely) and then peppering in random differences at rate ϵ into two distinct copies. The two perturbed copies were then given to the wave algorithm with seed point $(0,0)$. For various settings of the trimming parameters and ϵ we ran 1000 trials and recorded (a) what fraction of the trials were *successful* in that the entire 1Mbp alignment between the two copies was reported (Table 2), (b) the average wave span (Table 3), and (c) the time taken.

Table 1. Perturbation versus Observed Correlation and Effective Perturbation

Perturbation (ϵ)	Observed Correlation	Effective Perturbation
15.0%	76.1%	12.45%
10.0%	82.8%	8.60%
5.0%	90.7%	4.35%
2.5%	95.2%	2.40%
1.0%	98.0%	1.00%

The first thing we observed was that the perturbed copies of a sequence actually aligned with much better correlation than $1 - 2\epsilon$ and the larger ϵ the larger the relative improvement. We thus define the *effective perturbation* as the value $\epsilon^\star$ such that $1 - 2\epsilon^\star$ equals the observed correlation. Table 1 gives the observed correlation and effective perturbation for a range of values of ϵ.

The success rate and wave span both increase monotonically as $\mathcal{L}$ increases and as $\mathcal{M}$ decreases. In Table 2, we observe that achieving a 100% success rate depends very crucially on $\mathcal{M}$ being small enough, e.g. $\mathcal{M}$ must be 55% or less when the perturbation is $\epsilon = 15\%$, 60% or less for $\epsilon = 10\%$, and so on. But one should further note in Table 3 that the average wave span is virtually independent of $\mathcal{M}$ and really depends only on $\mathcal{L}$, at least for the values of $\mathcal{M}$ that are required to have a 100% success rate. One might then think that only the lag threshold is important and trimming on $\mathcal{M}$ can be dropped, but one must remember that in the general case, when two sequences stop aligning, it is regional alignment quality that stops the extension beyond the end of the local alignment.

So we then investigated how quickly the wave's die off after the end of a local alignment with trials where two sequences completely random with respect to each other were generated and then the wave algorithm was called with seed

Table 2. Success Rate of Heuristic on 1Mbp Alignments

		$\mathcal{L}$							
$1-2\epsilon$	$\mathcal{M}$	15	20	25	30	35	40	45	50
	55%	0.68	0.97	1.00	1.00	1.00	1.00	1.00	1.00
70%	60%	0.27	0.66	0.74	0.74	0.74	0.74	0.73	0.72
	65%	0.00	0.00	0.00	0.00	0.00	0.00	0.00	0.00
	55%	0.99	1.00	1.00	1.00	1.00	1.00	1.00	1.00
80%	60%	0.96	1.00	1.00	1.00	1.00	1.00	1.00	1.00
	65%	0.86	0.93	0.94	0.92	0.94	0.95	0.95	0.94
	70%	0.00	0.00	0.00	0.00	0.00	0.00	0.00	0.00
	70%	1.00	1.00	1.00	1.00	1.00	1.00	1.00	1.00
90%	75%	0.91	0.92	0.92	0.92	0.94	0.94	0.94	0.94
	80%	0.00	0.00	0.00	0.00	0.00	0.00	0.00	0.00
95%	80%	1.00	1.00	1.00	1.00	1.00	1.00	1.00	0.99
	85%	0.25	0.25	0.27	0.28	0.27	0.26	0.27	0.27
98%	85%	1.00	1.00	1.00	1.00	1.00	1.00	1.00	1.00

Table 3. Average Wave Span While Finding An Alignment

		$\mathcal{L}$							
$1-2\epsilon$	$\mathcal{M}$	15	20	25	30	35	40	45	50
	55%	6.4	7.9	9.5	11.1	12.8	14.3	15.9	17.5
70%	60%	6.4	7.9	9.5	11.1	12.8	14.3	15.9	17.5
	65%	6.4	7.9	9.5	11.1	12.8	14.3	15.9	17.5
	55%	4.4	5.5	6.5	7.5	8.6	9.6	10.7	11.7
80%	60%	4.4	5.5	6.5	7.5	8.6	9.6	10.7	11.7
	65%	4.4	5.5	6.5	7.5	8.6	9.6	10.7	11.7
	70%	4.4	5.5	6.5	7.5	8.6	9.6	10.7	11.7
	70%	2.7	3.2	3.7	4.2	4.7	5.2	5.7	6.2
90%	75%	2.7	3.2	3.7	4.2	4.7	5.2	5.7	6.2
	80%	2.7	3.2	3.7	4.2	4.7	5.2	5.7	6.2
95%	80%	1.8	2.1	2.3	2.6	2.8	3.1	3.3	3.6
	85%	1.8	2.1	2.3	2.6	2.8	3.1	3.3	3.6
98%	85%	1.3	1.4	1.5	1.6	1.7	1.8	1.9	2.0

point $(0, 0)$. We recorded the number of waves traversed in each trial, the average span of the waves, and the total number of furthest reaching (f.r.) points computed all together before the algorithm quit. The results are presented in Table 4. Basically the total time to terminate grows quadratically in $\mathcal{M}$ for large values but as $\mathcal{M}$ moves towards the rate at which two random DNA sequences will align (i.e. 48%) the growth in time begins to become exponential going to infinity at 48%. One can begin to see this at $\mathcal{M} = 55\%$ in the table.

We timed the local alignment algorithm on 15 operating points in Tables 2 and 3 for which the success rate was 100% so that each measurement involved exactly 1billion aligned base pairs. The points covered ϵ from 1% to 15% and $\mathcal{L}$ from 20

Table 4. Termination Efficiency

$\mathcal{M}$	Average Number of Waves	Average Wave Span	Average Number of F.R. Points
55%	38	24	910
60%	26	24	620
65%	23	22	510
70%	20	20	400
75%	17	17	290
80%	14	14	200
85%	11	11	120

to 50. The structure of the algorithm implies that the time it takes should be a linear function of (a) the number of waves, D, (b) the number of f.r. points computed, $D\bar{W}$ where $\bar{W}$ is the average span of a wave, and (c) the number of non-random aligned bases followed in snakes, a. But $D = \epsilon^* N$ and we know that $i+d+2s+2a = 2N$ where i, d, and s are the number of insertions, deletions, and substitutions in the alignment found. The later implies $a = N(1 - (1+\sigma)/2\epsilon^*)$ were σ is the relative portion of the alignment that is substitutions versus indels. Thus it follows that the time for the algorithm should be the linear function:

$$N(\alpha + \beta \cdot \epsilon^* + \gamma \cdot \epsilon^* \bar{W}) \tag{3}$$

for some choice of α, β, and γ. A linear regression on our 15 timing values gave a correlation of .9995 with the fit:

$$N(61 + 185\epsilon^* + 32\epsilon^* \bar{W}) \text{ nano seconds} \tag{4}$$

For example, with $\mathcal{L} = 30$, the algorithm takes 194s for $\epsilon = 15\%$, 134s for $\epsilon = 10\%$, 91s for $\epsilon = 5\%$, 75s for $\epsilon = 2.5\%$, and 66s for $\epsilon = 1\%$.

To time and estimate the sensitivity of the filtration algorithm we generated 40X coverage of an 10Mbp synthetic genome. Every read was of length 10Kbp and perturbed by $\epsilon = 15\%$ and we sought overlaps of 1Kbp or longer. In Table 5 we present a number of statistics and timings for a few operating points around our preferred choice of $(14, 35, 6)$ for the parameters k, h, and s. The table reveals that (a) the algorithm is very sensitive missing 1 in 5000 overlaps at the standard operating point, (b) the false discovery rate is generally low but does not have a large effect on the time taken by the filtration step, (c) the major determiner of time taken is k, and (d) the time for the filter, e.g. 132 seconds, is small compared to the time taken to find local alignments which was roughly 860 seconds.

For all the runs in Table 5, the speedup with 4 threads was 3.88 on average, implying for example that the wall clock time for the standard operating point was 256 seconds, or 4.25 minutes for comparing two 400Mb blocks. The 40X synthetic data set constituted a single 400Mbp block in the trials, and when compared against itself produced 1.23 million overlaps between the 37,000 reads

Table 5. DALIGN performance on a synthetic 40X dataset as a function of k, h, and s

(k,h,s)	Sensitivity (TP/(FN+TP))	False Discovery (FP/(TP+FP))	Filter Time (sec.)	Memory (Gb)	Total Time (sec.)
⋆ (14,35,6)	.020%	7.02%	132		995
(14,32,6)	.014%	7.50%	132	11.92	1007
(14,30,6)	.010%	9.51%	132		1014
(14,28,6)	.006%	22.30%	139		1039
(14,35,5)	.037%	6.87%			994
⋆ (14,35,6)	.020%	7.02%	132	11.92	995
(14,35,7)	.015%	7.23%			996
(14,35,8)	.013%	7.84%			998
(13,35,6)	.004%	10.53%	341	12.85	1193
⋆ (14,35,6)	.020%	7.02%	132	11.92	995
(15,35,6)	.109%	6.23%	90	8.22	933

Table 6. DALIGN versus BLASR

Block Size	BLASR Sensitivity	BLASR Time (sec.)	DALIGN Sensitivity	DALIGN Time (sec.)
100	87%	2463	98.7%	109
200	86%	5678	97.5%	222
400	85%	15334	97.3%	393

in the data set. One should note carefully, that for much bigger projects, the time for alignment is considerably less. For example, a 40X dataset over a 1Gbp synthetic genome, would produce 100 400Mb blocks, but comparing each block against itself would typically find only 12.3 thousand overlaps. Another way to look at it is that there will be 100 times more overlaps found, but the filter has to be run on roughly 5000 block pairs.

Real genomes are highly repetitive, implying that the number of overlaps found in practical situations is much higher. For example for the 218Mbp, 31,700 read *E. coli* data set produced by PacBio found 1.44 million overlaps in 1256 total seconds (5.36 wall clock minutes). Moreover, to obtain this result overly frequent k-mers had to be suppressed and low-complexity intervals of reads had to be soft masked. So while the synthetic results above characterize performance in a well understood situation, performance on real data is harder to predict. As our last result, we show in Table 6 the results of timing BLASR and DALIGN on blocks of various sizes from the PacBio human data set. DALIGN was run with $(k,h,s) = (14, 35, 6)$ and k-mers occurring more than 20 times were suppressed. BLASR was run with the parameters used by the PacBio team for their human genome assembly (private communication, J. Chin) which were "−nCandidates 24 −minMatch 14 −maxLCPLength 15 −bestn 12−minPctIdentity 70.0−maxScore 1000−nproc 4 noSplitSubreads". Reads in the block were mapped to the human genome reference in order to obtain the sensitivity numbers. It is clear that DALIGN is much more sensitive (despite the k-mer suppression) and 22 to 39 times faster depending on the block size. In the introduction we gave our time, 15,600 core hours, for overlapping the 54X PacBio human genome dataset, which has been informally reported as 404,000 core hours on the Google "Exacycle" platform using `BLASR` with the parameters as above except $-$`bestn 1` and $-$`minPctIdentity 75.0`. This represents a substantial 25X reduction in compute time and returns the problem to a manageable scale.

Acknowledgments. I would like to acknowledge Sigfried Schloissnig, who is my partner in building a new assembler for long read data. Also Sigfried's postdoc Martin Pippel produced the timing numbers for the big runs on Arabidopsis and Human, and his Ph.D. student Philip Kämpher produced the statistics for Table 6.

References

1. Myers, E.W.: An O(ND) difference algorithm and its variations. Algorithmica 1, 251–266 (1986)
2. Eid, R., Fehr, A., ... (51 authors) ... Korlach, J, Turner, S.W.: Real-Time DNA Sequencing from Single Polymerase Molecules. Science 323(5910), 133–138
3. Lander, E.S., Waterman, M.S.: Genomic mapping by fingerprinting random clones: a mathematical analysis. Genomics 2(3), 231–239 (1988)
4. Churchill, G.A., Waterman, W.S.: The accuracy of DNA sequences: estimating sequence quality. Genomics 14(1), 89–98 (1992)
5. Chin, C.S., Alexander, D.H., Marks, P., Klammer, A.A., Drake, J., Heiner, C., Clum, A., Copeland, A., Huddleston, J., Eichler, E.E., Turner, S.W., Korlach, J.: Nonhybrid, finished microbial genome assemblies from long-read SMRT sequencing data. Nature Methods 10, 563–569 (2013)
6. Pevzner, P.A., Tang, H., Waterman, M.S.: An Eulerian path approach to DNA fragment assembly. PNAS 98(17), 9748–9753 (2001)
7. Kececioglu, J., Myers, E.W.: Combinatorial algorithms for DNA sequence assembly. Algorithmica 13, 7–51 (1995)
8. Chaisson, M.J., Tesler, G.: Mapping single molecule sequencing reads using basic local alignment with successive refinement (BLASR): application and theory. BMC Bioinformatics 13, 238–245 (2012)
9. Burrows, M., Wheeler, D.J.: A block sorting lossless data compression algorithm. Technical Report 124, Digital Equipment Corporation (1994)
10. `https://github.com/PacificBiosciences/DevNet/wiki/Datasets`
11. Manber, U., Myers, E.: Suffix Arrays: A New Method for On-Line String Searches. SIAM Journal on Computing 22, 935–948 (1993)
12. Cormen, T.H., Leiserson, C.E., Rivest, R.L., Stein, C.: Introduction to Algorithms (3rd, 3rd edn., pp. 197–204. MIT Press (2009)
13. Yuan, W.: `http://projects.csail.mit.edu/wiki/pub/SuperTech/` `ParallelRadixSort/Fast_Parallel_Radix_Sort_Algorithm.pdf`

Efficient Indexed Alignment
of Contigs to Optical Maps

Martin D. Muggli[1], Simon J. Puglisi[2], and Christina Boucher[1]

[1] Department of Computer Science,
Colorado State University, Fort Collins, CO
{muggli,cboucher}@cs.colostate.edu
[2] Department of Computer Science,
University of Helsinki, Finland
puglisi@cs.helsinki.fi

Abstract. Since its emergence almost 20 years ago (Schwartz et al., Science 1995), optical mapping has undergone a transition from laboratory technique to commercially available data generation method. In line with this transition, it is only relatively recently that optical mapping data has started to be used for scaffolding contigs and assembly validation in large-scale sequencing projects — for example, the goat (Dong et al., Nature Biotech. 2013) and amborella (Chamala et al., Science 2013) genomes. One major hurdle to the wider use of optical mapping data is the efficient alignment of *in silico* digested contigs to an optical map. We develop TWIN to tackle this very problem. TWIN is the first index-based method for aligning *in silico* digested contigs to an optical map. Our results demonstrate that TWIN is an order of magnitude faster than competing methods on the largest genome. Most importantly, it is specifically designed to be capable of dealing with large eukaryote genomes and thus is the only non-proprietary method capable of completing the alignment for the budgerigar genome in a reasonable amount of CPU time.

1   Introduction

With the cost of next generation sequencing (NGS) continuing to fall, the last decade has been witness to the production of draft whole genome sequences for dozens of species. However, *de novo* genome assembly, the process of reconstructing long contiguous sequences (*contigs*) from short sequence reads, still produces a substantial number of errors [25,1] and is easily misled by repetitive regions [26].

One way to improve the quality of assembly is to use secondary information (independent of the short sequence reads themselves) about the order and orientation of contigs. Optical mapping, which constructs ordered genome-wide high-resolution restriction maps, can provide such information. Optical mapping is a system that works as follows [4,10]: an ensemble of DNA molecules adhered to a charged glass plate are elongated by fluid flow. An enzyme is then used

D. Brown and B. Morgenstern (Eds.): WABI 2014, LNBI 8701, pp. 68–81, 2014.
© Springer-Verlag Berlin Heidelberg 2014

to cleave them into fragments at loci where the enzyme's recognition sequence occurs. Next, the remaining fragments are highlighted with fluorescent dye and digitally photographed under a microscope. Finally, these images are analyzed to estimate the fragment sizes, producing a molecular map. Since the fragments stay relatively stationary during the aforementioned process, the images captures their relative order and size [23]. Multiple copies of the genome undergo this process, and a consensus map is formed that consists of an ordered sequence of fragment sizes, each indicating the approximate number of bases between occurrences of the recognition sequence in the genome [2].

The raw optical mapping data identified by the image processing is an ordered sequence of fragment lengths. Hence, an optical map with x fragments can be denoted as $\ell = \{\ell_1, \ell_2, \ldots, \ell_x\}$, where ℓ_i is the length of the ith fragment in base pairs. This raw data can then be converted into a sequence of locations, each of which determines where a restriction site occurs. We denote the converted data as follows: $L(x) = \{L_0 < L_1 < \cdots < L_n\}$, where $\ell_i = L_i - L_{i-1}$ for $i = 1, \ldots, n$, and L_0 and L_n are defined by the original molecule as a segment of the whole genome by shearing. This latter representation is convenient for algorithmic descriptions. The approximate mean and standard deviation of the fragment size error rate for current data [31] are zero and 150 bp, respectively. See Figure 1 for an illustration of the data produced by this technique. Each restriction enzyme recognizes a specific nucleotide sequence so a unique optical map results from each enzyme, and multiple enzymes can be used in combination to derive denser optical maps. Optical maps have recently become commercially available for mammalian-sized genomes[1], allowing them to be used in a variety of applications.

Although optical mapping data has been used for structural variation detection [28], scaffolding and validating contigs for several large sequencing projects — including those for various prokaryote species [24,32,33], *Oryza sativa* (rice) [35], maize [34], mouse [9], goat [11], *Melopsittacus Undulatus* (budgerigar) [16], and *Amborella trichopoda* [8] — there exist few non-proprietary tools for analyzing this data. Furthermore, the currently available tools are extremely slow because most of them were specifically designed for smaller, prokaryote genomes.

Our Contribution. We present the first index-based method for aligning contigs to an optical map. We call our tool TWIN to illustrate the association between the assembly and optical map as two representations of the genome sequence. The first step of our procedure is to *in silico* digest the contigs with the set of restriction enzymes, computationally mimicking how each restriction enzyme would cleave the short segment of DNA defined by the contig. Thus, *in silico digested contigs* are miniature optical maps that can be aligned to the much longer (sometimes genome-wide) optical maps. The objective is to search and align the *in silico* digested contigs to the correct location in the optical map. By using a suitably-constructed FM-Index data structure [12] built on the optical map, we

---

[1] OpGen (http://www.opgen.com) and BioNano (http://www.bionanogenomics.com) are commercial producers of optical mapping data.

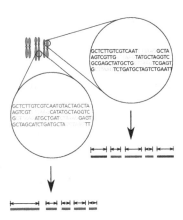

Fig. 1. An illustration of the data produced by optical mapping. Optical mapping locates and measures the distance between restriction sites. Analogous to sequence data, optical mapping data is produced for multiple copies of the same genome, and overlapping single molecular maps are analyzed to produce a map for each chromosome.

show that alignments between contigs and optical maps can be computed in time that is faster than competing methods by more than two orders of magnitude.

TWIN takes as input a set of contigs and an optical map, and produces a set of alignments. The alignments are output in Pattern Space Layout (PSL) format, allowing them to be visualized using any PSL visualization software, such as IGV [29]. TWIN is specifically designed to work on a wide range of genomes, anything from relatively small genomes, to large eukaryote genomes. Thus, we demonstrate the effectiveness of TWIN on *Yersinia kristensenii*, rice, and budgerigar genomes. Rice and budgerigar have genomes of total sizes 430 Mb and 1.2 Gb, respectively. *Yersinia kristensenii*, a bacteria with genome size of 4.6 Mb, is the smallest genome we considered. Short read sequence data was assembled for these genomes, and the resulting contigs were aligned to the respective optical map. We compared the performance of our tool with available competing methods; specifically, the method of Valouev et al. [30] and SOMA [22]. TWIN has superior performance on all datasets, and is demonstrated to be the only current method that is capable of completing the alignment for the budgerigar genome in a reasonable amount of CPU time; SOMA [22] required over 77 days of machine time to solve this problem, whereas, TWIN required just 35 minutes. Lastly, we verify our approach on simulated *E. coli* data by showing our alignment method found correct placements for the *in silico* digested contigs on a simulated optical map. TWIN is available for download at http://www.cs.colostate.edu/twin.

Roadmap. We review related tools for the problem in the remainder of this section. Section 2 then sets notation and formally lays the data structural tools we make use of. Section 3 gives details of our approach. We report our experimental results in Section 4. Finally, Section 5 offers reflections and some potentially fruitful avenues future work may take.

Related Work. The most recent tools to make use of optical mapping data in the context of assembly are AGORA [19] and SOMA [22]. AGORA [19] uses the optical map information to constrain de Bruijn graph construction with the aim of improving the resulting assembly. SOMA [22] is a scaffolding method that uses an optical map and is specifically designed for short-read assemblies. SOMA requires an alignment method for scaffolding and implements an $O(n^2 m^2)$-time dynamic programming algorithm. Gentig [2], and software developed by Valouev et al. [30] also use dynamic programming to address the closely related task of finding alignments between optical maps. Gentig is not available for download. BACop [34] also uses a dynamic programming algorithm and corresponding scoring scheme that gives more weight to contigs with higher fragment density. Antoniotti et al. [3] consider the unique problem of validating an optical map by using assembled contigs. This method assumes the contigs are error-free. Optical mapping data was produced for Assemblathon 2 [6].

2   Background

Strings. Throughout we consider a string $X = X[1..n] = X[1]X[2]\ldots X[n]$ of $|X| = n$ symbols drawn from the alphabet $[0..\sigma - 1]$. For $i = 1, \ldots, n$ we write $X[i..n]$ to denote the *suffix* of X of length $n - i + 1$, that is $X[i..n] = X[i]X[i + 1]\ldots X[n]$. Similarly, we write $X[1..i]$ to denote the *prefix* of X of length i. $X[i..j]$ is the *substring* $X[i]X[i + 1]\ldots X[j]$ of X that starts at position i and ends at j.

Optical Mapping. From a computational point of view, optical mapping is a process that takes two strings: a genome $A[1, n]$ and a restriction sequence $B[1, b]$, and produces an array (string) of integers $M[1, m]$, such that $M[i] = j$ if and only if $A[j..j + b] = B$ is the ith occurrence of B in A.

For example, if we let $B = act$ and

1	2	3	4	5	6	7	8	9	10	11	12	13	14	15	16	17	18	19	20	21	22
A	a	t	a	c	t	t	a	c	t	g	g	a	c	t	a	c	t	a	a	a	c	t

then we would have

$$M = 3, 7, 12, 15, 20.$$

It will also be convenient to view M slightly differently, as an array of fragment sizes, or distances between occurrences of B in A (equivalently differences between adjacent values in M). We denote this *fragment size domain* of M, as the array $F[1, m]$, defined such that $F[i] = (M[i] - M[i - 1])$, with $F[1] = M[1] - 1$. Continuing with the example above, we have

$$F = 2, 4, 5, 3, 5.$$

Suffix Arrays. The suffix array [20] SA_X (we drop subscripts when they are clear from the context) of a string X is an array $SA[1..n]$ which contains a permutation of the integers $[1..n]$ such that $X[SA[1]..n] < X[SA[2]..n] < \cdots < X[SA[n]..n]$. In other words, $SA[j] = i$ iff $X[i..n]$ is the jth suffix of X in lexicographical order.

SA Intervals. For a string Y, the Y-interval in the suffix array SA_X is the interval $SA[s..e]$ that contains all suffixes having Y as a prefix. The Y-interval is a representation of the occurrences of Y in X. For a character c and a string Y, the computation of cY-interval from Y-interval is called a *left extension*.

BWT and backward search. The Burrows-Wheeler Transform [7] $\mathsf{BWT}[1..n]$ is a permutation of X such that $\mathsf{BWT}[i] = \mathsf{X}[\mathsf{SA}[i] - 1]$ if $\mathsf{SA}[i] > 1$ and $ otherwise. We also define $\mathsf{LF}[i] = j$ iff $\mathsf{SA}[j] = \mathsf{SA}[i] - 1$, except when $\mathsf{SA}[i] = 1$, in which case $\mathsf{LF}[i] = I$, where $\mathsf{SA}[I] = n$.

Ferragina and Manzini [12] linked BWT and SA in the following way. Let $\mathsf{C}[c]$, for symbol c, be the number of symbols in X lexicographically smaller than c. The function $\mathsf{rank}(\mathsf{X}, c, i)$, for string X, symbol c, and integer i, returns the number of occurrences of c in $\mathsf{X}[1..i]$. It is well known that $\mathsf{LF}[i] = \mathsf{C}[\mathsf{BWT}[i]] + \mathsf{rank}(\mathsf{BWT}, \mathsf{BWT}[i], i)$. Furthermore, we can compute the left extension using C and rank. If $\mathsf{SA}[s..e]$ is the Y-interval, then $\mathsf{SA}[\mathsf{C}[c] + \mathsf{rank}(\mathsf{BWT}, c, s), \mathsf{C}[c] + \mathsf{rank}(\mathsf{BWT}, c, e)]$ is the $c\mathsf{Y}$-interval. This is called *backward search* [12], and a data structure supporting it is called an *FM-index*.

3   Methods

We find alignments in four steps. First, we convert contigs from the sequence domain to the optical map domain through the process of *in silico* digestion. Second, an FM-index is built from the sequence of optical map fragment sizes. Third, we execute a modified version of the FM-index backward search algorithm described in Section 2 that allows inexact matches. As a result of allowing inexact matches, there may be multiple fragments in an optical map that could each be a reasonable match for an *in silico* digested fragment, and in order to include all of these as candidate matches, backtracking becomes necessary in the backward search. For every backward search path that maintains a non-empty interval for the entire query contig, we emit the alignments denoted by the final interval.

3.1   Converting Contigs to the Optical Map Domain

In order to find alignments for contigs relative to the optical map, we must first convert the strings of bases into the domain of optical maps, that is, strings of fragment sizes. We do this by performing an *in silico* digest of each contig, which is performing a linear search over its bases, searching for occurrences of the enzyme recognition sequence and then computing the distances between adjacent restriction sites. These distances are taken to be equivalent to the fragment sizes that would result if the contig's genomic region underwent digestion in a lab. Additionally, the end fragments of the *in silico* digested contig are removed, as the outside ends are most likely not a result of the optical map restriction enzyme digestion, but rather an artifact of the sequencing and assembly process.

3.2   Building an FM-index from Optical Mapping Data

We construct the FM-index for ℓ, the string of fragment sizes. The particular FM-index implementation we use is the SDSL-Lite[2] [14] library's *compressed suffix array with integer wavelet tree* data structure[3].

---

[2] https://github.com/simongog/sdsl-lite.
[3] The exact revision we used was commit ae42592099707bc59cd1e74997e635324b210115.

In preparation for finding alignments, we also keep two auxiliary data structures. The first is the suffix array, SA_F, corresponding to our FM-index, which we use to report the positions in ℓ where alignments of a contig occur. While we could decode the relevant entries of SA on demand with the FM-index in $O(p)$ time, where p is the so-called sample period of the FM-index, storing SA explicitly significantly improves runtime at the cost of a modest increase in memory usage. The second data structure we store is M, which allows us to map from positions in ℓ to positions in the original genome in constant time.

3.3   Alignment of Contigs Using the FM-index

After constructing the FM-index of the optical map, we find alignments between the optical map and the *in silico* digested contigs.

Specifically, we try to find substrings of the optical map fragment sequence ℓ that are similar to the string of each *in silico* digested contig's non-end fragments F satisfying an alignment goodness metric suggested by Nagarajan et al. [22] [4]:

$$\left| \sum_{i=s}^{t} F_i - \sum_{j=u}^{v} \ell_j \right| \leq F_\sigma \sqrt{\sum_{j=u}^{v} \sigma_j^2},$$

where a parameter F_σ will affect the precision/recall tradeoff.

This computation is carried out using a modified FM-index backward search. A simplified, recursive version of our algorithm for finding alignments is shown in Algorithm 1. The original FM-index backward search proceeds by finding a succession of intervals in the suffix array of the original text that progressively match longer and longer suffixes of the query string, starting from the rightmost symbol of the query. Each additional symbol in the query string is matched in a process taking two arguments: 1) a suffix array interval, the Y-interval, corresponding to the suffixes in the text, ℓ, whose prefix matches a suffix of the query string, and 2) an extension symbol c. The process returns a new interval, the cY-interval, where a prefix of each text suffix corresponding to the new interval is a left extension of the previous query suffix. This process is preserved in TWIN, and is represented by the function *BackwardSearchOneSymbol* in the TWIN algorithm, displayed in Algorithm 1.

Since the optical map fragments include error from the measurement process, it cannot be assumed an *in silico* fragment size will exactly match the optical map fragment size from the same locus in the genome. To accomodate these differences, we determine a set of distinct candidate match fragment sizes, D, each similar in size to the next fragment to be matched in our query. These candidates are drawn from the interval of the BWT currently active in our backward search. We do this by a wavelet tree traversal function provided by SDSL-Lite, which implements the algorithm described in [13] and takes $O(|D| \log(f/\Delta))$ time. This

---

[4] N.B. Alternative goodness metrics could be substituted. They must satisfy the property that pairs of strings considered to align well are composed of substrings that are also considered to align well would also work.

is represented by the function *RestrictedUniqueRangeValues* in Algorithm 1. We emphasise that, due to the large alphabet of ℓ, the wavelet tree's ability to list unique values in a range efficiently is vital to overall performance. Unlike in other applications where the FM-index is used for approximate pattern matching (e.g. read alignment), we cannot afford a bruteforce enumeration of the alphabet at each step in the backward search.

These candidates are chosen to be within a reasonable noise tolerance, t, based on assumptions about the distribution of optical measurement error around the true fragment length. Since there may be multiple match candidates in the BWT interval of the optical map for a query fragment, we extend the backward search with backtracking so each candidate size computed from the wavelet tree is evaluated. That is, for a given *in silico* fragment size (i.e. symbol) c, every possible candidate fragment size, c', that can be found in the optical map in the range $c - t \ldots c + t$ and in the interval $s \ldots e$ (of the BWT) for some tolerance t is used as a substitute in the backward search. Each of these candidates is then checked to ensure that a left extension would still satify the goodness metric, and then used as the extension symbol in the backward search. So it is actually a set of c'Y-intervals that is computed as the left extension in TWIN. Additionally, small DNA fragments may not adhere sufficiently to the glass surface and can be lost in the optical mapping process, so we also branch the backtracking search both with and without small *in silico* fragments to accomodate the uncertainty.

Each time the backward search algorithm successfully progresses throughout the entire query (i.e. it finds some approximate match in the optical map for each fragment in the contig query), we take the contents of the resulting interval in the SA as representing a set of likely alignments.

3.4   Output of Alignments in PSL Format

For each *in silico* digested contig that has an approximate match in the optical map, we emit the alignment, converting positions in the fragment string ℓ to positions in the genome using the M table. We provide a script to convert the human readable output into PSL format.

4   Results

We evaluated the performance of TWIN against the best competing methods on *Yersinia kristensenii*, rice and budgerigar. These three genomes were chosen because they have available sequence and optical mapping data and are diverse in size. For each dataset, we compared the runtime, peak memory usage, and the number of contigs for which at least one alignment was found for TWIN, SOMA [22], and the software of Valouev et al. [30]. Peak memory was measured as the maximum resident set size as reported by the operating system. Runtime is the user process time, also reported by the operating system. SOMA [22] v2.0 was run with example parameters provided with the tool and the software of Valouev et al. [30] was run with its scoring parameters object constructed with

Algorithm 1. MATCH(s, e, q, h) Provided a suffix array start index s and end index e, query string q, and rightmost unmatched query string index h (initially $s = 1$, $e = m$, $h = |q| - 1$), emit alignments of an *in silico* digested contig to an optical map

---

procedure MATCH(s,e,q,h)
   if $h = -1$ **then**
      ▷ Recursion base case. Suffix array indexes $s..e$ denote original query matches.
      $Emit(s, e)$
   else
      ▷ The next symbol to match, c, is the last symbol in the query string.
      $c \leftarrow q[h]$
      ▷ Find the approximately matching values in $\mathsf{BWT}[s \ldots e]$, within tolerance t.
      $D \leftarrow RestrictedUniqueRangeValues(\text{s, e, } c + t, c - t)$
      ▷ Let c' be one possible substitute for c drawn from D
      for all $c' \in D$ **do**
         ▷ If Equation 1 is still satisfied with c' and c, ...
         if $\left| \sum_{i=0}^{|q|-h} \mathsf{SA}[s]_i + c' - \sum_{j=h}^{|q|-1} q_j - c \right| \le F_\sigma \sqrt{\sum_{j=0}^{|q|-h} \sigma_j^2}$ **then**
            ▷ ... determine the suffix array range of the left extension of c'.
            $s', e' \leftarrow BackwardSearchOneSymbol(s, e, c')$
            ▷ Recurse to attempt to match the currently unmatched prefix.
            MATCH($s', e', q, h - 1$)

---

arguments (0.2, 2, 1, 5, 17.43, 0.579, 0.005, 0.999, 3, 1). TWIN was run with $D_\sigma = 4$, $t = 1000$, and $[250 \ldots 1000]$ for the range of small fragments. Gentig [2] and BACop [34] were not available for download so we did not test the data using these approaches.

The sequence data was assembled for *Yersinia kristensenii*, rice and budgerigar by using various assemblers. The relevant assembly statistics are given in Table 1. An important statistic in this table is the number of contigs that have at least two restriction sites, since contigs with fewer than two are unable to be aligned meaningfully by any method, including TWIN. This statistic was computed to reveal cases of ambiguity in placement from lack of information. Indeed, Assemblathon 2 required there to be nine restriction sites present in a contig to align it to the optical mapping data [6]. All experiments were performed on Intel x86-64 workstations with sufficient RAM to avoid paging, running 64-bit Linux.

The experiments for *Yersinia kristensenii*, rice and budgerigar illustrate how each of the programs' running time scale as the size of the genome increases. However, due to the possibility of mis-assemblies in these draft genomes, comparing the actual alignments could possibly lead to erroneous conclusions. Therefore, we will verify the alignments using simulated *E. coli* data. See Subsection 4.4 for this experiment.

Table 1. Assembly and genome statistics for *Yersinia kristensenii*, rice and budgerigar. The assembly statistics were obtained from Quast. [15].

Genome	N50	Genome Size	No. of Contigs with $\geq$ 2 restriction sties
Y. kristensenii	30,719	4.6 Mb	92
Rice	5,299	430 Mb	3,103
Budgerigar	77,556	1.2 Gb	10,019

4.1 Performance on *Yersinia kristensenii*

The sequence and optical map data for *Yersinia kristensenii* are described by Nagarajan *et al.* [22]. The *Yersinia kristensenii* ATCC 33638 reads were generated using 454 GS 20 sequencing and assembled using SPAdes version 3.0.0 [5] using default parameters. Contigs from this assembly were aligned against an optical map of the bacterial strain generated by OpGen using the AflII restriction enzyme. There are approximately 1.4 million single-end reads for this dataset, and they were obtained from the NCBI Short Read Archive (accession SRX013205). Of the 92 contigs that could be aligned to the optical map, the software of Valouev et al. aligned 91 contigs, SOMA aligned 54 contigs, and TWIN aligned 61 contigs. Thus, TWIN found more alignments than SOMA, and did so faster. It should be noted that, for this dataset, all three tools had reasonable runtimes. However, while the software of Valouev et al. found more alignments, our validation experiments (below) suggest these results may favor recall over precision, and many of the additional alignments may not be credibled.

4.2 Performance on Rice Genome

The second dataset consists of approximately 134 million 76 bp paired-end reads from *Oryza sativa Japonica* rice, generated by Illumina, Inc. on the Genome Analayzer (GA) IIx platform, as described by Kawahara *et al.* [17]. These reads were obtained from the NCBI Short Read Archive (accession SRX032913) and assembled using SPAdes version 3.0.0 [5] using default parameters. The optical map for rice was constructed by Zhou *et al.* [35] using SwaI as the restriction enzyme. This optical map was assembled from single molecule restriction maps into 14 optical map contigs, labeled as 12 chromosomes, with chromosome labels 6 and 11 both containing two optical map contigs.

Again, TWIN found alignments for more contigs than SOMA on the rice genome. SOMA and TWIN found alignments for 2,434, and 3,098 contigs, respectively, out of 3,103 contigs that could be aligned to the optical map. However, while SOMA required over 29 minutes to run, TWIN required less than one minute. The software of Valouev executed faster than SOMA (taking around 3 minutes), though still several times slower than TWIN on this modest sized genome.

4.3   Performance on Budgerigar Genome

The sequence and optical map data for the budgerigar genome were generated for the Assemblathon 2 project of Bradnam et al. [6]. Sequence data consists of a combination of Roche 454, Illumina, and Pacific Biosciences reads, providing 16x, 285x, and 10x coverage (respectively) of the genome. All sequence reads are available at the NCBI Short Read Archive (accession ERP002324). For our analysis we consider the assembly generated using Celera [21], which was completed by the CBCB team (Koren and Phillippy) as part of Assemblathon 2 [6]. The optical mapping data was created by Zhou, Goldstein, Place, Schwartz, and Bechner using the SwaI restriction enzyme and consists of 92 separate pieces.

As with the two previous data sets, TWIN found alignments for more contigs than SOMA on the budgerigar genome. SOMA and TWIN found alignments for 9,668, and 9,826 contigs, respectively, out of 10,019 contigs that could be aligned to the optical map. However, SOMA required over 77 days of CPU time and TWIN required 35 minutes. The software of Valouev et al. returned 9,814 alignments and required over an order of magnitude (6.5 hours) of CPU time. Hence, TWIN was the only method that efficiently aligned the *in silico* digested budgerigar genome contigs to the optical map. It should be kept in mind that the competing methods were developed for prokaryote genomes and so we are repurposing them at a scale for which they were not designed. Lastly, the amount of memory used by all the methods on all experiments was low enough for them to run on a standard workstation.

We were forced to parallelize SOMA due to the enormous amount of CPU time SOMA required for this dataset. To accomplish this task, the FASTA file containing the contigs was split into 300 different files, and then IPython Parallel library was used to invoke up to two instances of SOMA on each machine from a set of 150 machines. Thus, when using a cluster with up to 300 jobs concurrently, the alignment for the budgerigar genome took about a day of wall clock time. In contrast, we ran the software of Valouev et al. and TWIN with a single thread running on a single core. However, it should be noted that the same parallelization could have been accomplished for both these software methods too. Also, even with parallelization of SOMA, TWIN is still an order of magnitude faster than it.

4.4   Alignment Verification

We compared the alignments given by TWIN against the alignments of the contigs of an *E. coli* assembly to the *E. Coli* (str. K-12 substr. MG1655) reference genome. Our prior experiments involved species for which the reference genome may have regions that are mis-asssembled and therefore, contig alignments to the reference genome may be inaccurate and cannot be used for comparison and verification of the *in silico* digested contig alignment. The *E. coli* reference genome is likely to contain the fewest errors and thus, is the one we used for assembly verification. The sequence data consists of approximately 27 million paired-end 100 bp reads from *E. coli* (str. K-12 substr. MG1655) generated by

Table 2. Comparsion of the alignment results for TWIN **and competing method.** The performance of TWIN was compared against SOMA [22] and the method of Valouev et al. [30] using the assembly and optical mapping data for *Yersinia Kristensenii*, rice, and budgerigar. Various assemblers were used to assemble the data for these species. The relevant statistics and information concerning these assemblies and genomes can be found in Table 1. The peak memory is given in megabytes (mb). The running time is reported in seconds (s), minutes (m), hours (h), and days.

Genome	Program	Memory	Time	Aligned Contigs
Y. Kristensenii				
	Valouev *et al.*	1.81	.17 s	91
	SOMA	1.71	7.32 s	54
	TWIN	18	.06 s	65
Rice				
	Valouev *et al.*	11.25	2 m 57 s	2,676
	SOMA	7.94	29 m 38 s	2,434
	TWIN	18.25	50 s	3,098
Budgerigar				
	Valouev *et al.*	390	6.5 h	9,814
	SOMA	380.95	77.2 d	9,668
	TWIN	127.112	35 m	9,826

Illumina, Inc. on the Genome Analayzer (GA) IIx platform, and was obtained from the NCBI Short Read Archive (accession ERA000206), and was assembled using SPAdes version 3.0.0 [5] using default parameters. This assembly consists of 160 contigs; 50 of which contain two restriction sites, the minimum required for any possible optical alignment, and complete alignments with minimal (<800 bp) total in/dels relative to the reference genome.

We simulated an optical map using the reference genome for *E. coli* (str. K-12 substr. MG1655) since there is no publicly available one for this genome.

The 50 contigs that contained more than two restriction sites were aligned to the reference genome using BLAT [18]. These same contigs were then *in silico* digested and aligned to the optical map using TWIN. The resulting PSL files were then compared. TWIN found alignment positions within 10% of those found by BLAT for all 50 contigs, justifying that our method is finding correct alignments. We repeated this verification approach with both SOMA and the software from Valouev. All of SOMA's reported alignments had matching BLAT alignments, while of the 49 alignments the software from Valuoev reported, only 18 could be matched with alignments from BLAT.

5   Discussion and Conclusions

We demonstrated that TWIN, an index-based algorithm for aligning *in silico* digested contigs to an optical map, gave over an order of magnitude improvement to runtime without sacrificing alignment quality. Our results show that we are able to handle genomes at least as large as the budgerigar genome directly, whereas SOMA cannot feasibly complete the alignment for this genome in a reasonable amount of time without significant parallelization, and even then is orders of magnitude slower than TWIN. Indeed, given its performance on the budgerigar genome, and its $O(m^2 n^2)$ time complexity, larger genomes seem beyond SOMA. For example, the loblolly pine tree genome, which is approximately 20 Gb [36], would take SOMA approximately 84 machine years, which, even with parallelization, is prohibitively long.

Lastly, optical mapping is a relatively new technology, and thus, with so few algorithms available for working with this data, we feel there remains good opportunities for developing more efficient and flexible methods. Dynamic programming optical map alignment approaches are still important today, as the assembly of the consensus optical maps from the individually imaged molecules often has to deal with missing or spurious restriction sites in the single molecule maps when enzymes fail to digest a recognition sequence or the molecule breaks. Though coverage is high (e.g. about 1,241 Gb of optical data was collected for the 2.66 Gb goat genome), there may be cases where missing restriction site errors are not resolved by the assembly process. In these rare cases (only 1% of alignments reported by SOMA on parrot contain such errors) they will inhibit TWIN's ability to find correct alignments. In essence, TWIN is trading a small degree of sensitivity for a huge speed increase, just as other index based aligners have done for sequence data. Sirén et al. [27] recently extended the Burrows-Wheeler transform (BWT) from strings to acyclic directed labeled graphs and to support path queries. In future work, an adaptation of this method for optical map alignment may allow for the efficient handling of missing or spurious restriction sites.

Acknowledgements. The authors would like to thank David C. Schwartz and Shiguo Zhou from the University of Wisconsin-Madison, Mihai Pop and Lee Mendelowitz from the University of Maryland, and Erich Jarvis and Jason Howard from Duke University for providing insightful comments and providing data for this project. The authors would also like to thank the reviewers for their insightful comments. MM and CB were funded by the Colorado Clinical and Translational Sciences Institute which is funded by National Institutes of Health (NIH-NCATS,UL1TR001082, TL1TR001081, KL2TR001080). SJP was supported by the the Helsinki Institute of Information Technology (HIIT) and by Academy of Finland through grants 258308 and 250345 (CoECGR). All authors greatly appreciate the funding provided for this project.

References

1. Alkan, C., Sajjadian, S., Eichler, E.: Limitations of next-generation genome sequence assembly. Nat. Methods 8(1), 61–65 (2010)
2. Anantharaman, T., Mishra, B.: A probabilistic analysis of false positives in optical map alignment and validation. In: Proc. of WABI, pp. 27–40 (2001)

3. Antoniotti, M., Anantharaman, T., Paxia, S., Mishra, B.: Genomics via optical mapping iv: sequence validation via optical map matching. Technical report, New York University (2001)
4. Aston, C., Schwartz, D.: Optical mapping in genomic analysis. John Wiley and Sons, Ltd. (2006)
5. Bankevich, A., et al.: others. SPAdes: a new Genome assembly algorithm and its applications to single-cell sequencing. J. Comp. Biol. 19(5), 455–477 (2012)
6. Bradnam, K.R., et al.: Assemblathon 2: evaluating *de novo* methods of genome assembly in three vertebrate species. GigaScience 2(1), 1–31 (2013)
7. Burrows, M., Wheeler, D.: A block sorting lossless data compression algorithm. Technical Report 124, Digital Equipment Corporation, Palo Alto, California (1994)
8. Chamala, S., et al.: Assembly and validation of the genome of the nonmodel basal angiosperm *amborella*. Science 342(6165), 1516–1517 (2013)
9. Church, D.M., et al.: Lineage-specific biology revealed by a finished genome assembly of the mouse. PLoS Biology 7(5), e1000112+ (2009)
10. Dimalanta, et al.: A microfluidic system for large dna molecule arrays. Anal. Chem. 76(18), 5293–5301 (2004)
11. Dong, Y., et al.: Sequencing and automated whole-genome optical mapping of the genome of a domestic goat (*capra hircus*). Nat. Biotechnol. 31(2), 136–141 (2013)
12. Ferragina, P., Manzini, G.: Indexing compressed text. J. ACM 52(4), 552–581 (2005)
13. Gagie, T., Navarro, G., Puglisi, S.J.: New algorithms on wavelet trees and applications to information retrieval. Theor. Comput Sci. 426-427, 25–41 (2012)
14. Gog, S., Petri, M.: Optimized succinct data structures for massive data. Software Pract. Expr. (to appear)
15. Gurevich, A., Saveliev, V., Vyahhi, N., Tesler, G.: QUAST: quality assessment tool for genome assemblies. Bioinformatics 29(8), 1072–1075 (2013)
16. Howard, J.T., et al.: De Novo high-coverage sequencing and annotated assemblies of the budgerigar genome (2013)
17. Kawahara, Y., et al.: Improvement of the *oryza sativa nipponbare* reference genome using next generation sequence and optical map data. Rice 6(4), 1–10 (2013)
18. Kent, J.: BLAT–The BLAST-Like Alignment Tool. Genome Res. 12(4), 656–664 (2002)
19. Lin, H., et al.: AGORA: Assembly Guided by Optical Restriction Alignment. BMC Bioinformatics 12, 189 (2012)
20. Manber, U., Myers, G.W.: Suffix arrays: A new method for on-line string searches. SIAM J. Sci. Comput. 22(5), 935–948 (1993)
21. Miller, J.R., et al.: Aggressive assembly of pyrosequencing reads with mates. Bioinformatics 24, 2818–2824 (2008)
22. Nagarajan, N., Read, T.D., Pop, M.: Scaffolding and validation of bacterial genome assemblies using optical restriction maps. Bioinformatics 24(10), 1229–1235 (2008)
23. Neely, R.K., Deen, J., Hofkens, J.: Optical mapping of DNA: single-molecule-based methods for mapping genome. Biopolymers 95(5), 298–311 (2011)
24. Reslewic, S., et al.: Whole-genome shotgun optical mapping of *rhodospirillum rubrum*. Appl. Environ. Microbiol. 71(9), 5511–5522 (2005)
25. Ronen, R., Boucher, C., Chitsaz, H., Pevzner, P.: SEQuel: Improving the Accuracy of Genome Assemblies. Bioinformatics 28(12), i188–i196 (2012)
26. Salzberg, S.: Beware of mis-assembled genomes. Bioinformatics 21(24), 4320–4321 (2005)

27. Sirén, J., Välimäki, N., Mäkinen, V.: Indexing graphs for path queries with applications in genome research. IEEE/ACM Trans. Comput. Biol. Bioinform. (to appear, 2014)
28. Teague, B., et al.: High-resolution human genome structure by single-molecule analysis. Proc. Natl. Acad. Sci. 107(24), 10848–10853 (2010)
29. Thorvaldsdòttir, H., Robinson, J.T., Mesirov, J.P.: Integrative Genomics Viewer (IGV): High-performance Genomics Data Visualization and Exploration. Brief. Bioinform. 14(2), 178–192 (2013)
30. Valouev, A., et al.: Alignment of optical maps. J. Comp. Biol. 13(2), 442–462 (2006)
31. VanSteenHouse, H. personal communication (2013)
32. Zhou, S., et al.: A whole-genome shotgun optical map of *yersinia pestis* strain KIM. Appl. Environ. Microbiol. 68(12), 6321–6331 (2002)
33. Zhou, S., et al.: Shotgun optical mapping of the entire *leishmania major* Friedlin genome. Mol. Biochem. Parasitol. 138(1), 97–106 (2004)
34. Zhou, S., et al.: A single molecule scaffold for the maize genome. PLoS Genet. 5(11), e1000711 (2009)
35. Zhou, S., et al.: Validation of rice genome sequence by optical mapping. BMC Genomics 8(1), 278 (2007)
36. Zimin, A., et al.: Sequencing and assembly of the 22-gb loblolly pine genome. Genetics 196(3), 875–890 (2014)

Navigating in a Sea of Repeats in RNA-seq without Drowning

Gustavo Sacomoto[1,2], Blerina Sinaimeri[1,2], Camille Marchet[1,2],
Vincent Miele[2], Marie-France Sagot[1,2], and Vincent Lacroix[1,2]

[1] INRIA Grenoble Rhône-Alpes, France
[2] UMR CNRS 5558 - LBBE, Université Lyon 1, France

Abstract. The main challenge in *de novo* assembly of NGS data is
certainly to deal with repeats that are longer than the reads. This is
particularly true for RNA-seq data, since coverage information cannot
be used to flag repeated sequences, of which transposable elements are
one of the main examples. Most transcriptome assemblers are based on de
Bruijn graphs and have no clear and explicit model for repeats in RNA-
seq data, relying instead on heuristics to deal with them. The results of
this work are twofold. First, we introduce a formal model for representing
high copy-number repeats in RNA-seq data and exploit its properties to
infer a combinatorial characteristic of repeat-associated subgraphs. We
show that the problem of identifying in a de Bruijn graph a subgraph
with this characteristic is NP-complete. In a second step, we show that in
the specific case of a local assembly of alternative splicing (AS) events,
using our combinatorial characterization we can *implicitly* avoid such
subgraphs. In particular, we designed and implemented an algorithm to
efficiently identify AS events that are not included in repeated regions.
Finally, we validate our results using synthetic data. We also give an
indication of the usefulness of our method on real data.

1 Introduction

Transcriptomes can now be studied through sequencing. However, in the absence of a reference genome, de novo assembly remains a challenging task. The
main difficulty certainly comes from the fact that sequencing reads are short,
and repeated sequences within transcriptomes could be longer than the reads.
This short read / long repeat issue is of course not specific to transcriptome
sequencing. It is an old problem that has been around since the first algorithms
for genome assembly. In this latter case, the problem is somehow easier because
coverage can be used to discriminate contigs that correspond to repeats, *e.g.*
using Myer's A-statistics [8] or [9]. In transcriptome assembly, this idea does
not apply, since the coverage of a gene does not only reflect its copy-number
in the genome, but also and mostly its expression level. Some genes are highly
expressed and therefore highly covered, while most genes are poorly expressed
and therefore poorly covered.

Initially, it was thought that repeats would not be a major issue in RNA-
seq, since they are mostly in introns and intergenic regions. However, the truth

D. Brown and B. Morgenstern (Eds.): WABI 2014, LNBI 8701, pp. 82–96, 2014.
© Springer-Verlag Berlin Heidelberg 2014

is that many regions which are thought to be intergenic are transcribed [3] and introns are not always already spliced out when mRNA is collected to be sequenced. Repeats, especially transposable elements, are therefore very present in real samples and cause major problems in transcriptome assembly.

Most, if not all current short-read transcriptome assemblers are based on de Bruijn graphs. Among the best known are OASES [14], TRINITY [4], and to a lesser degree TRANS-ABYSS [11] and IDBA-TRAN [10]. Common to all of them is the lack of a clear and explicit model for repeats in RNA-seq data. Heuristics are thus used to try and cope efficiently with repeats. For instance, in OASES short nodes are thought to correspond to repeats and are therefore not used for assembling genes. They are added in a second step, which hopefully causes genes sharing repeats not to be assembled together. In TRINITY, there is no attempt to deal with repeats explicitly. The first module of TRINITY, Inchworm, will try and assemble the most covered contig which hopefully corresponds to the most abundant alternative transcript. Then alternative exons are glued to this major transcript to form a splicing graph. The last step is to enumerate all alternative transcripts. If repeats are present, their high coverage may be interpreted as a highly expressed link between two unrelated transcripts. Overall, assembled transcripts may be chimeric or spliced into many sub-transcripts.

In the method we developed, KISSPLICE, which is a local transcriptome assembler [12], repeats may be less problematic, since the goal is not to assemble full-length transcripts. KISSPLICE instead aims at finding variations expressed at the transcriptome level (SNPs, indels and alternative splicings). However, as we previously reported in [12], KISSPLICE is not able to deal with large portions of a de Bruijn graph containing subgraphs associated to highly repeated sequences, e.g. transposable elements, the so-called complex BCCs.

Here, we try and achieve two goals: (i) give a clear formalization of the notion of repeats with high copy-number in RNA-seq data, and (ii) based on it, give a practical way to enumerate bubbles that are lost because of such repeats. Recall that we are in a *de novo* context, so we assume that neither a reference genome/transcriptome nor a database of known repeats, e.g. REPEAT-MASKER [15], are available.

First, we formally introduce a model for representing high copy-number repeats and exploit its properties to infer a parameter characterizing repeat-associated subgraphs in a de Bruijn graph. We prove its relevance but we also show that the problem of identifying, in a de Bruijn graph, a subgraph corresponding to repeats according to such characterization is NP-complete. Hence, a polynomial time algorithm is unlikely. We then show that in the specific case of a local assembly of alternative splicing (AS) events, by using a strategy based on that parameter, we can *implicitly* avoid such subgraphs. More precisely, it is possible to find the structures (*i.e.* bubbles) corresponding to AS events in a de Bruijn graph that are not contained in a repeat-associated subgraph. Finally, using simulated RNA-seq data, we show that the new algorithm improves by a factor of up to 2 the sensitivity of KISSPLICE, while also *improving* its precision. For the specific tasks of calling AS events, we further show that our algorithm more sensitive, by a factor of 2, than TRINITY, while also being slightly more precise. Finally, we give an indication of the usefulness of our method on real data.

2   Preliminaries

Let Σ be an alphabet of fixed size σ. Here we always assume $\Sigma = \{A, C, T, G\}$. Given a sequence (string) $s \in \Sigma^*$, let $|s|$ denote its length, $s[i]$ the ith element of s, and $s[i, j]$ the substring $s[i]s[i+1]\dots s[j]$ for any $1 \le i < j \le |s|$.

A k-mer is a sequence $s \in \Sigma^k$. Given an integer k and a set S of sequences each of length $n \ge k$, we define $span(S, k)$ as the set of all distinct k-mers that appear as a substring in S.

Definition 1. *Given a set of sequences (reads) $R \subseteq \Sigma^*$ and an integer k, we define the directed de Bruijn graph $G_k(R) = (V, A)$ where $V = span(R, k)$ and $A = span(R, k+1)$.*

Given a directed graph $G = (V, A)$ and a vertex $v \in V$, we denote its *out-neighborhood* (resp. *in-neighborhood*) by $N^+(v) = \{u \in V \mid (v, u) \in A\}$ (resp. $N^-(v) = \{u \in V \mid (u, v) \in A\}$), and its out-degree (resp. in-degree) by $d^+(v) = |N^+(v)|$ $(d^-(v) = |N^-(v)|)$. A (simple) *path* $\pi = s \leadsto t$ in G is a sequence of distinct vertices $s = v_0, \dots, v_l = t$ such that, for each $0 \le i < l$, (v_i, v_{i+1}) is an arc of G. If the graph is weighted, *i.e.* there is a function $w : A \to Q_{\ge 0}$ associating a weight to every arc in the graph, then the *length* of a path π is the sum of the weights of the traversed arcs, and is denoted by $|\pi|$.

An arc $(u, v) \in A$ is called *compressible* if $d^+(u) = 1$ and $d^-(v) = 1$. The intuition behind this definition comes from the fact that every path passing through u should also pass through v. It should therefore be possible to "compress" or contract this arc without losing any information. Note that the compressed de Bruijn graph [4,14] commonly used by transcriptomic assemblers is obtained from a de Bruijn graph by replacing, for each compressible arc (u, v), the vertices u, v by a new vertex x, where $N^-(x) = N^-(u)$, $N^+(x) = N^+(v)$ and the label is the concatenation of the k-mer of u and the k-mer of v without the overlapping part (see Fig. 1).

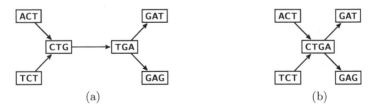

(a)   (b)

Fig. 1. (a) The arc (CTG, TGA) is the only compressible arc in the given de Bruijn graph $(k = 3)$. (b) The corresponding compressed de Bruijn graph.

3   Repeats in de Bruijn Graphs

Given a de Bruijn graph $G_k(R)$ generated by a set of reads R for which we do not have any prior information, our goal is to identify whether there are subgraphs of $G_k(R)$ that correspond each to a set of high copy-number repeats in R. To this end, we identify and then exploit some of the topological properties of the subgraphs that are induced by repeats. Starting with a formal model for representing repeats with high-copy number, we show that the number of compressible arcs, which we denote by γ, is a relevant parameter for such a characterization. This parameter will play an important role in the algorithm of Section 4. However, we also prove that, for an arbitrary de Bruijn graph, identifying a subgraph G' with bounded $\gamma(G')$ is NP-complete.

3.1   Simple Uniform Model for Repeats

We now present the model we adopted for representing high copy-number repeats, *e.g.* transposable elements, in a genome or transcriptome. Basically, our model consists of several "similar" sequences, each generated by uniformly mutating a fixed initial sequence. This model is a simple one and as such should be seen as only a first approximation of what may happen in reality. It is important to point out however that such model is realistic enough in some real cases. In particular, it enables to model well recent invasions of transposable elements which often involve high copy-number and low divergence rate (*i.e.* divergence from their consensus sequence). Consider indeed as an example the recent sub-families AluYa5 and AluYb8 with 2640 and 1852 copies respectively, which both present a divergence rate below 1% [2] (see [6] for other subfamilies with high copy-number and low divergence).

The model is as follows. First, due to mutations, the sequences $s_1, \ldots, s_m$ that represent the repeats are not identical. However, provided that the number of such mutations is not high (otherwise the concept of repeats would not apply), the repeats are considered "similar" in the sense of having a small pairwise Hamming distance between them. We recall that, given two equal length sequences s and s' in Σ^n, their *Hamming distance*, denoted by $d_H(s, s')$, is the number of positions i for which $s[i] \neq s'[i]$. Indels are thus not consider in this model. Mathematically, it is more convenient to consider substitutions only, but this is not a crucial part of the model.

The model has then the following parameters: Σ, the length n of the repeat, the number m of copies of the repeat, an integer k (for the length of the k-mers considered), and the mutation rate, α, *i.e.* the probability that a mutation happens in a particular position. The sequences $s_1, \ldots, s_m$ are then generated by the following process. We first choose uniformly at random a sequence $s_0 \in \Sigma^n$. At step $i \leq m$, we create a sequence s_i as follows: for each position j, $s_i[j] = s_0[j]$ with probability $1 - \alpha$, whereas with probability α a value different from $s[j]$ is chosen uniformly at random for $s_i[j]$. We repeat the whole process m times and thus create a set $S(m, n, \alpha)$ of m such sequences from s_0 (see Fig. 2 for a small example). The generated sequences thus have an expected Hamming distance of αn from s_0.

$$
\begin{matrix}
c_1 & c_2 & c_3 & c_4 & c_5 & c_6 & c_7 & c_8 & c_9 & c_{10} \\
\end{matrix}
$$

$$
\left(
\begin{matrix}
A & A & C & T & G & T & A & T & C & C \\
A & C & C & T & G & T & A & G & C & C \\
G & A & C & T & C & A & A & T & C & C \\
A & A & C & T & C & T & A & T & C & C \\
A & A & C & A & G & T & A & T & C & A \\
A & A & T & T & G & T & A & G & C & C \\
A & G & C & T & G & T & A & T & C & A \\
\vdots & \vdots & \vdots & \vdots & \vdots & \vdots & \vdots & \vdots & \vdots & \vdots \\
A & A & G & T & G & A & A & T & C & C \\
\end{matrix}
\right)
\begin{matrix}
s_0 \\ s_1 \\ s_2 \\ s_3 \\ s_4 \\ s_5 \\ s_6 \\ \\ s_{20}
\end{matrix}
$$

Fig. 2. An example of a set of repeats $S(20, 10, 0.1)$

3.2 Topological Characterization of the Subgraphs Generated by Repeats

Given a de Bruijn graph $G_k(R)$, if a is a compressible arc labeled by the sequence $s = s_1 \ldots s_{k+1}$, then by definition, a is the only outgoing arc of the vertex labeled by the sequence $s[1, k]$ and the only incoming arc of the vertex labeled by the sequence $s[2, k + 1]$. Hence the $(k - 1)$-mer $s[2, k]$ appears as a substring in R, always preceded by the symbol $s[1]$ and followed by the symbol $s[k + 1]$. We refer to such $(k - 1)$-mers as being *boundary rigid*. It is not difficult to see that the set of compressible arcs in a de Bruijn graph $G_k(R)$ stands in a one-to-one correspondence with the set of boundary rigid $(k - 1)$-mers in R.

We now calculate and compare among them the expected number of compressible arcs in $G = G_k(R)$ when R corresponds to a set of sequences that are generated: (i) uniformly at random, and (ii) according to our model. We show that γ is "small" in the cases where the induced graph corresponds to similar sequences, which provides evidence for the relevance of this parameter.

Claim. Let R be a set of m sequences randomly chosen from Σ^n. Then the expected number of compressible arcs in $G_k(R)$ is $\Theta(mn)$.

Proof. The probability that a sequence of length $k - 1$ occurs in a fixed position in a randomly chosen sequence of length n is $(1/4)^{k-1}$. Thus the expected number of appearances of a sequence of length $k - 1$ in a set of m randomly chosen sequences of length n is given by $m(n - k + 2)(1/4)^{k-1}$. If $m(n - k + 2) \leq 4^k$, then this value is upper bounded by 1, and all the sequences of length $k - 1$ are boundary rigid (as a sequence appears once). The claim follows by observing that there are $m(n - k + 1)$ different k-mers.   □

We consider now $\gamma(G_k(R))$ for $R = S(m, n, \alpha)$. We upper bound the expected number of compressible arcs by upper bounding the number of boundary rigid $(k - 1)$-mers.

Theorem 1. *Given integers k, n, m with $k < n$ and a real number $0 \leq \alpha \leq 3/4$, the de Bruijn graph $G_k(S(m, n, \alpha))$ has $o(nm)$ expected compressible arcs.*

Proof. Let s_0 be a sequence chosen randomly from Σ^n. Let $S(m, n, \alpha)$ be the set $\{s_1, \ldots, s_m\}$ of m repeats generated according to our model starting from s_0. Consider now the de Bruijn graph $G = G_k(S(m, n, \alpha))$. Recall that the number of compressible arcs in this graph is equal to the number of boundary rigid $(k-1)$-mers in $S(m, n, \alpha)$. Let X be a random variable representing the number of boundary rigid $(k-1)$-mers in G. Consider the repeats in $S(m, n, \alpha)$ in a matrix-like ordering as in Fig.2 and observe that the mutations from one column to another are independent. Due to the symmetry and the linearity of expectation, $E[X]$ is given by $m(n-k-1)$ (the total number of $(k-1)$-mers) multiplied by the probability that a given $(k-1)$-mer is boundary rigid.

The probability that the $(k-1)$-mer $\hat{s} = s[i, i+k-2]$ is boundary rigid clearly depends on the distance from the starting sequence $\hat{s}_0 = s_0[i, i+k-2]$. Let d be the distance $d_H(\hat{s}, \hat{s}_0)$.

Observe that if the $(k-1)$-mer $s[i] \ldots s[k-1]$ is not boundary rigid then there exists a sequence y in $S(m, n, \alpha)$ such that $y[j] = s[j]$ for all $i \leq j \leq i+k-2$ and either $y[i+k-1] \neq s[i+k-1]$ or $y[i-1] \neq s[i-1]$. It is not difficult to see that the probability that this happens is lower bounded by $(2\alpha - 4/3\alpha^2)(1-\alpha)^{k-1-d}(\alpha/3)^d$. Hence we have:

$$Pr[\hat{s} \text{ is boundary rigid}|d_H(\hat{s}, \hat{s}_0) = d] \leq \left(1 - (2\alpha - 4/3\alpha^2)(1-\alpha)^{k-1-d}(\alpha/3)^d\right)^{m-1}$$

By approximating the above expression we therefore have that,

$$E[X] \leq (n-k-1)m \sum_{d=0}^{k-1} Pr[\hat{s} \text{ is boundary rigid}|d_H(\hat{s}, \hat{s}_0) = d] \qquad (1)$$

$$\leq (n-k-1)me^{-(m-1)(2\alpha-4/3\alpha^2)/(\frac{\alpha}{3})^{k-1}}$$

For a sufficiently large number of copies (*e.g.* $m = \binom{k}{\alpha k}$) and using the fact that $\binom{k}{\alpha k} \geq (1/\alpha)^{\alpha k}$, we have that $E[X]$ is $o(mn)$. This concludes the proof. $\square$

The previous result shows that the number of compressible arcs is a good parameter for characterizing a repeat-associated subgraph.

3.3 Identifying a Repeat-Associated Subgraph

As we showed, a subgraph due to repeated elements has a distinctive feature: it contains few compressible arcs. Based on this, a natural formulation to the repeat identification problem in RNA-seq data is to search for large enough subgraphs that do not contain many compressible arcs. This is formally stated in Problem 1. In order to disregard trivial solutions, it is necessary to require a large enough *connected* subgraph, otherwise any set of disconnected vertices or any small subgraph would be a solution. Unfortunately, we show that this problem is NP-complete, so an efficient algorithm for the repeat identification problem based on this formulation is unlikely.

Problem 1 (Repeat Subgraph).
   INSTANCE: A directed graph G and two positive integers m, t.
   DECIDE: If there exists a connected subgraph $G' = (V', E')$, with $|V'| \geq m$ and having at most t compressible arcs.

In Theorem 2, we prove that this problem is NP-complete for all directed graphs with (total) degree, *i.e.* sum of in and out-degree, bounded by 3. The reduction is from the Steiner tree problem which requires finding a minimum weight subgraph spanning a given subset of vertices. It remains NP-hard even when all arc weights are 1 or 2 (see [1]). This version of the problem is denoted by STEINER$(1,2)$. More formally, given a complete undirected graph $G = (V, E)$ with arc weights in $\{1, 2\}$, a set of *terminal* vertices $N \subseteq V$ and an integer B, it is NP-complete to decide if there exists a subgraph of G spanning N with weight at most B, *i.e.* a connected subgraph of G containing all vertices of N.

We specify next a family of directed graphs that we use in the reduction. Given an integer x we define the directed graph $R(x)$ as a cycle on $2x$ vertices numbered in a clockwise order and where the arcs have alternating directions, *i.e.* for any $i \leq x$, (v_{2i}, v_{2i+1}) is an arc. Note that in $R(x)$ all vertices in even positions, *i.e.* all vertices v_{2i}, have out-degree 2 and in-degree 0, while all vertices v_{2i+1}, have out-degree 0 and in-degree 2. Clearly, none of the arcs of $R(x)$ is compressible.

Theorem 2. *The* Repeat Subgraph Problem *is NP-complete even for directed graphs with degree bounded by d, for any $d \geq 3$.*

Proof. Given a complete graph $G = (V, E)$, a set of terminal vertices N and an upper bound B, *i.e.* an instance of STEINER$(1,2)$, we transform it into an instance of *Repeat Subgraph Problem* for a graph G' with degree bounded by 3. Let us first build the graph $G' = (V', E')$. For each vertex v in $V \setminus N$, add a corresponding subgraph $r(v) = R(|V|)$ in G' and for each vertex v in N, add a corresponding subgraph $r(v) = R(|E| + |V|^2 + 1)$ in G'. For each arc (u, v) in E with weight $w \in \{1, 2\}$, add a simple directed path composed by w compressible arcs connecting $r(u)$ to $r(v)$ in G'; these are the subgraphs corresponding to u and v. The first vertex of the path should be in a sink of $r(u)$ and the last vertex in a source of $r(v)$. By construction, there are at least $|V|$ vertices with in-degree 2 and out-degree 0 (sink) and $|V|$ vertices with out-degree 2 and in-degree 0 (source) in both $r(v)$ and $r(u)$. It is clear that G' has degree bounded by 3. Moreover, the size of G' is polynomial in the size of G and it can be constructed in polynomial time.

In this way, the graph G' has one subgraph for each vertex of G and a path with one or two (depending on the weight of the corresponding arc) compressible arcs for each arc of G. Thus, there exists a subgraph spanning N in G with weight at most B if and only if there exists a subgraph in G' with at least $m = 2|N| + 2|E||N| + 2|V|^2|N|$ vertices and at most $t = |B|$ compressible arcs.

This follows from the fact that any subgraph of G' with at least m vertices necessarily contains all the subgraphs $r(v)$, where $v \in N$, since the number of vertices in all $r(v)$, with $v \in V \setminus N$, is at most $|E| + 2|V|^2$ and the only compressible arcs of G' are in the paths corresponding to the arcs of G.   □

We can obtain the same result for the specific case of de Bruijn graphs. The reduction is very similar but uses a different graph family.

Theorem 3. *The* Repeat Subgraph Problem *is NP-complete even for subgraphs of de Bruijn graphs on* $|\Sigma| = 4$ *symbols.*

4   Bubbles "Drowned" in Repeats

In the previous section, we showed that an efficient algorithm to *directly* identify the subgraphs of a de Bruijn graph corresponding to repeated elements, according to our model (*i.e.* containing few compressible arcs), is unlikely to exist since the problem is NP-complete. However, in this section we show that in the specific case of a local assembly of alternative splicing (AS) events, based on the compressible-arc characterization of Section 3.2, we can *implicitly* avoid such subgraphs. More precisely, it is possible to find the structures (*i.e.* bubbles) corresponding to AS events in a de Bruijn graph that are not contained in a repeat-associated subgraph, thus answering to the main open question of [12].

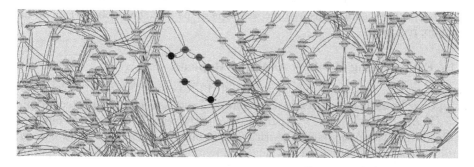

Fig. 3. An alternative splicing event in the SCN5A gene (human) trapped inside a complex region, likely containing repeat-associated subgraphs, in a de Bruijn graph. The alternative isoforms correspond to a pair of paths shown in red and blue.

KISSPLICE [12] is a method for *de novo* calling of AS events through the enumeration of so-called *bubbles*, that correspond to pairs of vertex-disjoint paths in a de Bruijn graph. The bubble enumeration algorithm proposed in [12] was later improved in [13]. However, even the improved algorithm is not able to enumerate all bubbles corresponding to AS events in a de Bruijn graph. There are certain complex regions in the graph, likely containing repeat-associated subgraphs but also real AS events [12], where both algorithms take a huge amount of time. See

Fig. 3 for an example of a complex region with a bubble corresponding to an AS event. The enumeration is therefore halted after a given timeout. The bubbles *drowned* (or trapped) inside these regions are thus missed by KISSPLICE.

In Section 3, the repeat-associated subgraphs are characterized by the presence of few compressible arcs. This suggests that in order to avoid repeat-associated subgraphs, we should restrict the search to bubbles containing many compressible arcs. Equivalently, in a compressed de Bruijn graph (see Section 2), we should restrict the search to bubbles with few branching vertices. Indeed, in a compressed de Bruijn graph, given a fixed sequence length, the number of branching vertices in a path is inversely proportional to the number of compressible arcs of the corresponding path in the non-compressed de Bruijn graph. We thus modify the definition of $(s, t, \alpha_1, \alpha_2)$-bubbles in compressed de Bruijn graphs (Def. 1 in [13]) by adding the extra constraint that each path should have at most b branching vertices.

Definition 2 ($(s, t, \alpha_1, \alpha_2, b)$-bubbles). *Given a weighted directed graph $G = (V, E)$ and two vertices $s, t \in V$, an $(s, t, \alpha_1, \alpha_2, b)$-bubble is a pair of vertex-disjoint st-paths π_1, π_2 with lengths bounded by α_1, α_2, each containing at most b branching vertices.*

By restricting the search to bubbles with few branching vertices, we are able to enumerate them in complex regions implicitly avoiding repeat-associated subgraphs. Indeed, in Section 5 we show that by considering bubbles with at most b branching vertices in KISSPLICE, we increase both its sensitivity and precision. This supports our claim that by focusing on $(s, t, \alpha_1, \alpha_2, b)$-bubbles, we avoid repeat-associated subgraphs and recover at least part of the bubbles trapped in complex regions.

4.1   Enumerating Bubbles Avoiding Repeats

In this section, we modify the algorithm of [13] to enumerate all bubbles with at most b branching vertices in each path. Given a weighted directed graph $G = (V, E)$ and a vertex $s \in V$, let $\mathcal{B}_s(G)$ denote the set of $(s, *, \alpha_1, \alpha_2, b)$-bubbles of G. The algorithm recursively partitions the solution space $\mathcal{B}_s(G)$ at every call until the considered subspace is a singleton (contains only one solution), and in that case it outputs the corresponding solution. In order to avoid unnecessary recursive calls, it maintains the invariant that the current partition contains at least one solution. The algorithm proceeds as follows.

Invariant: At a generic recursive step on vertices u_1, u_2 (initially, $u_1 = u_2 = s$), let $\pi_1 = s \rightsquigarrow u_1, \pi_2 = s \rightsquigarrow u_2$ be the paths discovered so far (initially, π_1, π_2 are empty). Let G' be the current graph (initially, $G' := G$). More precisely, G' is defined as follows: remove from G all the vertices in π_1 and π_2 but u_1 and u_2. Moreover, we also maintain the following invariant $(*)$: there exists at least one pair of paths $\bar{\pi}_1$ and $\bar{\pi}_2$ in G' that extends π_1 and π_2 so that $\pi_1 \cdot \bar{\pi}_1$ and $\pi_2 \cdot \bar{\pi}_2$ belong to $\mathcal{B}_s(G)$.

Base case: When $u_1 = u_2 = u$, output the $(s, u, \alpha_1, \alpha_2, b)$-bubble given by π_1 and π_2.

Recursive rule: Let $\mathcal{B}_s(\pi_1, \pi_2, G')$ denote the set of $(s, *, \alpha_1, \alpha_2, b)$-bubbles to be listed by the current recursive call, *i.e.* the subset of $\mathcal{B}_s(G)$ with prefixes π_1, π_2. It is the union of the following disjoint sets[1].

- The bubbles of $\mathcal{B}_s(\pi_1, \pi_2, G')$ that use e, for each arc $e = (u_1, v)$ outgoing from u_1, that is $\mathcal{B}_s(\pi_1 \cdot e, \pi_2, G' - u_1)$, where $G' - u_1$ is the subgraph of G' after the removal of u_1 and all its incident arcs.
- The bubbles that do not use any arc from u_1, that is $\mathcal{B}_s(\pi_1, \pi_2, G'')$, where G'' is the subgraph of G' after the removal of all arcs outgoing from u_1.

In order to maintain the invariant $(*)$, we only perform the recursive calls when $\mathcal{B}_s(\pi_1 \cdot e, \pi_2, G' - u)$ or $\mathcal{B}_s(\pi_1, \pi_2, G'')$ are non-empty. In both cases, we have to decide if there exist a pair of (internally) vertex-disjoint paths $\bar{\pi}_1 = u_1 \rightsquigarrow t_1$ and $\bar{\pi}_2 = u_2 \rightsquigarrow t_2$, such that $|\bar{\pi}_1| \leq \alpha'_1$, $|\bar{\pi}_2| \leq \alpha'_2$, and $\bar{\pi}_1, \bar{\pi}_2$ have at most b_1, b_2 branching vertices, respectively. Since both the length and the number of branching vertices are monotonic properties, *i.e.* the length and the number of branching vertices of a path prefix is smaller than this number for the full path, we can drop the vertex-disjoint condition. Indeed, let $\bar{\pi}_1$ and $\bar{\pi}_2$ be a pair of paths satisfying all conditions but the vertex-disjointness one. The prefixes $\bar{\pi}_1^* = u_1 \rightsquigarrow t^*$ and $\bar{\pi}_2^* = u_2 \rightsquigarrow t^*$, where t^* is the first intersection of the paths, satisfy all conditions and are internally vertex-disjoint. Moreover, using a dynamic programming algorithm, we can obtain the following result.

Lemma 1. *Given a non-negatively weighted directed graph $G = (V, E)$ and a source $s \in V$, we can compute the shortest paths from s using at most b branching vertices in $O(b|V||E|)$ time.*

As a corollary, we can decide if $\mathcal{B}_s(\pi_1, \pi_2, G)$ is non-empty in $O(b|V||E|)$ time. Now, using an argument similar to [13], *i.e.* leaves of the recursion tree and solutions are in one-to-one correspondence and the height of the recursion tree is bounded by $2n$, we obtain the following theorem.

Theorem 4. *The $(s, *, \alpha_1, \alpha_2, b)$-bubbles can be enumerated in $O(b|V|^3|E||\mathcal{B}_s(G)|)$ time. Moreover, the time elapsed between the output of any two consecutive solutions (i.e. the delay) is $O(b|V|^3|E|)$.*

5   Experimental Results

5.1   Experimental Setup

To evaluate the performance of our method, we simulated RNA-seq data using the FLUXSIMULATOR version 1.2.1 [5]. We generated 100 million reads of 75 bp using its the default error model. We used the RefSeq annotated Human transcriptome (hg19 coordinates) as a reference and we performed a two-step pipeline to obtain a mixture of mRNA and pre-mRNA (*i.e.* with introns not

---

[1] The same holds for u_2 instead of u_1.

yet spliced). To achieve this, we first ran the FLUXSIMULATOR with the Refseq annotations. We then modified the annotations to include the introns and re-ran it on this modified version. In this second run, we additionally constrained the expression values of the pre-mRNAs to be correlated to the expression values of their corresponding mRNAs, as simulated in the first run. Finally, we mixed the two sets of reads to obtain a total of 100M reads. We tested two values: 5% and 15% for the proportion of reads from pre-mRNAs. Those values were chosen so as to correspond to realistic ones as observed in a cytoplasmic mRNA extraction (5%) and a total (cytoplasmic + nuclear) mRNA extraction (15%) [16].

On these simulated datasets, we ran KISSPLICE [12] versions 2.1.0 (KSOLD) and 2.2.0 (KSNEW, with a maximum number of branching vertices set to 5) and obtained lists of detected bubbles that are putative alternative splicing (AS) events. We also ran the full-length transcriptome assembler TRINITY version r2013_08_14 on both datasets, obtaining a list of predicted transcripts, from which we then extracted a list of putative AS events.

In order to assess the precision and the sensitivity of our method, we compared our set of *found* AS events to the set of *true* AS events. Following the definition of ASTALAVISTA, an AS event is composed of two sets of transcripts, the inclusion/exclusion isoforms respectively. An AS event is said to be *true* if at least one transcript among the inclusion isoforms and one among the exclusion isoforms is present in the simulated dataset with at least one read. We stress that this definition is very permissive and includes AS events with very low coverage. This means that our ground truth, *i.e.* the set of *true* AS events, contains some events that are very hard, or even impossible, to detect. We chose to proceed in this way as it reflects what happens in real data.

To compare the results of KISSPLICE with the *true* AS events, we propose that a true AS event is a *true positive* (TP) if there is a bubble such that one path matches the inclusion isoform and the other the exclusion isoform. If there is no such bubble among the results of KISSPLICE, the event is counted as a *false negative* (FN). If a bubble does not correspond to any *true* AS event, it is counted as a *false positive* (FP). To align the paths of the bubbles to transcript sequences, we used the BLAT aligner [7] with 95% identity and a constraint of 95% of each bubble path length to be aligned (to account for the sequencing errors simulated by FLUXSIMULATOR). We computed the sensitivity TP/(TP+FN) and precision TP/(TP+FP) for each simulation case and we report their values for various classes of expression of the minor isoform. Expression values are measured in reads per kilobase (RPK).

5.2   KSNEW vs KSOLD

The plots for the sensitivity of each version on the two simulated datasets are shown in Fig. 4. On the one hand, both versions of KISSPLICE have similar sensitivity in the 5% pre-mRNA dataset, with KSNEW performing slightly better, especially for highly expressed variants. The overall sensitivity in this dataset is 32% and 37% for KSOLD and KSNEW, respectively. On the other hand, the sensitivity of the new version is considerably better over all expression levels in

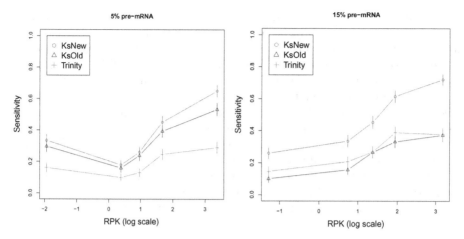

Fig. 4. Sensitivity of KsNew, KsOld and Trinity for several classes of expression of the minor isoform. Each class (*i.e.* point in the graph) contains the same number of AS events (250). It is therefore an average sensitivity on a potentially broad class of expression.

the 15% pre-mRNA dataset. In this case, the sensitivity for KsNew and KsOld are 24% and 48%, respectively. This represents an improvement of 100% over the old version. The results reflect the fact that the most problematic repeats are in intronic regions. A small unspliced mRNA rate leads to few repeat-associated subgraphs, so there are not many AS events drowned in them (which are then missed by KsOld). In this case, the advantage of using KsNew is less obvious, whereas a large proportion of pre-mRNA leads to more AS events drowned in repeat-associated subgraphs which are identified by KsNew and missed by KsOld.

Clearly, any improvement in the sensitivity is meaningless if there is also a significant decrease in precision. This is not the case here. In both datasets, KsNew *improves* the precision of KsOld. It increases from 95% to 98% and from 90% to 99%, in the 5% and 15% datasets, respectively. The high precision we obtain indicates that very few FP bubbles, including the ones generated by repeats, are mistakenly identified as AS events. Moreover, both running times and memory consumption are very similar for the two versions.

5.3   KsNew vs Trinity

The plots for the sensitivity of Trinity on the two simulated datasets are also shown in Fig. 4. In both cases, KsNew performs considerably better than Trinity over all expression levels, with a larger gap for highly expressed variants. The overall sensitivity of Trinity for the 5% and 15% pre-mRNA datasets is 18% and 28%, whereas for KsNew we have 37% and 48%, respectively. Similarly to both KsNew and KsOld, the specificity of Trinity improved from the

5% pre-mRNA to the 15% pre-mRNA dataset. However, this improvement was coupled with a *decrease* of precision from 94% to 75%. This drop in precision is actually mostly due to the prediction of a large number of intron retention, since TRINITY assembles both the mRNA and pre-mRNA. KISSPLICE does not have this problem because most of these apparent intron retentions are bubbles with more than 5 branches (KSNEW) or drowned in complex regions of the graph (KSOLD). To summarize, KSNEW is almost a factor of 2 more sensitive than TRINITY, while also being slightly more precise.

As it was already reported in [12], KISSPLICE (*i.e.* both KSNEW and KSOLD) is faster and uses considerably less memory than TRINITY. For instance, on these datasets, KISSPLICE uses around 5GB of RAM, while TRINITY uses more than 20GB. However, it should be noted that TRINITY tries to solve a more general problem than KISSPLICE, that is reconstructing the full-length transcripts.

5.4   On the Usefulness of KSNEW

In order to give an indication of the usefulness of our repeat-avoiding bubble enumeration algorithm with real data, we also ran KSNEW and KSOLD on the SK-N-SH Human neuroblastoma cell line RNA-seq dataset (wgEncodeEH000169, total RNA). In Fig. 5, we have an example of a *non-annotated* exon skipping event not found by KSOLD. Observe that the intronic region contains several transposable elements (many of which are Alu sequences), while the exons contain none. This is a good example of a bubble (exon skipping event) drowned in a complex region of the de Bruijn graph. The bubble (composed by the two alternative paths) itself contains no repeated elements, but it is surrounded by them. In other words, this is a bubble with few branching vertices that is surrounded by repeat-associated subgraphs. Since KSOLD is unable to differentiate between repeat-associated subgraphs and the bubble, it spends a prohibitive amount of time in the repeat-associated subgraph and fails to find the bubble.

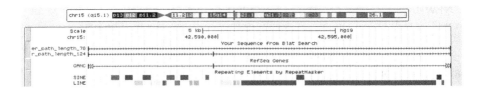

Fig. 5. One of the bubbles found only by KSNEW with the corresponding sequences mapped to the reference human genome and visualized using the UCSC Genome Browser. The first two lines correspond to the sequences of, respectively, the shortest (exon exclusion variant) and longest paths (exon inclusion variant) of the bubble mapped to the genome. The blue line is the Refseq annotation. The last line shows the annotated SINE and LINE sequences (transposable elements).

6   Conclusion

Although transcriptome assemblers are now commonly used, their way to handle repeats is not satisfactory, arguably because the presence of repeats in transcriptomes has been underestimated so far. Given that most RNA-seq datasets correspond to total mRNA extractions, many introns are still present in the data and their repeat content cannot be simply ignored. In this paper, we first proposed a simple formal model for representing high copy-number repeats in RNA-seq data. Exploiting the properties of this model we established that the number of compressible arcs is a relevant quantitative characteristic of repeat-associated subgraphs. We proved that the problem of identifying in a de Bruijn graph a subgraph with this characteristic is NP-complete. However, this characteristic drove the design of an algorithm for efficiently identifying AS events that are not included in repeated regions. The new algorithm was implemented in KISSPLICE (KSNEW), and by using simulated RNA-seq data, we showed that it improves by a factor of up to 2 the sensitivity of the previous version of KISSPLICE, while also improving its precision. In addition, we compared our algorithm with TRINITY and showed that for the specific tasks of calling AS events, our algorithm is more sensitive, by a factor of 2, while also being slightly more precise. Finally, we gave an indication of the usefulness of our method on real data.

Clearly our model could be improved, for instance by using a tree-like structure to take into account the evolutionary nature of repeat (sub)families. Indeed, many TE families are composed by different subfamilies that can be divergent from each other. Consider for instance the human ALU family of TEs that contains at least 7 high copy-number subfamilies with intra-family divergence less than 1% and substantially higher inter-family divergence [6]. In this model, the repeats are generated through a branching process on binary trees. Starting from the root to which we associate a sequence s_0, the tree generation process follows recursively the following rule: each node has probability γ to give birth to two children and $1 - \gamma$ to give birth to a single child. In each case the node is associated to a sequence obtained by independently mutating each symbol of the parent sequence with probability α. In this way, the height of the tree reflects the passing of the time. Hence, the maximum height of the tree would correspond to the time passed since the appearance of the first element of this repeat family. The leaves will be associated to the set of repetitions of s_0 in a genome. Beside representing in a more realistic way the generation of copies of transposable elements, this would also allow to model subfamilies of repeats. Indeed, sequences corresponding to leaves of the same subtree are more similar between them then to sequences belonging to leaves outside the subtree.

However, a formal mathematical analysis on this model seems more difficult to obtain. Observe that in the case α is sufficiently small, such model would converge to the one presented in this paper.

Finally, an interesting open problem remains on how to efficiently enumerate AS events for which their variable region (*i.e.* the skipped exon) is itself a high copy number and low divergence repeat.

Acknowledgments. We thank Alexandre Trindade for implementing an early prototype of the model. This work was supported by the European Research Council under the European Community's Seventh Framework Programme (FP7/2007-2013) / ERC grant agreement no. [247073]10, and the French project ANR-12-BS02-0008 (Colib'read). Part of this work was supported by the ABS4NGS ANR project (ANR-11-BINF-0001-06) and Action n3.6 Plan Cancer 2009-2013.

References

1. Bern, M., Plassmann, P.: The steiner problem with edge lengths 1 and 2. Information Processing Letters (1989)
2. Carroll, M.L., Roy-Engel, A.M., Nguyen, S.V., Salem, A.-H., et al.: Large-scale analysis of the alu ya5 and yb8 subfamilies and their contribution to human genomic diversity. Journal of Molecular Biology 311(1), 17–40 (2001)
3. Djebali, S., Davis, C., Merkel, A., Dobin, A., et al.: Landscape of transcription in human cells. Nature (2012)
4. Grabherr, M., Haas, B., Yassour, M., Levin, J., et al.: Full-length transcriptome assembly from RNA-Seq data without a reference genome. Nat. Biot. (2011)
5. Griebel, T., Zacher, B., Ribeca, P., Raineri, E., et al.: Modelling and simulating generic RNA-Seq experiments with the flux simulator. Nucleic Acids Res. (2012)
6. Jurka, J., Bao, W., Kojima, K.: Families of transposable elements, population structure and the origin of species. Biology Direct 6(1), 44 (2011)
7. Kent, W.J.: BLAT–the BLAST-like alignment tool. Genome Res. 12 (2002)
8. Myers, E., Sutton, G., Delcher, A., Dew, I., et al.: A whole-genome assembly of drosophila. Science 287(5461), 2196–2204 (2000)
9. Novák, P., Neumann, P., Macas, J.: Graph-based clustering and characterization of repetitive sequences in next-generation sequencing data. BMC Bioinf. (2010)
10. Peng, Y., Leung, H., Yiu, S.-M., Lv, M.-J., et al.: IDBA-tran: a more robust de novo de bruijn graph assembler for transcriptomes with uneven expression levels. Bioinf. 29(13) (2013)
11. Robertson, G., Schein, J., Chiu, R., Corbett, R., et al.: De novo assembly and analysis of RNA-seq data. Nat. Met. 7(11), 909–912 (2010)
12. Sacomoto, G., Kielbassa, J., Chikhi, R., Uricaru, R., et al.: KISSPLICE: denovo calling alternative splicing events from RNA-seq data. BMC Bioinformatics 13(Suppl 6), S5 (2012)
13. Sacomoto, G., Lacroix, V., Sagot, M.-F.: A polynomial delay algorithm for the enumeration of bubbles with length constraints in directed graphs and its application to the detection of alternative splicing in RNA-seq data. In: Darling, A., Stoye, J. (eds.) WABI 2013. LNCS, vol. 8126, pp. 99–111. Springer, Heidelberg (2013)
14. Schulz, M., Zerbino, D., Vingron, M., Birney, E.: Oases: robust de novo RNA-seq assembly across the dynamic range of expression levels. Bioinf. (2012)
15. Smit, A.F.A., Hubley, R., Green, P.: RepeatMasker Open-3.0, 1996-2004
16. Tilgner, H., Knowles, D., Johnson, R., Davis, C., et al.: Deep sequencing of subcellular RNA fractions shows splicing to be predominantly co-transcriptional in the human genome but inefficient for lncRNAs. Genome Res. (2012)

Linearization of Median Genomes under DCJ

Shuai Jiang and Max A. Alekseyev

Computational Biology Institute, The George Washington University,
Ashburn, VA, U.S.A.
{jiangs89,maxal}@gwu.edu

Abstract. Reconstruction of the median genome consisting of linear chromosomes from three given genomes is known to be intractable. There exist efficient methods for solving a relaxed version of this problem, where the median genome is allowed to have circular chromosomes. We propose a method for construction of an approximate solution to the original problem from a solution to the relaxed problem and prove a bound on its approximation accuracy. Our method also provides insights into the combinatorial structure of genome transformations with respect to appearance of circular chromosomes.

Keywords: DCJ, median genome, circular chromosome.

1 Introduction

One of the key computational problems in comparative genomics is the *genome median problem* (GMP), which asks to reconstruct a *median* genome M from three given genomes such that the total number of genome rearrangements between M and the given genomes is minimized. The GMP represents a particular case of the more general *ancestral genome reconstruction problem* (AGRP) and is often used as a building block for AGRP solvers [1–5]. The GMP is NP-hard under several models of genome rearrangements, such as reversals only [6] and DCJs [7]. While Double-Cut-and-Join (DCJ) operations [8] (also known as 2-breaks [9]) mimic most common genome rearrangements (i.e., reversals, translocations, fissions, and fusions) and simplify their analysis, they do not take into account linearity of genome chromosomes. As a result, a solution to the GMP under DCJ may contain circular chromosomes even if the given genomes are *linear* (i.e., consist only of linear chromosomes). We will therefore distinguish between *DCJ genome median problem* (DCJ-GMP) and *linear genome median problem* (L-GMP), where the latter is restricted to linear genomes.

There exist some advanced DCJ-GMP solvers [10–12], which allow the median genome to have circular chromosomes. To the best of our knowledge, there exist no solvers for L-GMP, so we pose the problem of using the solution for DCJ-GMP to obtain a linear genome approximating the solution to L-GPM. In the present study, we propose an algorithm that linearizes chromosomes of the given DCJ-GMP solution in some optimal way. Our method also provides insights into the combinatorial structure of genome transformations with DCJs with respect to appearance of circular chromosomes.

D. Brown and B. Morgenstern (Eds.): WABI 2014, LNBI 8701, pp. 97–106, 2014.
© Springer-Verlag Berlin Heidelberg 2014

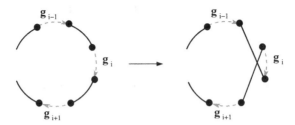

Fig. 1. Graph representation of a gene sequence g_{i-1}, g_i, g_{i+1} and a DCJ that inverses the gene g_i

We remark that a similar *Linearization Problem* appears in adjacency-based reconstructions of median genomes and is known to be intractable [13], forcing the existing approaches [13–16] to solve its relaxation and allow the resulting median genome to contain circular chromosomes.

2   Genome Median Problem under Various Models

Below we briefly describe the concepts of genome graphs and DCJs (2-breaks), for further details we refer the reader to [9, 17].

We represent a circular chromosome on n genes as a cycle with n directed edges (encoding genes and their strands) alternating with n undirected edges connecting adjacent genes. A linear chromosome on n genes is represented as a path of n directed edges alternating with $n-1$ undirected edges (Fig 1). In addition, we introduce an vertex ∞ and connect it to the chromosomal ends (telomeres) with undirected edges. A genome is thus represented as a collection of such paths (starting and ending at vertex ∞) and cycles. Since this does not cause any confusion, we will not distinguish between a genome and its genome graph.

In the genome graph, DCJ corresponds to replacement of a pair of undirected edges with a different pair of undirected edges on the same set of four vertices (Fig 1). A *transformation* from genome P into genome Q is a sequence of DCJs that starts with P and results in Q. Transformations between the same two genomes are called *equivalent*. We denote by $|T|$ the length of the transformation T (i.e., the number of DCJs in T). We define the *DCJ distance* $d_{DCJ}(P,Q)$ between genomes P and Q as the minimum length of transformations between them. DCJ-GMP asks to construct a genome M from three given genomes G_1, G_2, G_3 such that the *DCJ median score* $\sum\limits_{i=1}^{3} d_{DCJ}(M, G_i)$ is minimized.

Similarly, the *genomic distance* $d_g(P,Q)$ between genomes P and Q is defined as the minimum number of genome rearrangements (reversals, translocations, fissions, and fusions) required to transform P into Q. Since each genome rearrangement can be modelled by a DCJ, we trivially have $d_g(P,Q) \geq d_{DCJ}(P,Q)$. However, in contrast to transformations with DCJs between linear genomes that may produce intermediate genomes with circular chromosomes, actual genome rearrangements preserve linearity of genomes and may sometimes require more steps than DCJs (i.e., resulting in $d_g(P,Q) > d_{DCJ}(P,Q)$). For given genomes G_1, G_2, G_3, L-GMP asks to construct a linear median genome M' with the minimum *genomic median score* $\sum_{i=1}^{3} d_g(M', G_i)$. While we are not aware of efficient algorithms for solving L-GMP (let alone, software solvers), we pose the problem of constructing an approximate solution for L-GMP from the given solution for DCJ-GMP as follows.

Given linear genomes G_1, G_2, G_3 and their DCJ median genome M (which may contain circular chromosomes), construct a linear genome M' such that $\sum_{i=1}^{3} d_{DCJ}(M', G_i)$ is minimal. While genome M' may not necessarily represent a solution to L-GMP, the DCJ distance gives a good approximation for genomic distance [17] and simplifies the genome rearrangements analysis. As soon as the genome M' is constructed, its transformations into the given genomes (or vice versa) with genome rearrangements (preserving linearity along all intermediate genomes) can be obtained with GRIMM [18].

We will measure the accuracy of constructed approximate solution M' by the difference $\sum_{i=1}^{3} d_{DCJ}(M', G_i) - \sum_{i=1}^{3} d_{DCJ}(M, G_i)$. We remark that an attempt to arbitrarily cut some gene adjacency (in other words, apply arbitrary fissions) in each circular chromosome of M to obtain genome M' may increase each of the three distances $d_{DCJ}(G_i, M)$ by $c(M)$ and thus the median score by up to $3 \cdot c(M)$, where $c(\cdot)$ denotes the number of circular chromosomes. We will show that we can get much better result, namely increase the median score by at most $c(M)$.

3   Construction of Approximate Solution for L-GMP

Suppose we are given linear genomes G_1, G_2, G_3 and their DCJ median genome M as well as shortest transformations[1] $M \xrightarrow{T_i} G_i (i = 1, 2, 3)$. Our algorithm will modify one of the transformations, say T_1, to produce a transformation T_1' of the form:

$$M \xrightarrow{t_0} M' \xrightarrow{t_1} G_1,$$

where M' is a linear genome, $|t_0| = c(M)$ (i.e., each DCJ in t_0 decreases the number of circular chromosomes) and $|t_0| + |t_1| = |T_1|$ (Fig. 2).

Below we will show that such genome M' suites our purposes and explain details of its construction.

---

[1] Transformations to the median genome may be produced by a DCJ-GMP solver or constructed directly from the median genome and the given genomes, since finding a shortest transformation between two genomes is polynomially solvable [19].

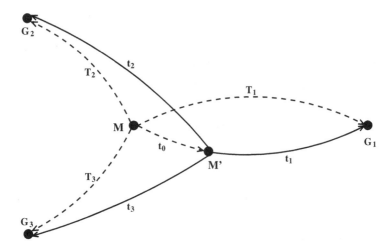

Fig. 2. Linear genomes G_1, G_2, G_3 and their median genome M represented as vertices and the corresponding shortest transformations T_1, T_2, T_3 represented as directed dashed edges. Under the assumption that M contains circular chromosomes, we construct another shortest transformation from M to G_1 (i.e., equivalent to T_1) composed of t_0 and t_1 such that t_0 results in a linear genome M' and $|t_0| = c(M)$. The corresponding shortest transformations from M' to G_2 and G_3 are represented as bold directed edges and denoted by t_2 and t_3.

3.1   Accuracy Estimation

We remark that $d_{DCJ}(G_1, M') = |t_1| = |T_1| - c(M)$. Clearly, t_1 represents a shortest transformation from M' to G_1. Let t_2 and t_3 be shortest transformations from M' to G_2 and G_3, respectively (Fig. 2). By the triangle inequality, $|t_i| = d_{DCJ}(G_i, M') \leq d_{DCJ}(G_i, M) + d_{DCJ}(M, M') = |T_i| + c(M)$ for $i = 2, 3$. Therefore, we have

$$\sum_{i=1}^{3} d_{DCJ}(M', G_i) - \sum_{i=1}^{3} d_{DCJ}(M, G_i) = \sum_{i=1}^{3} |t_i| - \sum_{i=1}^{3} |T_i|$$
$$\leq (|T_1| + |T_2| + |T_3| + c(M)) - (|T_1| + |T_2| + |T_3|) = c(M).$$

3.2   Construction of Transformation T_1'

Our construction of the transformation T_1' relies on the following theorem.

Theorem 1. *Let $P \xrightarrow{z} Q$ be a transformation between genomes P and Q with $c(P) > c(Q)$. Then there exists a transformation $P \xrightarrow{r} P' \xrightarrow{z'} Q$ such that r is a single DCJ, $c(P') = c(P) - 1$, and $|z'| = |z| - 1$.*

Applying Theorem 1 $c(P) - c(Q)$ times, we easily get the following statement.

Corollary 1. *Let $P \xrightarrow{z} Q$ be a transformation between genomes P and Q with $c(P) > c(Q)$. Then there exists a transformation $P \xrightarrow{r_1} P' \xrightarrow{r_2} P'' \xrightarrow{r_3} \cdots \xrightarrow{r_k} P^{(k)} \to Q$ of the same total length $|z|$, where $k = c(P) - c(Q)$, $r_1, r_2, \ldots, r_k$ are DCJs, and $c(P^{(k)}) = c(Q)$.*

We apply this corollary for $z = T_1$ (with $P = M$ and $Q = G_1$) to obtain $M' = P^{(k)}$ with $c(M') = c(G_1) = 0$ chromosomes (i.e., M' is a linear genome). The rest of this section is devoted to the proof of Theorem 1.

We find it convenient to view each DCJ as an operation that removes and creates edges in the genome graph. Two adjacent DCJs α and β in a transformation are called *independent* if β removes edges that were not created by α. Otherwise, when β removes edge(s) created by α, we say that β *depends* on α.

Lemma 1. *Let ϑ be a DCJ that transforms genome P into genome Q. Then $c(P) > c(Q)$ if and only if the two edges removed by ϑ in P belong to distinct chromosomes, at least one of which is circular.*

Proof. If $c(P) > c(Q)$, ϑ must either destroy one circular chromosome in the genome graph P or combine two circular chromosomes into a new one. In either case, the two edges removed by ϑ must belong to different chromosomes in P and at least one of them is circular.

If the two edges removed by ϑ in P belong to distinct chromosomes, one of which is circular, then ϑ destroys this circular chromosome. Thus, $c(Q) < c(P)$ unless ϑ creates a new circular chromosome in Q. However, in the latter case ϑ must also destroy another circular chromosome in P (i.e., ϑ is a fusion on circular chromosomes), implying that $c(Q) < c(P)$. □

Theorem 2. *Let $P \xrightarrow{\vartheta_1} Q \xrightarrow{\vartheta_2} R$ be a transformation between genomes P, Q, R such that ϑ_1, ϑ_2 are independent DCJs and $c(P) \geq c(Q) > c(R)$. Then in the transformation $P \xrightarrow{\vartheta_2} Q' \xrightarrow{\vartheta_1} R$, we have $c(P) > c(Q')$.*

Proof. Let a and b be the edges removed by ϑ_2. Since ϑ_1 and ϑ_2 are independent, the edges a and b are present in both P and Q. By Lemma 1, $c(Q) > c(R)$ implies that in Q one edge, say a, belongs to a circular chromosome, which does not contain b.

Suppose that a belongs to a circular chromosome C in Q. Genome P can be obtained from genome Q by a DCJ ϑ_1^{-1} that reverses ϑ_1 (i.e., ϑ_1^{-1} replaces the edges created by ϑ_1 with the edges removed by it). We consider two cases, depending on whether $c(P) > c(Q)$ or $c(P) = c(Q)$.

If $c(P) > c(Q)$, ϑ_1^{-1} must either split one circular chromosome in Q into two or create a new circular chromosome from linear chromosome(s). In the former case, the edge a belongs to a circular chromosome C' in P even if ϑ_1^{-1} splits the chromosome C. The set of vertices of C' is a subset of the vertices of C and thus does not contain b. In the latter case, C is not affected, while b remains outside it.

If $c(P) = c(Q)$, C is either not affected by ϑ_1^{-1} or remains circular after ϑ_1^{-1} reverses its segment.

In either case, we get that a belongs to a circular chromosome in P, and this chromosome does not contain b. By Lemma 1, ϑ_2 will decrease the number of circular chromosomes in P, i.e., $c(P) > c(Q')$.   □

Suppose that DCJ β depends on DCJ α. Let $k \in \{1, 2\}$ be the number of edges created by α and removed by β. We say that β *strongly depends* on α if $k = 2$ and *weakly depends* on α if $k = 1$. We remark that adjacent pair of strongly dependent DCJs may not appear in shortest transformations between genomes, since such pair can be replaced by an equivalent single DCJ, decreasing the transformation length.

In a genome graph, a pair of dependent DCJs replaces three edges with three other edges on the same six vertices (this operation is known as 3-break [9]). It is easy to see that for a pair of weakly dependent DCJs, there exist exactly two other equivalent pairs of weakly dependent DCJs (Fig. 3):

Lemma 2. *A triple of edges $\{(x_1, x_2), (y_1, y_2), (w_1, w_2)\}$ can be transformed into a triple of edges $\{(x_1, w_2), (y_1, w_2), (w_1, y_2)\}$ with two weakly dependent DCJs in exactly three different ways:*

1. $\{(x_1, x_2), (y_1, y_2), (w_1, w_2)\} \xrightarrow{r_1} \{(x_1, y_2), (y_1, x_2), (w_1, w_2)\} \xrightarrow{r_2} \{(x_1, w_2), (y_1, w_2), (w_1, y_2)\}$;
2. $\{(x_1, x_2), (y_1, y_2), (w_1, w_2)\} \xrightarrow{r_3} \{(x_1, w_2), (y_1, y_2), (w_1, x_2)\} \xrightarrow{r_4} \{(x_1, w_2), (y_1, w_2), (w_1, y_2)\}$;
3. $\{(x_1, x_2), (y_1, y_2), (w_1, w_2)\} \xrightarrow{r_5} \{(x_1, x_2), (y_1, w_2), (w_1, y_2)\} \xrightarrow{r_6} \{(x_1, w_2), (y_1, w_2), (w_1, y_2)\}$.

Theorem 3. *Let $P \xrightarrow{\vartheta_1} Q \xrightarrow{\vartheta_2} R$ be a transformation between genomes P, Q, R such that ϑ_2 depends on ϑ_1 and $c(P) \geq c(Q) > c(R)$. Then there exists a transformation: $P \xrightarrow{\vartheta_3} Q' \xrightarrow{\vartheta_4} R$, where ϑ_3 and ϑ_4 are DCJs and $c(P) > c(Q')$.*

Proof. If DCJ ϑ_2 strongly depends on ϑ_1, then $(\vartheta_1, \vartheta_2)$ is equivalent to a single DCJ ϑ' between genomes P and R. We let $\vartheta_3 = \vartheta'$ and ϑ_4 be any identity DCJ (which removes and adds the same edges and thus does not change the genome) in $Q' = R$ to complete the proof in this case.

For the rest of the proof we assume that ϑ_2 weakly depends on ϑ_1.

By Lemma 2, for a pair of DCJs $(\vartheta_1, \vartheta_2)$, there exists another equivalent pair $(\vartheta_3, \vartheta_4)$, which also transforms P into R. Let Q' be a genome resulting from ϑ_3 in P. We will use Lemma 1 to prove $c(P) > c(Q')$ and show that the two edges removed by ϑ_3 belong to distinct chromosomes, one of which is circular.

Let a, b be the edges removed by ϑ_1 in P and c, d be the edges removed by ϑ_2 in Q. Since $c(Q) > c(R)$, by Lemma 1, one of the edges removed by ϑ_2, say c, belongs to a circular chromosome C in Q, and C does not contain d. Since ϑ_2 weakly depends on ϑ_1, either c or d is created by ϑ_1. We consider these two cases below.

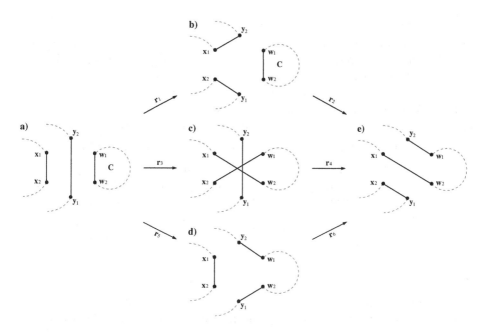

Fig. 3. Illustration of Lemma 2. **a)** Initial genome graph, where the dashed edges represent some gene sequences. The dashed edge and black undirected edge between w_1 and w_2 form a circular chromosome C. **b-d)** The intermediate genomes after first DCJs in the three equivalent pairs of weakly dependent DCJs. **e)** The resulting genome graph after the equivalent pairs of DCJs, where C is destroyed. Namely, C is destroyed by DCJs r_2, r_3, and r_5.

If d is created by ϑ_1, then edge c exists in P and is removed by ϑ_3. Let e be the other edge created by ϑ_1 in Q. If e belongs to circular chromosome C, then by Lemma 1 ϑ_1^{-1} (transforming Q into P) decreases the number of circular chromosomes, i.e., $c(P) < c(Q)$, a contradiction. Therefore, neither of d, e belong to C, implying that C also exists in P and does not contain edges a, b. Then the edges removed by ϑ_3 (i.e., c and one of a, b) belong to circular chromosome C and some other chromosome, implying by Lemma 1 that $c(P) > c(Q')$.

If c is created by ϑ_1, then edge d exists in P and is removed by ϑ_3. We can easily see that the other edge created by ϑ_1 must also belong to C, since otherwise by Lemma 1 ϑ_1^{-1} (transforming Q into P) would decrease the number of circular chromosomes, i.e., $c(P) < c(Q)$, a contradiction. Thus, ϑ_1^{-1} replaces two edges in C with edges a, b, resulting in either two circular chromosomes or a single circular chromosome in P. In either case, edges a and b in P belong to one or two circular chromosomes that do not contain edge d. Since ϑ_3 operates on d and one of a, b, by Lemma 1, $c(P) > c(Q')$.

$\square$

Now we are ready to prove Theorem 1.

Proof (of Theorem 1). Let $P \xrightarrow{z} Q$ be a transformation between genomes P and Q with $c(P) > c(Q)$. Suppose that $z = (q_1, q_2, \ldots, q_n)$, where q_i are DCJs, i.e., z has the form: $P = P_0 \xrightarrow{q_1} P_1 \xrightarrow{q_2} \cdots \xrightarrow{q_n} Q$.

If $c(P_0) > c(P_1)$, then we simply let $r = q_1$, $P' = P_1$, and $z' = (q_2, q_3, \ldots, q_n)$.

If $c(P_0) \leq c(P_1)$, we find the smallest index $j = j(z) \in \{2, 3, \ldots, n\}$ such that $c(P_{j-2}) \geq c(P_{j-1}) > c(P_j)$. Using Theorem 2 or 3, we can obtain a new transformation z_1 from z by replacing the pair of adjacent DCJs $P_{j-2} \xrightarrow{q_{j-1}} P_{j-1} \xrightarrow{q_j} P_j$ with an equivalent pair $P_{j-2} \xrightarrow{q'_{j-1}} P'_{j-1} \xrightarrow{q'_j} P_j$ such that $c(P_{j-2}) > c(P'_{j-1})$ in z_j. We remark that $j(z_1) = j(z) - 1$. Similarly, from transformation z_1 we construct a transformation z_2. After $j(z) - 1$ such steps, we get a transformation

$z_{j-1} : P \xrightarrow{q'_1} P'_1 \xrightarrow{q'_2} \cdots \xrightarrow{q'_{j-1}} P'_{j-1} \xrightarrow{q'_j} P_j \xrightarrow{q_{j+1}} \cdots \xrightarrow{q_n} Q$ such that $c(P) > c(P'_1)$.

Now we let $r = q'_1$ and $z' = (q'_2, \ldots, q'_j, q_{j+1}, \ldots, q_n)$ to complete the proof.   □

4   Discussion

For given three linear genomes G_1, G_2, G_3 and their DCJ median genome M (which may contain circular chromosomes), we described an algorithm that constructs a linear genome M' such that the approximation accuracy of M' (i.e., the difference in the DCJ median scores of M' and M) is bounded by $c(M)$, the number of circular chromosomes in M. In the Appendix we give an example, where $c(M)$ also represents the lower bound for the accuracy of any linearization of M and thus our algorithm achieves the best possible accuracy in this case. It was earlier observed by Xu [11] on simulated data that the number of circular chromosomes produced by their DCJ-GMP solver is typically very small, implying that the approximation accuracy of M' would be very close to 0.

We remark that the proposed algorithm relies on a transformation between M and one of the genomes G_1, G_2, G_3. For presentation purposes, we chose it to be G_1 but other choices may sometimes result in better approximation accuracy. It therefore makes sense to apply the algorithm for each of the three transformations from M to G_i and obtain three corresponding linear genomes M'_i, among which select the genome M' with the minimum DCJ median score. At the same time, we remark that the linear genomes M'_i may be quite distant from each other. In the Appendix, we show that the pairwise DCJ distances between the linear genomes M'_i may be as large as $2/3 \cdot N$, where $N = |G_1| = |G_2| = |G_3|$ is the number of genes in the given genomes.

The proposed algorithm can be viewed as $c(M)$ iterative applications of Theorem 1, each of which takes at most $d_{DCJ}(G_1, M) < N$ steps. Therefore, the overall time complexity is $O(c(M) \cdot N)$ elementary (in sense of Theorems 2 and 3) operations on DCJs. The algorithm is implemented in the AGRP solver MGRA [20, 21].

Acknowledgments. The work was supported by the National Science Foundation under the grant No. IIS-1253614.

References

1. Sankoff, D., Cedergren, R.J., Lapalme, G.: Frequency of insertion-deletion, transversion, and transition in the evolution of 5S ribosomal RNA. Journal of Molecular Evolution 7(2), 133–149 (1976)
2. Kováč, J., Brejová, B., Vinař, T.: A practical algorithm for ancestral rearrangement reconstruction. In: Przytycka, T.M., Sagot, M.-F. (eds.) WABI 2011. LNCS, vol. 6833, pp. 163–174. Springer, Heidelberg (2011)
3. Gao, N., Yang, N., Tang, J.: Ancestral genome inference using a genetic algorithm approach. PLoS One 8(5), e62156 (2013)
4. Moret, B.M., Wyman, S., Bader, D.A., Warnow, T., Yan, M.: A New Implementation and Detailed Study of Breakpoint Analysis. In: Pacific Symposium on Biocomputing, vol. 6, pp. 583–594 (2001)
5. Bourque, G., Pevzner, P.A.: Genome-scale evolution: reconstructing gene orders in the ancestral species. Genome Research 12(1), 26–36 (2002)
6. Caprara, A.: The reversal median problem. INFORMS Journal on Computing 15(1), 93–113 (2003)
7. Tannier, E., Zheng, C., Sankoff, D.: Multichromosomal median and halving problems under different genomic distances. BMC Bioinformatics 10(1), 120 (2009)
8. Yancopoulos, S., Attie, O., Friedberg, R.: Efficient sorting of genomic permutations by translocation, inversion and block interchange. Bioinformatics 21(16), 3340–3346 (2005)
9. Alekseyev, M.A., Pevzner, P.A.: Multi-Break Rearrangements and Chromosomal Evolution. Theoretical Computer Science 395(2-3), 193–202 (2008)
10. Xu, A.W.: A fast and exact algorithm for the median of three problem: A graph decomposition approach. Journal of Computational Biology 16(10), 1369–1381 (2009)
11. Xu, A.W.: DCJ median problems on linear multichromosomal genomes: Graph representation and fast exact solutions. In: Ciccarelli, F.D., Miklós, I. (eds.) RECOMB-CG 2009. LNCS, vol. 5817, pp. 70–83. Springer, Heidelberg (2009)
12. Zhang, M., Arndt, W., Tang, J.: An exact solver for the DCJ median problem. In: Pacific Symposium on Biocomputing, vol. 14, pp. 138–149 (2009)
13. Maňuch, J., Patterson, M., Wittler, R., Chauve, C., Tannier, E.: Linearization of ancestral multichromosomal genomes. BMC Bioinformatics 13(suppl. 19), S11 (2012)
14. Ma, J., Zhang, L., Suh, B.B., Raney, B.J., Burhans, R.C., Kent, W.J., Blanchette, M., Haussler, D., Miller, W.: Reconstructing contiguous regions of an ancestral genome. Genome Research 16(12), 1557–1565 (2006)
15. Muffato, M., Louis, A., Poisnel, C.-E., Crollius, H.R.: Genomicus: a database and a browser to study gene synteny in modern and ancestral genomes. Bioinformatics 26(8), 1119–1121 (2010)
16. Ma, J., Ratan, A., Raney, B.J., Suh, B.B., Zhang, L., Miller, W., Haussler, D.: Dupcar: reconstructing contiguous ancestral regions with duplications. Journal of Computational Biology 15(8), 1007–1027 (2008)
17. Alekseyev, M.A.: Multi-break rearrangements and breakpoint re-uses: from circular to linear genomes. Journal of Computational Biology 15(8), 1117–1131 (2008)
18. Tesler, G.: Efficient algorithms for multichromosomal genome rearrangements. Journal of Computer and System Sciences 65(3), 587–609 (2002)
19. Bergeron, A., Mixtacki, J., Stoye, J.: A unifying view of genome rearrangements. In: Bücher, P., Moret, B.M.E. (eds.) WABI 2006. LNCS (LNBI), vol. 4175, pp. 163–173. Springer, Heidelberg (2006)

20. Alekseyev, M.A., Pevzner, P.A.: Breakpoint graphs and ancestral genome reconstructions. Genome Research 19(5), 943–957 (2009)
21. Jiang, S., Avdeyev, P., Hu, F., Alekseyev, M.A.: Reconstruction of ancestral genomes in presence of gene gain and loss (2014) (submitted)

Appendix. An Extremal Example

In Fig. 4, each of the three linear genomes M'_i on the same genes a, b, c can be viewed as a result of a single fission in their circular unichromosomal median genome M, while all the pairwise DCJ distances between M'_i equal 2. If for each $i = 1, 2, 3$, a genome G^k_i consists of k copies of M'_i (on different triples of genes), then their DCJ median genome M^k consists of k corresponding copies of M and has $c(M^k) = k$ circular chromosomes. We claim (and will prove elsewhere) that the DCJ median score of M^k is $3k$, while any linearization of M^k has the DCJ median score at least $4k$, implying that our algorithm on such genomes G^k_i achieves the best possible accuracy equal $c(M^k) = k$. We also notice that the three linearizations $M^{k'}_i$ of M^k have pairwise DCJ distances equal $2k$.

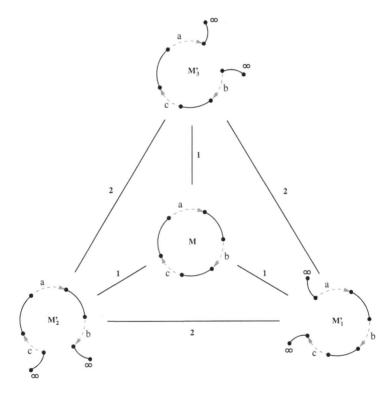

Fig. 4. A circular median genome M of three unichromosomal linear genomes M'_1, M'_2, M'_3 on genes a, b, c with specified pairwise DCJ distances

An LP-Rounding Algorithm for Degenerate Primer Design*

Yu-Ting Huang and Marek Chrobak

Department of Computer Science, University of California, Riverside

Abstract. In applications where a collection of similar sequences needs to be amplified through PCR, *degenerate primers* can be used to improve the efficiency and accuracy of amplification. Conceptually, a degenerate primer is a sequence in which some bases are *ambiguous*, in the sense that they can bind to more than one nucleotide. These ambiguous bases allow degenerate primers to bind to multiple target sequences. When designing degenerate primers, it is essential to find a good balance between high coverage (the number of amplified target sequences) and low degeneracy. In this paper, we propose a new heuristic, called RRD2P, for computing a pair of forward and reverse primers with near-optimal coverage, under the specified degeneracy threshold. The fundamental idea of our algorithm is to represent computing optimal primers as an integer linear program, solve its fractional relaxation, and then apply randomized rounding to compute an integral solution. We tested Algorithm RRD2P on three biological data sets, and our experiments confirmed that it produces primer pairs with good coverage, comparing favorably with a similar tool called HYDEN.

1 Introduction

Polymerase Chain Reaction (PCR) is an amplification technique widely used in molecular biology to generate multiple copies of a desired region of a given DNA sequence. In a PCR process, two small pieces of synthetic DNA sequences called *primers*, typically of length 15-30 bases, are required to identify the boundary of amplification. This pair of primers, referred to as *forward* and *reverse primers*, are obtained from the 5' end of the target sequences and their opposite strand, respectively. Each primer hybridizes to the 3' end of another strand and starts to amplify toward the 5' end.

In applications where a collection of similar sequences needs to be amplified through PCR, *degenerate primers* can be used to improve the efficiency and accuracy of amplification. Degenerate primers [1] can be thought of, conceptually, as having *ambiguous bases* at certain positions, that is bases that represent several different nucleotides. This enables degenerate primers to bind to several different sequences at once, thus allowing amplification of multiple sequences in a single PCR experiment. Degenerate primers are represented as strings formed

---

* Research supported by NSF grants CCF-1217314 and NIH 1R01AI078885.

D. Brown and B. Morgenstern (Eds.): WABI 2014, LNBI 8701, pp. 107–121, 2014.
© Springer-Verlag Berlin Heidelberg 2014

from IUPAC codes, where each code represents multiple possible alternatives for each position in a primer sequence (see Table 1).

The *degeneracy* deg(p) of a primer p is the number of distinct non-degenerate primers that it represents. For example, the degeneracy of primer $p =$ ACMCM is 4, because it represents the following four non-degenerate primers: ACACA, ACACC, ACCCA, and ACCCC.

Table 1. IUPAC nucleotide code table for ambiguous bases

IUPAC nucleotide code	M	R	W	S	Y	K	V	H	D	B	N
represented bases	A	A	A				A	A	A		A
	C			C	C		C	C		C	C
		G		G		G	G		G	G	G
			T		T	T		T	T	T	T

Figure 1 shows a simple example of a pair of primers binding to a target DNA sequence. The forward primer TRTAWTGATY matches the substring TGGACTGATT of the target sequence in all but two positions, illustrating that, in practice, binding can occur even in the presence of a small number of mismatched bases. The reverse primer AGAAAAGTCM matches the target sequence (or, more precisely, its reverse complement) perfectly. This primer pair can produce copies of the region ACCGATGACT of the target sequence, as well as its reverse complement.

```
                              ‹ε--MƆƬƆⱯⱯⱯⱯⱯƆⱯ--‹S  ⇒ reverse primer
                                 ||||||||||
      5'--GATGGACTGATTACCGATGACTGGACTTTTCTG--3'  ⇒ target sequence
      ‹ε--ƆƬⱯƆƆƬƆⱯƆƆƬⱯⱯƬƆƆƆƬⱯƆƬƆⱯƆƆƆƬƆⱯⱯⱯⱯƆⱯƆ--‹S
                || | |||||
forward primer ⇐   5'--TRTAWTGATY--3'
```

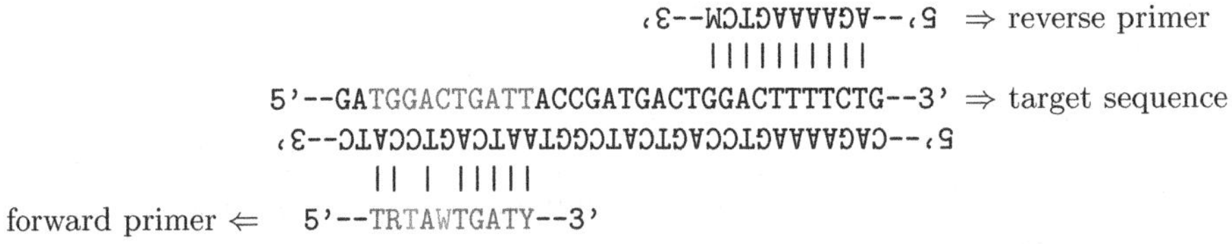

Fig. 1. A symbolic illustration of a pair of primers binding to a DNA sequence

Quite obviously, primers with higher degeneracy can cover more target sequences, but in practice high degeneracy can also negatively impact the quality and quantity of amplification. This is because, in reality, degenerate primers are just appropriate mixtures of regular primers, and including too many primers in the mixture could lead to problems such as mis-priming, where unrelated sequences may be amplified, or primer cross-hybridization, where primers may hybridize to each other. Thus, when designing degenerate primers, it is essential to find a good balance between high coverage and low degeneracy.

PCR experiments involving degenerate primers are useful in studying the composition of microbial communities that typically include many different but similar organisms (see, for example, [2,3]). This variant of PCR is sometimes referred to as *Multiplex PCR (MP-PCR)* [4], although in the literature the term MP-PCR is also used in the context of applications where non-similar sequences are amplified, in which case using degenerate primers may not be beneficial. Designing (non-degenerate) primers for MP-PCR applications also leads to interesting algorithmic problems – see, for example, [5] and references therein.

For the purpose of designing primers we can assume that our target sequences are single-strain DNA sequences. Thus from now on target sequences will be represented by strings of symbols A, C, T, and G.

We say that a (degenerate) primer p *covers* a target sequence s if at least one of the non-degenerate primers represented by p occurs in s as a substring. In practice, a primer can often hybridize to the target sequence even if it only approximately matches the sequence. Formally, we will say that p *covers s with at most m mismatches* if there exists a sub-string s' of s of length $|p|$ such some non-degenerate primer represented by p matches s' on at least $|p| - m$ positions. We refer to m as *mismatch allowance*.

Following the approach in [6], we model the task as an optimization problem that can be formulated as follows: *given a collection of target sequences, a desired primer length, and bounds on the degeneracy and mismatch allowance, we want to find a pair of degenerate primers that meet these criteria and maximize the number of covered target sequences.* We add, however, that, as discussed later in Section 2, there are other alternative approaches that emphasize other aspects of primer design, for example biological properties of primers.

As in [6], using heuristics taking advantage of properties of DNA sequences, the above task can be reduced to the following problem, which, though conceptually simpler, still captures the core difficulty of degenerate primer design:

Problem: MCDPD^{mis}.
> *Instance:* A set of n target strings $A = \{a^1, a^2, \ldots, a^n\}$ over alphabet Σ, each of length k, integers d (degeneracy threshold) and m (mismatch allowance);
> *Objective:* Find a degenerate primer p of length k and degeneracy at most d that covers the maximum number of strings in A with up to m mismatches.

This reduction involves computing the left and right primers separately, as well as using local alignment of target sequences to extract target strings that have the same length as the desired primer. There may be many collections of such target strings (see Section 4), and only those likely to produce good primer candidates need to be considered. Once we solve the instance of **MCDPD^{mis}** for each collection, obtaining a number of forward and reverse primer candidates, we select the final primer pair that optimizes the joint coverage, either through exhaustive search or using heuristic approaches.

The main contribution of this paper is a new algorithm for **MCDPD^{mis}**, called SRR$_{dna}$, based on LP-rounding. We show that **MCDPD^{mis}** can be formulated as an integer linear program. (This linear program actually solves a slightly modified version of **MCDPD^{mis}** – see Section 3 for details.) Algorithm SRR$_{dna}$ computes the optimal fractional solution of this linear program, and then uses an appropriate randomized rounding strategy to convert this fractional solution into an integral one, which represents a degenerate primer.

Using the framework outlined above, we then use Algorithm SRR$_{dna}$ to design a new heuristic algorithm, called RRD2P, that computes pairs of primers for a given collection of target DNA sequences. We implemented Algorithm RRD2P and tested it on three biological data sets. The results were compared to those generated by the existing state-of-the-art tool HYDEN, developed by Linhart and Shamir [7]. Our experiments show that Algorithm RRD2P is able to find primer pairs with better coverage than HYDEN.

2   Related Work

The problem of designing high-quality primers for PCR experiments has been extensively studied and has a vast literature. Much less is known about designing *degenerate primers*. The work most relevant to ours is by Linhart and Shamir [7], who introduced the **MCDPD**^{mis} model, proved that the problem is NP-hard, and gave some efficient approximation algorithms.

In their paper [7], the ideas behind their approximation algorithms were incorporated into a heuristic algorithm HYDEN for designing degenerate primers. HYDEN uses an efficient heuristic approach to design degenerate primers [7] with good coverage. It constructs primers of specified length and with specified degeneracy threshold. HYDEN consists of three phases. It first uses a non-gap local alignment algorithm to find best-conserved regions among target sequences. These regions are called *alignments*. The degree to which an alignment A is conserved is measured by its entropy score:

$$H_A = - \sum_{j=1}^{k} \sum_{\sigma \in \Sigma} \frac{D_A(\sigma,j)}{n} \cdot \log_2 \frac{D_A(\sigma,j)}{n},$$

where k is the length of A, n is the number of target sequences, and $D_A(\sigma, j)$ is the number of sequences in A that have symbol σ at the j_{th} position. Matrix $D_A()$ is called the *column distribution matrix*.

Then, HYDEN designs degenerate primers for these regions using two heuristic algorithms called CONTRACTION and EXPANSION. Finally, it chooses a certain number of best primer candidates, from which it computes a pair of primers with good coverage using a hill-climbing heuristic.

HYDEN has a number of parameters that users can specify, including the desired primer length, the preferred binding regions, the degeneracy threshold, and the mismatch allowance. HYDEN has been tested in a real biological experiment with 127 human olfactory receptor (OR) genes, showing that it produces fairly good primer pairs [7].

Linhart and Shamir [7] also introduced another variant of degenerate primer design problem, called **MDDPD**, where the objective is to find a degenerate primer that covers all given target sequences and has minimum degeneracy. This problem is also NP-hard. An extension of this model where multiple primers are sought was studied by Souvenir *et al.* [8]. See also [9,10] for more related work.

We briefly mention two other software packages for designing degenerate primers. (See the full paper for a more comprehensive survey.) PrimerHunter [11,12] is a software tool that accepts both target and non-target sequences on input, to ensure that the selected primers can efficiently amplify target sequences but avoid the amplification of non-target sequences. This feature allows Primer-Hunter to distinguish closely related subtypes. PrimerHunter allows users to set biological parameters. However, it does not provide the feature to directly control the primer degeneracy. Instead, it uses a degeneracy mask that specifies the positions at which fully degenerate nucleotides are allowed.

iCODEHOP is a web application which designs degenerate primers at the amino acid level [13,14]. This means that during the primer design process, it will reverse-translate the amino acid sequences to DNA, using a user-specified codon usage table. iCODEHOP does not explicitly attempt to optimize the coverage.

3   Randomized Rounding

We now present our randomized rounding approach to solving the $\mathbf{MCDPD}^{mis}$ problem defined in the introduction. Recall that in this problem we are given a collection A of strings over an alphabet Σ, each of the same length k, a degeneracy threshold d, and a mismatch allowance m, and the objective is to compute a degenerate primer p of length k and degeneracy at most d, that covers the maximum number of strings in A with at most m mismatches.

An optimal primer p covers at least one target string $a^i \in A$ with at most m mismatches. In other words, p can be obtained from a^i by (i) changing at most m bases in a^i to different bases, and (ii) changing some bases in a^i to ambiguous bases that match the original bases, without exceeding the degeneracy limit d. Let $Tmpl_m(A)$ denote the set of all strings of length k that can be obtained from some target string $a^i \in A$ by operation (i), namely changing up to m bases in a^i. By trying all strings in $Tmpl_m(A)$, we can reduce $\mathbf{MCDPD}^{mis}$ to its variant where p is required to cover a given template string (without mismatches). Formally, this new optimization problem is:

Problem: $\mathbf{MCDPD}^{mis}_{tmpl}$.
   Instance: A set of n strings $A = \{a^1, a^2, \dots, a^n\}$, each of length k, a template string $\hat{p}$, and integers d (degeneracy threshold) and m (mismatch allowance);
   Objective: Find a degenerate primer p of length k, with $\deg(p) \leq d$ that covers $\hat{p}$ and covers the maximum number of sequences in A with mismatch allowance m.

We remark that our algorithm for $\mathbf{MCDPD}^{mis}$ will not actually try all possible templates from $Tmpl_m(A)$ – there are simply too many of these, if m is large. Instead, we randomly sample templates from $Tmpl_m(A)$ and apply the algorithm for $\mathbf{MCDPD}^{mis}_{tmpl}$ only to those sampled templates. The number of samples affects the running time and accuracy (see Section 5).

We present our algorithm for $\mathbf{MCDPD}^{mis}_{tmpl}$ in two steps. In Section 3.1 that follows, we explain the fundamental idea of our approach, by presenting the linear program and our randomized rounding algorithm for the case of binary strings, where $\Sigma = \{0, 1\}$. The extension to DNA strings is somewhat complicated due to the presence of several ambiguous bases. We present our linear program formulation and the algorithm for DNA strings in Section 3.2.

3.1   Randomized Rounding for Binary Strings

In this section we focus on the case of the binary alphabet $\Sigma = \{0, 1\}$. For this alphabet we only have one ambiguous base, denoted by N, which can represent either 0 or 1. We first demonstrate the integer linear program representation of $\mathbf{MCDPD}^{mis}_{tmpl}$ for the binary alphabet and then we give a randomized rounding algorithm for this case, called $\mathrm{SRR}_{\mathrm{bin}}$. The idea of $\mathrm{SRR}_{\mathrm{bin}}$ is to compute an optimal fractional solution of this linear program and then round it to a feasible integral solution.

Let $\hat{p} = \hat{p}_1\hat{p}_2 \cdots \hat{p}_k$ be the template string from the given instance of $\mathbf{MCDPD}^{mis}_{tmpl}$. It is convenient to think of the objective of $\mathbf{MCDPD}^{mis}_{tmpl}$ as converting $\hat{p}$ into a degenerate primer p by changing up to $\log d$ symbols in $\hat{p}$ to N. For each target string $a^i = a^i_1 a^i_2 \cdots a^i_k$, we use a binary variable x^i to indicate if a^i is covered by p. For each position j, a binary variable n_j is used to indicate whether $\hat{p}_j$ will be changed to N. To take mismatch allowance into consideration, we also use binary variables μ^i_j, which indicate if we allow a mismatch between p and a^i on position j, that is, whether or not $a^i_j \not\subseteq p_j$.

With the above variables, the objective is to maximize the sum of all x^i. Next, we need to specify the constraints. One constraint involves the mismatch allowance m; for a string a^i, the number of mismatches $\sum_j \mu^i_j$ should not exceed m. Next, we have the bound on the degeneracy. In the binary case, the degeneracy of p can be written as $\deg(p) = \prod_j 2^{n_j}$, and we require that $\deg(p) \leq d$. To convert this inequality into a linear constraint, we take the logarithms of both sides. The last group of constraints are the covering constraints. For each j, if p covers a^i and $\hat{p}_j \neq a^i_j$, then either $p_j = \mathrm{N}$ or p_j contributes to the number of mismatches. This can be expressed by inequalities $x^i \leq n_j + \mu^i_j$, for all i, j such that $a^i_j \neq \hat{p}_j$. Then the complete linear program is:

$$
\begin{array}{lll}
\text{maximize} & \sum_i x^i & \qquad\qquad (1) \\
\text{subject to} \sum_j \mu^i_j \leq m & \forall i \\
\sum_j n_j \leq \log_2 d & \\
x^i \leq n_j + \mu^i_j, & \forall i, j : a^i_j \neq \hat{p}_j \\
x^i, n_j, \mu^i_j \in \{0, 1\} & \forall i, j
\end{array}
$$

The pseudo-code of our Algorithm $\mathrm{SRR}_{\mathrm{bin}}$ is given below in Pseudocode 1. The algorithm starts with $p = \hat{p}$ and gradually changes some symbols in p to N,

solving a linear program at each step. At each iteration, the size of the linear
program can be reduced by discarding strings that are too different from the
current p, and by ignoring strings that are already matched by p. More precisely,
any a^i which differs from the current p on more than $m + \log_2 d$ positions cannot
be covered by any degenerate primer obtained from p, so this a^i can be discarded.
On the other hand, if a^i differs from p on at most m positions then it will
always be covered, in which case we can set $x^i = 1$ and we can also remove it
from A. This pruning process in Algorithm SRR_{bin} is implemented by function
FILTEROUT.

Pseudocode 1. Algorithm $\text{SRR}_{\text{bin}}(\hat{p}, A, d, m)$

1: $p \leftarrow \hat{p}$
2: **while** $\deg(p) < d$ **do**
3:     FILTEROUT(p, A, d, m)
4:     **if** $A = \emptyset$ **then break**   ▷ updates A
5:     $LP \leftarrow$ GENLINPROGRAM(p, A, d, m)
6:     $FracSol \leftarrow$ SOLVELINPROGRAM(LP)
7:     RANDROUNDING$_{bin}(p, FracSol, d)$ ▷ updates p and d
8: **return** p

If no sequences are left in A then we are done; we can output p. Otherwise,
we construct the linear program for the remaining strings. This linear program
is essentially the same as the one above, with $\hat{p}$ replaced by p, and with the
constraint $x^i \leq n_j + \mu_j^i$ included only if $p_j \neq \text{N}$. Additional constraints are added
to take into account the rounded positions in p, namely we add the constraint
$n_j = 1$ for all p_j already replaced by N.

We then consider the relaxation of the above integer program, where all in-
tegral constraints $x^i, n_j, \mu_j^i \in \{0, 1\}$ are replaced by $x^i, n_j, \mu_j^i \in [0, 1]$, that is,
all variables are allowed to take fractional values. After solving this relaxation,
we call Procedure RANDROUNDING$_{bin}$, which chooses one fractional variable n_j,
with probability proportional to its value, and rounds it up to 1. (It is sufficient to round only the n_j variables, since all other variables are uniquely determined from the n_j's.) To do so, let J be the set of all j for which $n_j \neq 1$ and
$\pi = \sum_{j \in J} n_j$. The interval $[0, \pi]$ can be split into consecutive $|J|$ intervals, with
the interval corresponding to $j \in J$ having length n_j. Thus we can randomly
(uniformly) choose a value c from $[0, \pi]$, and if c is in the interval corresponding
to $j \in J$ then we round n_j to 1.

If the degeneracy of p is still below the threshold, Algorithm SRR_{bin} executes
the next iteration: it correspondingly adjusts the constraints of the linear program, which produces a new linear program, and so on. The process stops when
the degeneracy allowance is exhausted.

3.2 Randomized Rounding for DNA Sequences Data

We now present our randomized rounding scheme for $\mathbf{MCDPD}_{tmpl}^{mis}$ when the input consists of DNA sequences.

We start with the description of the integer linear program for $\mathbf{MCDPD}_{tmpl}^{mis}$ with $\Sigma = \{\mathtt{A}, \mathtt{C}, \mathtt{G}, \mathtt{T}\}$. Degenerate primers for DNA sequences, in addition to four nucleotide symbols $\mathtt{A}$, $\mathtt{C}$, $\mathtt{G}$ and $\mathtt{T}$, can use eleven symbols corresponding to ambiguous positions, described by their IUPAC codes $\mathtt{M}$, $\mathtt{R}$, $\mathtt{W}$, $\mathtt{S}$, $\mathtt{Y}$, $\mathtt{K}$, $\mathtt{V}$, $\mathtt{H}$, $\mathtt{D}$, $\mathtt{B}$, and $\mathtt{N}$. The interpretation of these codes was given in Table 1 in Section 1. Let Λ denote the set of these fifteen symbols. We think of each $\lambda \in \Lambda$ as representing a subset of Σ, and we write $|\lambda|$ for the cardinality of this subset. For example, we have $|\mathtt{C}| = 1$, $|\mathtt{H}| = 3$ and $|\mathtt{N}| = 4$.

The complete linear program is given below. As for binary sequences, x^i indicates whether the i-th target sequence a^i is covered. Then the objective of the linear program is to maximize the primer coverage, that is $\sum_i x^i$.

$$\text{maximize} \quad \sum_i x^i$$

$$m_j + r_j + w_j + s_j + y_j + k_j + v_j + h_j + d_j + b_j + n_j \leq 1 \qquad \forall j$$

$$\sum_j \mu_j^i \leq m \qquad \forall i$$

$$\sum_j \left[(m_j + r_j + w_j + s_j + y_j + k_j) + \log 3 \cdot (v_j + h_j + d_j + b_j) + 2 \cdot n_j \right] \leq \log d$$

$$x^i \leq m_j + v_j + h_j + n_j + \mu_j^i \qquad \forall i, j : (\hat{p}_j = \mathtt{A}, a_{ij} = \mathtt{C}) \vee (\hat{p}_j = \mathtt{C}, a_{ij} = \mathtt{A})$$

$$x^i \leq r_j + v_j + d_j + n_j + \mu_j^i \qquad \forall i, j : (\hat{p}_j = \mathtt{A}, a_{ij} = \mathtt{G}) \vee (\hat{p}_j = \mathtt{G}, a_{ij} = \mathtt{A})$$

$$x^i \leq w_j + h_j + d_j + n_j + \mu_j^i \qquad \forall i, j : (\hat{p}_j = \mathtt{A}, a_{ij} = \mathtt{T}) \vee (\hat{p}_j = \mathtt{T}, a_{ij} = \mathtt{A})$$

$$x^i \leq s_j + v_j + b_j + n_j + \mu_j^i \qquad \forall i, j : (\hat{p}_j = \mathtt{C}, a_{ij} = \mathtt{G}) \vee (\hat{p}_j = \mathtt{G}, a_{ij} = \mathtt{C})$$

$$x^i \leq y_j + h_j + b_j + n_j + \mu_j^i \qquad \forall i, j : (\hat{p}_j = \mathtt{C}, a_{ij} = \mathtt{T}) \vee (\hat{p}_j = \mathtt{T}, a_{ij} = \mathtt{C})$$

$$x^i \leq k_j + d_j + b_j + n_j + \mu_j^i \qquad \forall i, j : (\hat{p}_j = \mathtt{G}, a_{ij} = \mathtt{T}) \vee (\hat{p}_j = \mathtt{T}, a_{ij} = \mathtt{G})$$

$$x^i, m_j, r_j, w_j, s_j, y_j, k_j, v_j, h_j, d_j, b_j, n_j, \mu_j^i \in \{0, 1\} \qquad \forall i, j$$

To specify the constraints, we now have eleven variables representing the presence of ambiguous bases in the degenerate primer, namely m_j, r_j, w_j, s_j, y_j, k_j, v_j, h_j, d_j, b_j, and n_j, denoted using letters corresponding to the ambiguous symbols. Specifically, for each position j and for each symbol $\lambda \in \Lambda$, the corresponding variable λ_j indicates whether $\hat{p}_j$ is changed to this symbol in the computed degenerate primer p. For example, r_j represents the absence or presence of $\mathtt{R}$ in position j. For each j, at most one of these variables can be 1, which can be represented by the constraint that their sum is at most 1.

Variables μ_j^i indicate a mismatch between p and a^i on position j. Then the bound on the number of mismatches can be written as $\sum_j \mu_j^i \leq m$, for each i. The bound on the degeneracy of the primer p can be written as

$$\deg(p) = \prod_j 2^{(m_j + r_j + w_j + s_j + y_j + k_j)} \times 3^{(v_j + h_j + d_j + b_j)} \times 4^{n_j} \leq d,$$

which after taking logarithms of both sides gives us another linear constraint.

In order for a^i to be covered (that is, when $x^i = 1$), for each position j for which $a^i_j \neq \hat{p}_j$, we must either have a mismatch at position j or we need $a^i_j \subseteq p_j$. Expressing this with linear constraints can be done by considering cases corresponding to different values of $\hat{p}_j$ and a^i_j. For example, when $\hat{p}_j = \mathtt{A}$ and $a^i_j = \mathtt{C}$ (or vice versa), then either we have a mismatch at position j (that is, $\mu^i_j = 1$) or p_j must be one of ambiguous symbols that match $\mathtt{A}$ and $\mathtt{C}$ (that is $\mathtt{M}$, $\mathtt{V}$, $\mathtt{H}$, or $\mathtt{N}$). This can be expressed by the constraint $x^i \leq m_j + v_j + h_j + n_j + \mu^i_j$. We will have one such case for any two different choices of $\hat{p}_j$ and a^i_j, giving us six groups of such constraints.

We then extend our randomized rounding approach from the previous section to this new linear program. From the linear program, we can see that the integral solution can be determined from the values of all variables λ_j, for $\lambda \in \Lambda$. In the fractional solution, a higher value of λ_j indicates that p_j is more likely to be the ambiguous symbol λ. We thus determine ambiguous bases in p one at a time by rounding the corresponding variables.

As for binary strings, Algorithm $\mathrm{SRR_{dna}}$ will start with $p = \hat{p}$ and gradually change some bases in p to ambiguous bases, solving a linear program at each step. At each iteration we first call function FILTEROUT that filters out target sequences that are either too different from the template $\hat{p}$, so that they cannot be matched, or too similar, in which case they are guaranteed to be matched. The pseudocode of Algorithm $\mathrm{SRR_{dna}}$ is the same as in Pseudocode 1 except that the procedure RANDROUNDING$_{bin}$ is replaced by the corresponding procedure RANDROUNDING$_{dna}$ for DNA strings.

If no sequences are left in A then we output p and halt. Otherwise, we construct a linear program for the remaining sequences. This linear program is a slight modification of the one above, with $\hat{p}$ replaced by p. Each base p_j that was rounded to an ambiguous symbol is essentially removed from consideration and will not be changed in the future. Specifically, the constraints on x^i associated with this position j will be dropped from the linear program (because these constraints apply only to positions where $p_j \in \{\mathtt{A}, \mathtt{C}, \mathtt{G}, \mathtt{T}\}$). For each position j that was already rounded, we appropriately modify the corresponding variables. If $p_j = \lambda$, for some $\lambda \in \Lambda - \Sigma$, then the corresponding variable λ_j is set to 1 and all other variables λ'_j are set to 0. If $a^i_j \in p_j$, that is, a^i_j is already matched, then we set $\mu^i_j = 0$, and if $a^i_j \notin p_j$ then we set $\mu^i_j = 1$, which effectively reduces the mismatch allowance for a^i in the remaining linear program.

Next, Algorithm $\mathrm{SRR_{dna}}$ solves the fractional relaxation of such constructed integer program, obtaining a fractional solution *FracSol*. Finally, the algorithm calls function RANDROUNDING$_{dna}$ that will round one fractional variable λ_j to 1. (This represents setting p_j to λ.) To choose j and the symbol λ for p_j, we randomly choose a fractional variable λ_j proportionally to their values among undetermined positions. This is done similarly as in the binary case, by summing up fractional values corresponding to different symbols and positions, and choosing uniformly a random number c between 0 and this sum. This c determines which variable should be rounded up to 1.

3.3 Experimental Approximation Ratio

To examine the quality of primers generated by algorithm SRR_{dna}, we compared the coverage of these primers to the optimal coverage. In our experiments we used the human OR gene [7] data set consisting of 50 sequences, each of length approximately 1Kbps. For this dataset we computed 15 alignments (regions) of length 25 with highest entropy scores (representing sequence similarity, see Section 2). Thus each obtained alignment consists of 50 target strings of length 25. Then, for each of these alignments A, we use each target string in A as a template to run SRR_{dna}, which gives us 50 candidate primers, from which we choose the best one. We then compared this selected primer to a primer computed with Cplex using a similar process, namely computing an optimal integral solution for each template and choosing the best solution.

Table 2. Algorithm SRR_{dna} versus the integral solution obtained with Cplex. The numbers represent coverage values for the fifteen alignments.

	$d = 10000, m = 0$													$d = 625, m = 2$												
A_i	1	2	3	4	5	6	7	8	9	10	11	12	13	1	2	3	4	5	6	7	8	9	10	11	12	13
Opt	26	24	24	24	26	26	24	24	24	26	24	24	24	43	42	42	42	43	43	42	42	42	43	42	42	42
SRR	26	24	23	23	26	26	23	24	23	26	24	23	23	42	40	42	42	43	43	40	41	42	43	42	42	40

This experiment was repeated for two different settings for m (the mismatch allowance) and d (the degeneracy threshold), namely for $(m, d) = (0, 1000), (2, 625)$. The results are shown in Table 2. As can be seen from this table, Algorithm SRR_{dna} computes degenerate primers that are very close, and often equal, to the values obtained from the integer program. Note that for $m = 0$ the value obtained with the integer program represents the true optimal solution for the instance of $\mathbf{MCDPD}^{mis}$, because we try all target strings as templates. For $m = 2$, to compute the optimal solution we would have to try all template strings in $Tmpl_2(A_h)$, which is not feasible; thus the values in the first row are only close approximations to the optimum.

The linear programs we construct are very sparse. This is because for any given a^i and position j, the corresponding constraint on x^i is generated only when p and a^i differ on position j (see Section 3.2), and our data sets are very conserved. Thus, for sufficently small data sets one could simply use integral solutions from Cplex instead of rounding the fractional solution. For example, the initial linear programs in the above instances had typically around 150 constraints, and computing each integral solution took only about 5 times longer than for the fractional solution (roughly, $0.7s$ versus $0.15s$). For larger datasets, however, computing the optimal integral solution becomes quickly infeasible.

4   RRD2P – Complete Primer Design Algorithm

To assess the effectiveness of our randomized rounding approach, we have extended Algorithm SRR_{dna} to a complete primer design algorithm, called RRD2P, and we tested it experimentally on real data sets. In this section we describe Algorithm RRD2P; the experimental evaluation is given in the next section.

Algorithm RRD2P (see Pseudocode 2) has two parameters: S_{fvd} and S_{rev}, which are, respectively, two sets of target sequences, one for forward and the other for reverse primers. They are provided by the user and represent desired binding regions for the two primers. The algorithm finds candidates for forward primers and reverse primers separately. Then, from among these candidates, it iterates over all primer pairs to choose primer pairs with the best joint coverage.

Pseudocode 2. Algorithm RRD2P($S_{fvd}, S_{rev}, k, d, m$)

1: $PrimerList_{fvd} \leftarrow$ DESIGNPRIMERS(S_{fvd}, k, d, m)
2: $PrimerList_{rev} \leftarrow$ DESIGNPRIMERS(S_{rev}, k, d, m)
3: CHOOSEBESTPAIRS($PrimerList_{fvd}, PrimerList_{rev}$)   ▷ Find best primer pairs (f, r)

For both types of primers, we call Algorithm DESIGNPRIMERS (Pseudocode 3), that consists of two parts. In the first part, the algorithm identifies conserved regions within target sequences (Line 1). As before, these regions are also called alignments, and they are denoted A_h. In the second part we design primers for these regions (Lines 2-7).

Pseudocode 3. Algorithm DESIGNPRIMERS($S = \{s^1, s^2, \cdots, s^n\}, k, d, m$)

1: $A_1, A_2, \cdots A_N \leftarrow$ FINDALIGNMENTS(S, k)
2: **for all** alignments $A_h, h = 1, \cdots N$ **do**
3:     $PL_h \leftarrow \emptyset$
4:     $T_h \leftarrow$ set of templates   ▷ see explanation in text
5:     **for all** $\hat{p} \in T_h$ **do**
6:         $p \leftarrow$ SRR$_{\mathrm{dna}}(\hat{p}, A_h, d, m)$
7:         Add p to PL_h
8: $PrimerList \leftarrow PL_1 \cup PL_2 \cdots \cup PL_N$
9: **return** $PrimerList$ (sorted according to coverage)

Finding alignments. Algorithm FINDALIGNMENTS for locating conserved regions (Pseudocode 4) follows the strategy from [7]. It enumerates over all sub-strings of length k of the target sequences. For each k-mer, K, we align it against every target sequence s^i without gaps, to find the best match a^i of length k, i.e, a^i has the smallest Hamming distance with K. The resulting set $A = \{a^1, a^2, \cdots, a^n\}$ of the n best matches, one for each target string, is a conserved region (alignment). Intuitively, more conserved alignments are preferred, since they are more likely to generate low-degeneracy primers. In order to identify how well-conserved an alignment A is, the entropy score is applied.

Pseudocode 4. Algorithm FINDALIGNMENTS($S = \{s^1, s^2 \cdots, s^n\}, k$)

1: $AlignmentList \leftarrow \emptyset$
2: **for all** k-mers, K, in S **do**
3:    $A \leftarrow \emptyset$
4:    **for all** $s^i \in S$ **do**
5:      $a^i \leftarrow$ substring of s^i that is the best match for K
6:      Add a^i to A
7:    Add A to $AlignmentList$
8: **return** $AlignmentList$ (sorted according to entropy)

Computing primers. In the second part (Lines 2-7), the algorithm considers all alignments A_h computed by Algorithm FINDALIGNMENTS. For each A_h, we use the list T_h of template strings (see below), and for each $\hat{p} \in T_h$ we call $\text{SRR}_{\text{dna}}(\hat{p}, A_h, d, m)$ to compute a primer p that is added to the list of primers PL_h. All lists PL_h are then combined into the final list of candidate primers.

It remains to explain how to choose the set T_h of templates. If the set $Tmpl_m(A_h)$ of all candidate templates is small then one can take T_h to be the whole set $Tmpl_m(A_h)$. (For instance, when $m = 0$ then $Tmpl_0(A_h) = A_h$.) In general, we take T_h to be a random sample of r strings from $Tmpl_m(A_h)$, where the value of r is a parameter of the program, which can be used to optimize the tradeoff between the accuracy and the running time. Each $\hat{p} \in T_h$ is constructed as follows: (i) choose uniformly a random $a^i \in A_h$, (ii) choose uniformly a set of exactly m random positions in a^i, and (iii) for each chosen position j in a^i, set a^i_j to a randomly chosen base, where this base is selected with probability proportional to its frequency in position j in all sequences from A_h.

5   Experiments

We tested Algorithm RRD2P on three biological data sets, and we compared our results to those from Algorithm HYDEN [7].
1. The first data set is a set of 50 sequences of human olfactory receptor (OR) gene [7], of length around 1Kbps, provided along with the HYDEN program.
2. The second data set is from the NCBI flu database [15], from which we chose Human flu sequences of lengths 900-1000 bps (dated from November 2013). This set contains 229 flu sequences.
3. The third one contains 160 fungal ITS genes of various lengths, obtained from NCBI-INSD [16]. Sequence lengths vary from 400 to 2000 bps.

We run Algorithm RRD2P with the following parameters: (i) Primer length = 25. (ii) Primer degeneracy threshold (forward, reverse) : (625,3750), (1250,7500), (1875, 11250), (2500, 15000), (3750, 22500), (5000,30000), (7500,45000), (10000,60000). Note that the degeneracy values increase roughly exponentially, which corresponds to a linear increase in the number of ambiguous bases.The degeneracy of the reverse primer is six times larger than that of the forward primer (the default in HYDEN). (iii) Forward primer binding range : $0 \sim 300$, reverse primer binding range : $-1 \sim -350$. (iv) Mismatch allowance : $m = 0, 1, 2$ (m represents the mismatch allowance for each primer separately). (v) Number of alignments: $N = 50$. (vi) Number of template samples: $r = 5$.

We compare our algorithm to HYDEN in terms of the coverage of computed primers. To make this comparison meaningful, we designed our algorithm to have similar input parameters, which allows us to run HYDEN with the same settings. For the purpose of these experiments, we use the best primer pair from the list computed by Algorithm RRD2P.

The results are shown in Figures 2, 3 and 4, respectively. The x-axis represents the degeneracy of the forward primer; the degeneracy of the reverse primer is six times larger. The y-axis is the coverage of the computed primer pair. The results show that RRD2P is capable to find better degenerate primers than HYDEN, for different choices of parameters.

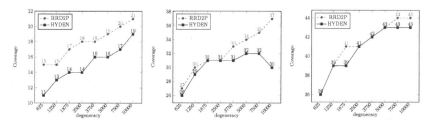

Fig. 2. Comparison of RRD2P and HYDEN on human OR genes for $m = 0$ (left), $m = 1$ (center) and $m = 2$ (right)

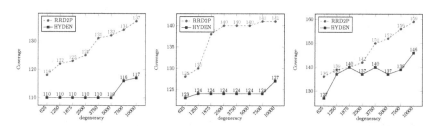

Fig. 3. Comparison of RRD2P and HYDEN on flu sequences for $m = 0$ (left), $m = 1$ (center) and $m = 2$ (right)

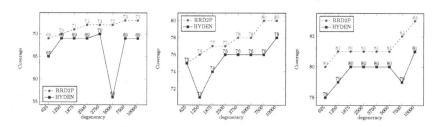

Fig. 4. Comparison of RRD2P and HYDEN on fungal sequences for $m = 0$ (left), $m = 1$ (center) and $m = 2$ (right)

Running time. The running time of Algorithm RRD2P is dominated by the module running Cplex to solve the linear program, and it depends, roughly linearly, on the number of times the LP solver is run. The above experiments were performed for $r = 5$. For the third dataset above and $m = 0$, the running times of Algorithm RRD2P varied from $110s$ for $d = 625$ to $164s$ for $d = 10000$ (on Windows 8 2.4 GHz CPU, 8.0 G memory). The respective run times of HYDEN were lower, between $25s$ and $28s$. The run time of Algorithm RRD2P can be adjusted by using smaller values of r. For example, for $r = 1, 2$, RRD2P is actually faster than HYDEN for small to moderate degeneracy values, and the loss of accuracy is not significant.

6   Discussion

We studied the problem of computing a pair of degenerate forward and reverse primers that maximizes the number of covered target sequences, assuming upper bounds on the primer degeneracy and the number of mismatches. We proposed an algorithm for this problem, called RRD2P, based on representing the problem as an integer linear program, solving its fractional relaxation, and then rounding the optimal fractional solution to integral values. We tested Algorithm RRD2P on three biological datasets. Our algorithm usually finds solutions that are near optimal or optimal, and it produces primer pairs with higher coverage than Algorithm HYDEN from [7], regardless of the parameters.

Our work focussed on optimizing the coverage of the sequence data by degenerate primers. Algorithm RRD2P does not consider biological parameters that affect the quality of the primers in laboratory PCR, including the melting temperature of the primers, GC content, secondary structure and other. In the future, we are planning to integrate Algorithm RRD2P into our software tool, called PRISE2 [17], that can be used to interactively design PCR primers based both on coverage and on a variety of biological parameters.

The integrality gap of the linear programs in Section 3.1 can be shown to be $\Omega(n(m + \log d)/k)$, where n is the number of target sequences, k is their length, d is the degeneracy bound and m is the mismatch allowance. An example with this integrality gap consists of n target binary sequences of length k such that each two differ from each other on more than $m + \log_2 d$ positions. When we choose any target sequence as template, the optimal coverage can only be 1. However, there is a fractional solution with value $n(m + \log d)/k$, obtained by setting $n_j = (\log d)/k$ and $\mu_j^i = m/k$, for all i, j.

Nevertheless, as we show, for real DNA datasets the solutions produced by rounding the fractional solution are very close to the optimum. Providing some analytical results that explain this phenomenon would be of considerable interest, both from the theoretical and practical standpoint, and will be a focus of our future work. This work would involve developing formal models for "conserved sequences" (or adapting existing ones) and establishing integrality gap results, both lower and upper bounds, for such datasets.

Acknowledgements. We would like to thank anonymous reviewers for pointing some deficiencies in the earlier version of the paper and for insightful comments.

References

1. Kwok, S., Chang, S., Sninsky, J., Wang, A.: A guide to the design and use of mismatched and degenerate primers. PCR Methods and Applications 47, S39–S47 (1994)
2. Hunt, D.E., Klepac-Ceraj, V., Acinas, S.G., Gautier, C., Bertilsson, S., Polz, M.F.: Evaluation of 23s rRNA PCR primers for use in phylogenetic studies of bacterial diversity. Applied Environmental Microbiology 72, 2221–2225 (2006)
3. Ihrmark, K., Bödeker, I.T., Cruz-Martinez, K., Friberg, H., Kubartova, A., Schenck, J., Strid, Y., Stenlid, J., Brandström-Durling, M., Clemmensen, K.E., Lindahl, B.D.: New primers to amplify the fungal its2 region–evaluation by 454-sequencing of artificial and natural communities. FEMS Microbiology Ecology 82, 666–677 (2012)
4. Chamberlain, J.S., Gibbs, R.A., Rainer, J.E., Nguyen, P.N., Casey, C.T.: Deletion screening of the duchenne muscular dystrophy locus via multiplex dna amplification. Nucleic Acid Research 16, 11141–11156 (1988)
5. Konwar, K.M., Mandoiu, I.I., Russell, A.C., Shvartsman, A.A.: Improved algorithms for multiplex PCR primer set selection with amplification length constraints. In: Proc. 3rd Asia-Pacific Bioinformatics Conference, pp. 41–50 (2005)
6. Linhart, C., Shamir, R.: The degenerate primer design problem: theory and applications. Journal of Computational Biology 12(4), 431–456 (2005)
7. Linhart, C., Shamir, R.: The degenerate primer design problem. Bioinformatics 180, S172–S180 (2002)
8. Souvenir, R., Buhler, J.P., Stormo, G., Zhang, W.: Selecting degenerate multiplex PCR primers. In: Benson, G., Page, R.D.M. (eds.) WABI 2003. LNCS (LNBI), vol. 2812, pp. 512–526. Springer, Heidelberg (2003)
9. Balla, S., Rajasekaran, S.: An efficient algorithm for minimum degeneracy primer selection. IEEE Transactions on NanoBioscience 6, 12–17 (2007)
10. Sharma, D., Balla, S., Rajasekaran, S., DiGirolamo, N.: Degenerate primer selection algorithms. Computational Intelligence in Bioinformatics and Computational Biology, 155–162 (2009)
11. Duitama, J., Kumar, D.M., Hemphill, E., Khan, M., Mandoiu, I.I., Nelson, C.E.: Primerhunter: a primer design tool for PCR-based virus subtype identification. Nucleic Acids Research 37, 2483–2492 (2009)
12. http://dna.engr.uconn.edu/software/PrimerHunter/primerhunter.php
13. Boyce, R., Chilana, P., Rose, T.M.: iCODEHOP: a new interactive program for designing COnsensus-DEgenerate Hybrid Oligonucleotide Primers from multiply aligned protein sequences. Nucleid Acids Research 37, 222–228 (2009)
14. http://blocks.fhcrc.org/codehop.html
15. http://www.ncbi.nlm.nih.gov/genomes/FLU/FLU.html
16. http://www.ncbi.nlm.nih.gov/genbank/collab/
17. Huang, Y.T., Yang, J.I., Chrobak, M., Borneman, J.: Prise2: Software for designing sequence-selective PCR primers and probes (2013) (in preparation)

GAML: Genome Assembly by Maximum Likelihood

Vladimír Boža, Broňa Brejová, and Tomáš Vinař

Faculty of Mathematics, Physics, and Informatics, Comenius University,
Mlynská dolina, 842 48 Bratislava, Slovakia

Abstract. The critical part of genome assembly is resolution of repeats and scaffolding of shorter contigs. Modern assemblers usually perform this step by heuristics, often tailored to a particular technology for producing paired reads or long reads. We propose a new framework that allows systematic combination of diverse sequencing datasets into a single assembly. We achieve this by searching for an assembly with maximum likelihood in a probabilistic model capturing error rate, insert lengths, and other characteristics of each sequencing technology.

We have implemented a prototype genome assembler GAML that can use any combination of insert sizes with Illumina or 454 reads, as well as PacBio reads. Our experiments show that we can assemble short genomes with N50 sizes and error rates comparable to ALLPATHS-LG or Cerulean. While ALLPATHS-LG and Cerulean require each a specific combination of datasets, GAML works on any combination.

Data and software is available at http://compbio.fmph.uniba.sk/gaml

1 Introduction

The second and third generation sequencing technologies have dramatically decreased the cost of sequencing. Nowadays, we have a surprising variety of sequencing technologies, each with its own strengths and weaknesses. For example, Illumina platforms are characteristic by low cost and high accuracy, but the reads are short. On the other hand, Pacific Biosciences offer long reads at the cost of quality and coverage. In the meantime, the cost of sequencing was brought down to the point, where it is no longer a sole domain of large sequencing centers; even small labs can experiment with cost-effective genome sequencing. In this setting, it is no longer possible to recommend a single protocol that should be used to sequence genomes of a particular size. In this paper, we propose a framework for genome assembly that allows flexible combination of datasets from different technologies in order to harness their individual strengths.

Modern genome assemblers are usually based either on the overlap–layout–consensus framework (e.g. Celera by Myers et al. (2000), SGA by Simpson and Durbin (2010)), or on de Bruijn graphs (e.g. Velvet by Zerbino and Birney (2008), ALLPATHS-LG by Gnerre et al. (2011)). Both approaches can be seen as special cases of a string graph (Myers, 2005), in which

D. Brown and B. Morgenstern (Eds.): WABI 2014, LNBI 8701, pp. 122–134, 2014.

we represent sequence fragments as vertices, while edges represent possible adjacencies of fragments in the assembly. A genome assembly is simply a set of walks through this graph. The main difference between the two frameworks is how we arrive at a string graph: through detecting long overlaps of reads (overlap–layout–consensus) or through construction of de Bruijn graphs based on k-mers.

However, neither of these frameworks is designed to systematically handle pair-end reads and additional heuristic steps are necessary to build larger scaffolds from assembled contigs. For example, ALLPATHS-LG (Gnerre et al., 2011) uses libraries with different insert lengths for scaffolding of contigs assembled without the use of paired read information, while Cerulean (Deshpande et al., 2013) uses Pacific Biosystems long reads for the same purpose. Recently, the techniques of paired de Bruijn graphs (Medvedev et al., 2011) and pathset graphs (Pham et al., 2013) were developed to address paired reads systematically, however these approaches cannot combine several libraries with different insert sizes.

Combination of sequencing technologies with complementary strengths can help to improve assembly quality. However, it is not feasible to design new algorithms for every possible combination of datasets. Often it is possible to supplement previously developed tools with additional heuristics for new types of data. For example, PBJelly (English et al., 2012) uses Pacific Biosystems reads solely to aid gap filling in draft assemblies. Assemblers like PacbioToCa (Koren et al., 2012) or Cerulean (Deshpande et al., 2013) use short reads to "upgrade" the quality of Pacific Biosystems reads so that they can be used within traditional assemblers. However, such approaches hardly use all information contained within the data sets.

We propose a new framework that allows a systematic combination of diverse datasets into a single assembly, without requiring a particular type of data for specific heuristic steps. Recently, probabilistic models have been used very successfully to evaluate the quality of genome assemblers (Rahman and Pachter, 2013; Clark et al., 2013; Ghodsi et al., 2013). In our work, we use likelihood of a genome assembly as an optimization criterion, with the goal of finding the highest likelihood genome assembly. Even though this may not be always feasible, we demonstrate that optimization based on simulated annealing can be very successful at finding high likelihood genome assemblies.

To evaluate likelihood, we use a relatively complex model adapted from Ghodsi et al. (2013), which can capture characteristics of each dataset, such as sequencing error rate, as well as length distribution and expected orientation of paired reads (Section 2). We can thus transparently combine information from multiple diverse datasets into a single score. Previously, there have been several works in this direction in much simpler models without sequencing errors (Medvedev and Brudno, 2009; Varma et al., 2011). These papers used likelihood to estimate repeat counts, without considering other problems, such as how exactly are repeats integrated within scaffolds.

To test our framework, we have implemented a prototype genome assembler GAML (Genome Assembly by Maximum Likelihood) that can use any combination of insert sizes with Illumina or 454 reads, as well as PacBio reads. The starting point of the assembly are short contigs derived from Velvet (Zerbino and Birney, 2008) with very conservative settings in order to avoid assembly errors. We then use simulated annealing to combine these short contigs into high likelihood assemblies (Section 3). We compare our assembler to existing tools on benchmark datasets (Section 4), demonstrating that we can assemble genomes of up to 10 MB long with N50 sizes and error rates comparable to ALLPATHS-LG or Cerulean. While ALLPATHS-LG and Cerulean each require a very specific combination of datasets, GAML works on any combination.

2   Probabilistic Model for Sequence Assembly

Recently, several probabilistic models were introduced as a measure of the assembly quality (Rahman and Pachter, 2013; Clark et al., 2013; Ghodsi et al., 2013). All of these authors have shown that the likelihood consistently favours higher quality assemblies. In general, the probabilistic model defines the probability $\Pr(R|A)$ that a set of sequencing reads R is observed assuming that assembly A is the correct assembly of the genome. Since the sequencing itself is a stochastic process, it is very natural to characterize concordance of reads and an assembly by giving a probability of observing a particular read. In our work, instead of evaluating the quality of a single assembly, we use the likelihood as an optimization criterion with the goal of finding high likelihood genome assemblies. We adapt the model of Ghodsi et al. (2013), which we describe in this section.

Basics of the likelihood model. The model assumes that individual reads are independently sampled, and thus the overall likelihood is the product of likelihoods of the reads: $\Pr(R|A) = \prod_{r \in R} \Pr(r|A)$. To make the resulting value independent of the number of reads in set R, we use as the main assembly score the log average probability of a read computed as follows: $\mathrm{LAP}(A|R) = (1/|R|) \sum_{r \in R} \log \Pr(r|A)$. Note that maximizing $\Pr(R|A)$ is equivalent to maximizing $\mathrm{LAP}(A|R)$.

If the reads were error-free and each position in the genome was sequenced equally likely, the probability of observing read r would simply be $\Pr(r|A) = n_r/(2L)$, where n_r is the number of occurrences of the read as a substring of the assembly A, L is the length of A, and thus $2L$ is the length of the two strands combined (Medvedev and Brudno, 2009). Ghodsi et al. (2013) have shown a dynamic programming computation of read probability for more complex models, accounting for sequencing errors. The algorithm marginalizes over all possible alignments of r and A, weighting each by the probability that a certain number of substitution and indel errors would happen during sequencing. In particular, the probability of a single alignment with m matching positions and s errors (substitution and indels) is defined as $R(s,m)/(2L)$, where $R(s,m) = \epsilon^s (1-\epsilon)^m$ and ϵ is the sequencing error rate.

However, full dynamic programming is too time consuming, and in practice only several best alignments contribute significantly to the overall probability. Thus Ghodsi et al. (2013) propose to approximate the probability of observing read r with an estimate based on a set S_r of a few best alignments of r to genome A, as obtained by a standard fast read alignment tool:

$$\Pr(r|A) \approx \frac{\sum_{j \in S_r} R(s_j, m_j)}{2L}, \tag{1}$$

where m_j is the number of matches in the j-th alignment, and s_j is the number of mismatches and indels implied by this alignment. The formula assumes the simplest possible error model, where insertions, deletions and substitutions have the same probability and ignores GC content bias. Of course, much more comprehensive read models are possible (see e.g. Clark et al. (2013)).

Paired reads. Many technologies provide paired reads produced from the opposite ends of a sequence insert of certain size. We assume that the insert size distribution in a set of reads R can be modeled by the normal distribution with known mean μ and standard deviation σ. The probability of observing paired reads r_1 and r_2 can be estimated from sets of alignments S_{r_1} and S_{r_2} as follows:

$$\Pr(r_1, r_2|A) \approx \frac{1}{2L} \sum_{j_1 \in S_{r_1}} \sum_{j_2 \in S_{r_2}} R(s_{j_1}, m_{j_1}) R(s_{j_2}, m_{j_2}) \Pr(d(j_1, j_2)|\mu, \sigma) \tag{2}$$

As before, m_{j_i} and s_{j_i} are the numbers of matches and sequencing errors in alignment j_i respectively, and $d(j_1, j_2)$ is the distance between the two alignments as observed in the assembly. If alignments j_1 and j_2 are in two different contigs, or on inconsistent strands, $\Pr(d(j_1, j_2)|\mu, \sigma)$ is zero.

Reads that have no good alignment to A. Some reads or read pairs do not align well to A, and as a result, their probability $\Pr(r|A)$ is very low; our approximation by a set of high-scoring alignments can even yield zero probability if set S_r is empty. Such extremely low probabilities then dominate the log likelihood score. Ghodsi et al. (2013) propose a method that assigns such a read a score approximating the situation when the read would be added as a new contig to the assembly. We modify their formulas for variable read length, and use score $e^{c+k\ell}$ for a single read of length ℓ or $e^{c+k(\ell_1+\ell_2)}$ for a pair of reads of lengths ℓ_1 and ℓ_2. Values k and c are scaling constants set similarly as in Ghodsi et al. (2013). These alternative scores are used instead of the read probability $\Pr(r|A)$ whenever the probability is lower than the score.

Multiple read sets. Our work is specifically targeted at a scenario, where we have multiple read sets obtained from different libraries with different insert lengths or even with different sequencing technologies. We use different model parameters for each set and compute the final score as a weighted combination of log average probabilities for individual read sets $R_1, \ldots, R_k$:

$$\text{LAP}(A|R_1, \ldots, R_k) = w_1 \text{LAP}(A|R_1) + \ldots + w_k \text{LAP}(A|R_k) \tag{3}$$

In our experiments we use weight $w_i = 1$ for most datasets, but we lower the weight for Pacific Biosciences reads, because otherwise they dominate the likelihood value due to their longer length. The user could also increase or decrease weights w_i of individual sets based on their reliability.

Penalizing spuriously joined contigs. The model of Ghodsi et al. (2013) does not penalize obvious misassemblies when two contigs are joined together without any evidence in the reads. We have observed that to make the likelihood function applicable as an optimization criterion for the best assembly, we need to introduce a penalty for such spurious connections. We say that a particular base j in the assembly is *connected* with respect to read set R if there is a read which covers base j and starts at least k bases before j, where k is a constant specific to the read set. In this setting, we treat a pair of reads as one long read. If the assembly contains d disconnected bases with respect to R, penalty αd is added to the LAP$(A|R)$ score (α is a scaling constant).

Properties of different sequencing technologies. Our model can be applied to different sequencing technologies by appropriate settings of model parameters. For example, Illumina technology typically produces reads of length 75-150bp with error rate below 1% (Quail et al., 2012). For smaller genomes, we often have a high coverage of Illumina reads. Using paired reads or mate pair technologies, it is possible to prepare libraries with different insert sizes ranging up to tens of kilobases, which are instrumental in resolving longer repeats (Gnerre et al., 2011). To align these reads to proposed assemblies, we use Bowtie2 (Langmead and Salzberg, 2012). Similarly, we can process reads by the Roche 454 technology, which are characteristic by higher read lengths (hundreds of bases).

Pacific Biosciences technology produces single reads of variable length, with median length reaching several kilobases, but the error rate exceeds 10% (Quail et al., 2012; Deshpande et al., 2013). Their length makes them ideal for resolving ambiguities in alignments, but the high error rate makes their use challenging. To align these reads, we use BLASR (Chaisson and Tesler, 2012). When we calculate the probability $\Pr(r|A)$, we consider not only the best alignments found by BLASR, but for each BLASR alignment, we also add probabilities of similar alignments in its neighborhood. More specifically, we run a banded version of the forward algorithm by Ghodsi et al. (2013), considering all alignments in a band of size 3 around a guide alignment produced by BLASR.

3   Finding a High Likelihood Assembly

Complex probabilistic models, like the one described in Section 2, were previously used to compare the quality of several assemblies (Ghodsi et al., 2013; Rahman and Pachter, 2013; Clark et al., 2013). In our work, we instead attempt to find the highest likelihood assembly directly. Of course, the search space is huge, and the objective function too complex to admit exact methods. Here, we describe an effective optimization routine based on the simulated annealing framework (Eglese, 1990).

Our algorithm for finding the maximum likelihood assembly consists of three main steps: preprocessing, optimization, and postprocessing. In *preprocessing*, we decrease the scale of the problem by creating an assembly graph, where vertices correspond to contigs and edges correspond to possible adjacencies between contigs supported by reads. In order to make the search viable, we will restrict our search to assemblies that can be represented as a set of walks in this graph. Therefore, the assembly graph should be built in a conservative way, where the goal is not to produce long contigs, but rather to avoid errors inside them. In the *optimization step*, we start with an initial assembly (a set of walks in the assembly graph), and iteratively propose changes in order to optimize the assembly likelihood. Finally, *postprocessing* examines the resulting walks and splits some of them into shorter contigs if there are multiple equally likely possibilities of resolving ambiguities. This happens, for example, when the genome contains long repeats that cannot be resolved by any of the datasets.

In the rest of this section, we discuss individual steps in more detail.

3.1 Optimization by Simulated Annealing

To find a high likelihood assembly, we use an iterative simulated annealing scheme. We start from an initial assembly A_0 in the assembly graph. In each iteration, we randomly choose a *move* that proposes a new assembly A' similar to the current assembly A. The next step depends on the likelihoods of the two assemblies A and A' as follows:

- If $\mathrm{LAP}(A'|R) \geq \mathrm{LAP}(A|R)$, the new assembly A' is accepted and the algorithm continues with the new assembly.
- If $\mathrm{LAP}(A'|R) < \mathrm{LAP}(A|R)$, the new assembly A' is accepted with probability $e^{(\mathrm{LAP}(A'|R)-\mathrm{LAP}(A|R))/T}$; otherwise A' is rejected and the algorithm retains the old assembly A for the next step.

Here, parameter T is called the temperature, and it changes over time. In general, the higher the temperature, the more aggressive moves are permitted. We use a simple cooling schedule, where $T = T_0/\ln(i)$ in the i-th iteration. The computation ends when there is no improvement in the likelihood for a certain amount of time. We select the assembly with the highest LAP score as the result.

To further reduce the complexity of the assembly problem, we classify all contigs as either *long* (more than 500bp) or *short* and concentrate on ordering the long contigs correctly. The short contigs are used to fill the gaps between the long contigs.

Recall that each assembly is a set of walks in the assembly graph. A contig can appear in more than one walk or can be present in a single walk multiple times. In all our experiments, the starting assembly simply contains each long contig as a separate walk. However, other assemblies (such as assemblies from other tools) can easily serve as a starting point as long as they can be mapped to the assembly graph.

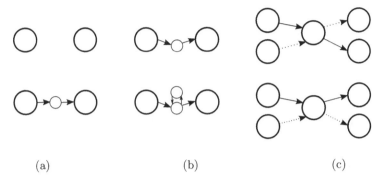

Fig. 1. Examples of proposal moves. (a) Walk extension joining two walks. (b) Local improvement by addition of a new loop. (c) Repeat interchange.

Proposals of new assemblies are created from the current assembly using the following moves:

- *Walk extension.* (Fig.1a) We start from one end of an existing walk and randomly walk through the graph, at every step uniformly choosing one of the edges outgoing from the current node. Each time we encounter the end of another walk, the two walks are considered for joining. We randomly (uniformly) decide whether we join the walks, end the current walk without joining, or continue walking.
- *Local improvement.* (Fig.1b) We optimize the part of some walk connecting two long contigs s and t. We first sample multiple random walks starting from contig s. In each walk, we only consider nodes from which contig t is reachable. Then we evaluate these random walks and choose the one that increases the likelihood the most. If the gap between contigs s and t is too big, we instead use a greedy strategy where in each step we explore multiple random extensions of the walk (of length around 200bp) and pick the one with the highest score.
- *Repeat optimization.* We optimize the copy number of short tandem repeats. We do this by removing or adding a loop to some walk. We precompute the list of all short loops (up to five nodes) in the graph and use it for adding loops.
- *Joining with advice.* We join two walks that are spanned by long reads or paired reads with long inserts. We fist select a starting walk, align all reads to the starting walk and randomly choose a read which has the other end outside the current walk. Then we find to which node this other end belongs to and join appropriate walks. If possible, we fill the gap between the two walks using the same procedure as in the local improvement move. Otherwise we introduce a gap filled with Ns.
- *Disconnecting.* We remove a path through short contigs connecting two long contigs in the same walk, resulting in two shorter walks.

- *Repeat interchange.* (Fig.1c) If a long contig has several incoming and outgoing walks, we optimize the pairing of incoming and outgoing edges. In particular, we evaluate all moves that exchange parts of two walks through this contig. If one of these changes improves the score, we accept it and repeat this step, until the score cannot be improved at this contig.

At the beginning of each annealing step, the type of the move is chosen randomly; each type of move has its own probability. We also choose randomly the contig at which we attempt to apply the move.

Note that some moves (e.g. local improvement) are very general, while other moves (e.g. joining with advice) are targeted at specific types of data. This does not contradict a general nature of our framework; it is possible to add new moves as new types of data emerge, leading to improvement when using specific data sets, while not affecting the performance when such data is unavailable.

3.2   Preprocessing and Postprocessing

To obtain the assembly graph, we use Velvet with basic error correction and unambiguous concatenation of k-mers. These settings will produce very short contigs, but will also give a much lower error rate than a regular Velvet run.

The resulting assembly obtained by the simulated annealing may contain positions with no evidence for a particular configuration of incoming and outgoing edges in the assembly graph (e.g., a repeat that is longer than the span of the longest paired read). Such arbitrary joining of walks may lead to assembly errors, since data give no indication which configuration of edges is correct. In the postprocessing step, we therefore apply the repeat interchange move at every possible location of the assembly. If the likelihood change resulting from such a move is negligible, we break corresponding walks into shorter contigs.

3.3   Fast Likelihood Evaluation

The most time consuming step in our algorithm is evaluation of the assembly likelihood, which we perform in each iteration of simulated annealing. This step involves alignment of a large number of reads to the assembly. However, we can significantly reduce required time by using the fact that only a small part of the assembly is changed in each annealing step.

To achieve this, we split walks into overlapping windows, each window containing several adjacent contigs of a walk. Windows should be as short as possible, but the adjacent windows should overlap by at least $2\ell_r$ bases, where ℓ_r is the length of the longest read. As a result, each alignment is completely contained in at least one window even in the presence of extensive indels.

We determine window boundaries by a simple greedy strategy, which starts at the first contig of a walk, and then extends the window by at least $2\ell_r$ bases beyond the boundary of the first contig. The next window always starts at the latest possible location that ensures a sufficient overlap and extends at least $2\ell_r$ bases beyond the end of the previous window.

For each window, we keep the position and edit distance of all alignments. In each annealing step, we identify which windows of the assembly were modified. We glue together overlapping windows and align reads against these sequences using a read mapping tool. Finally, we use alignments in all windows to calculate the probability of each read and combine them into the score of the whole assembly. This step requires careful implementation to ensure that we count each alignment exactly once.

To speed up read mapping even more, we use a simple prefiltering scheme, where we only align reads which contain some k-mer (usually $k = 13$) from the target sequence. In the current implementation, we store an index of all k-mers from all reads in a simple hash map. In each annealing step, we can therefore iterate over all k-mers in the target portion of the genome and retrieve reads that contain them. We use a slightly different filtering approach for PacBio reads. In particular, we take all reasonably long contigs (at least 100 bases) and align them to PacBio reads. Since BLASR can find alignments where a contig and a read overlap by only around 100 bases, we can use these alignments as a filter.

4   Experimental Evaluation

We have implemented the algorithm proposed in the previous section in a prototype assembler GAML (Genome Assembly by Maximum Likelihood). At this stage, GAML can assemble small genomes (approx. 10 Mbp) in a reasonable amount of time (approximately 4 days on a single CPU and using 50GB of memory). In future, we plan to explore efficient data structures to further speed up likelihood computation and to lower the memory requirements.

To evaluate the quality of our assembler, we have adopted the methodology of Salzberg et al. (2012) used for Genome Assembly Gold-Standard Evaluation, using metrics on scaffolds. We have used the same genomes and libraries as in Salzberg et al. (2012) (the *S. aureus* genome) and in Deshpande et al. (2013) (the *E. coli* genome); the overview of the data sets is shown in Tab.1. An additional dataset EC3 (long insert, low coverage) was simulated using the ART software (Huang et al., 2012). We have evaluated GAML in three different scenarios:

1. combination of fragment and short insert Illumina libraries (SA1, SA2),
2. combination of a fragment Illumina library and a long-read high-error-rate Pacific Biosciences library (EC1, EC2),
3. combination of a fragment Illumina library, a long-read high-error-rate Pacific Biosciences library, and a long jump Illumina library (EC1, EC2, EC3)

In each scenario, we use the short insert Illumina reads (SA1 or EC1) in Velvet with conservative settings to build the initial contigs and assembly graph. In the LAP score, we give Illumina datasets weight 1 and PacBio dataset weight 0.01.

The results are summarized in Tab.2. Note that none of the assemblers considered here can effectively run in all three of these scenarios, except for GAML.

Table 1. Properties of data sets used

ID	Source	Technology	Insert len. (bp)	Read len. (bp)	Coverage	Error rate
Staphylococus aureus (2.87Mbp)						
SA1	Salzberg et al. (2012)	Illumina	180bp	101bp	90	3%
SA2	Salzberg et al. (2012)	Illumina	3500bp	37bp	90	3%
Escherichia coli (4.64Mbp)						
EC1	Deshpande et al. (2013)	Illumina	300bp	151bp	400	0.75%
EC2	Deshpande et al. (2013)	PacBio		4000bp	30	13%
EC3	simulated	Illumina	37,000bp	75bp	0.5	4%

In the first scenario, GAML performance ranks third among zero-error assemblers in N50 length. The best N50 assembly is given by ALLPATHS-LG (Gnerre et al., 2011). A closer inspection of the assemblies indicates that GAML missed several possible joins. One such miss was caused by a 4.5 kbp repeat, while the longest insert size in this dataset is 3.5 kbp. Even though in such cases it is sometimes possible to reconstruct the correct assembly thanks to small differences in the repeated regions, the difference in likelihood between alternative repeat resolutions may be very small. Another missed join was caused by a sequence coverage gap penalized in our scoring function. Perhaps in both of these cases the manually set constants may have caused GAML to be overly conservative. Otherwise, the GAML assembly seems very similar to the one given by ALLPATHS-LG.

In the second scenario, Pacific Biosystems reads were employed instead of jump libraries. These reads pose a significant challenge due to their high error rate, but they are very useful due to their long length. Assemblers such as Cerulean (Deshpande et al., 2013) deploy special algorithms taylored to this technology. GAML, even though not explicitly tuned to handle Pacific Biosystems reads, builds an assembly with N50 size and the number of scaffolds very similar to that of Cerulean. In N50, both programs are outperformed by PacbioToCA (Koren et al., 2012), however, this is again due to a few very long repeats (approx. 5000 bp) in the reference genome which were not resolved by GAML or Cerulean. (Deshpande et al. (2013) also aim to be conservative in repeat resolution.) Note that in this case, simulated annealing failed to give the highest likelihood assembly among those that we examined, so perhaps our results can be improved by tuning the likelihood optimization.

Finally, the third scenario shows that the assembly quality can be hugely improved by including a long jump library, even if the coverage is really small (we used 0.5× coverage in this experiment). This requires a flexible genome assembler; in fact, only Celera (Myers et al., 2000) can process this data, but GAML assembly is clearly superior. We have attempted to also run ALLPATHS-LG, but the program could not process this combination of libraries. Compared to the previous scenario, GAML N50 size increased approximately 7 fold (or approx. 4 fold compared to the best N50 from the second scenario assemblies).

Table 2. Comparison of assembly accuracy in three experiments. For all
assemblies, N50 values are based on the actual genome size. All misjoins were considered
as errors and error-corrected values of N50 and contig sizes were obtained by breaking
each contig at each error (Salzberg et al., 2012). All assemblies except for GAML and
conservative Velvet were obtained from Salzberg et al. (2012) in the first experiment,
and from Deshpande et al. (2013) in the second experiment.

Assembler	Number of scaffolds	Longest scaffold (kb)	Longest scaffold corr. (kb)	N50 (kb)	Err.	N50 corr. (kb)	LAP
Staphylococus aureus, read sets SA1, SA2							
GAML	28	1191	1191	514	0	514	−23.45
Allpaths-LG	**12**	1435	**1435**	1092	0	1092	−25.02
SOAPdenovo	99	518	518	332	0	332	−25.03
Velvet	45	958	532	762	17	126	−25.34
Bambus2	17	1426	1426	1084	0	1084	−25.73
MSR-CA	17	**2411**	1343	**2414**	3	1022	−26.26
ABySS	246	125	125	34	1	28	−29.43
Cons. Velvet*	219	95	95	31	0	31	−30.82
SGA	456	286	286	208	1	208	−31.80
Escherichia coli, read sets EC1, EC2							
PacbioToCA	55	1533	1533	**957**	0	**957**	−33.86
GAML	29	1283	1283	653	0	653	−33.91
Cerulean	**21**	**1991**	**1991**	694	0	694	−34.18
AHA	54	477	477	213	5	194	−34.52
Cons. Velvet*	383	80	80	21	0	21	−36.02
Escherichia coli, read sets EC1, EC2, EC3							
GAML	4	**4662**	**4661**	**4662**	3	**4661**	**−60.38**
Celera	19	4635	2085	4635	19	2085	−61.47
Cons. Velvet*	383	80	80	21	0	21	−72.03

*: Velvet with conservative settings used to create the assembly graph in our method.

5   Conclusion

We have presented a new probabilistic approach to genome assembly, maximizing
likelihood in a model capturing essential characteristics of individual sequencing
technologies. It can be used on any combination of read datasets and can be
easily adapted to other technologies arising in the future.

Our work opens several avenues for future research. First, we plan to implement more sophisticated data structures to improve running time and memory
and to allow the use of our tool on larger genomes. Second, the simulated annealing procedure could be improved by optimizing probabilities of individual
moves or devising new types of moves. The tool could also be easily adapted to
improve existing assemblies after converting a given assembly to a set of walks.
Finally, it would be interesting to explore even more detailed probabilistic models, featuring coverage biases and various sources of experimental error.

Acknowledgements. This research was funded by VEGA grants 1/1085/12 (BB) and 1/0719/14 (TV). The authors would like to thank Viraj Deshpande for sharing his research data.

References

Chaisson, M.J., Tesler, G.: Mapping single molecule sequencing reads using basic local alignment with successive refinement (BLASR): application and theory. BMC Bioinformatics 13(1), 238 (2012)

Clark, S.C., Egan, R., Frazier, P.I., Wang, Z.: ALE: a generic assembly likelihood evaluation framework for assessing the accuracy of genome and metagenome assemblies. Bioinformatics 29(4), 435–443 (2013)

Deshpande, V., Fung, E.D.K., Pham, S., Bafna, V.: Cerulean: A hybrid assembly using high throughput short and long reads. In: Darling, A., Stoye, J. (eds.) WABI 2013. LNCS, vol. 8126, pp. 349–363. Springer, Heidelberg (2013)

Eglese, R.: Simulated annealing: a tool for operational research. European Journal of Operational Research 46(3), 271–281 (1990)

English, A.C., Richards, S., et al.: Mind the gap: upgrading genomes with Pacific Biosciences RS long-read sequencing technology. PLoS One 7(11), e47768 (2012)

Ghodsi, M., Hill, C.M., Astrovskaya, I., Lin, H., Sommer, D.D., Koren, S., Pop, M.: De novo likelihood-based measures for comparing genome assemblies. BMC Research Notes 6(1), 334 (2013)

Gnerre, S., MacCallum, I., et al.: High-quality draft assemblies of mammalian genomes from massively parallel sequence data. Proceedings of the National Academy of Sciences 108(4), 1513–1518 (2011)

Huang, W., Li, L., Myers, J.R., Marth, G.T.: ART: a next-generation sequencing read simulator. Bioinformatics 28(4), 593–594 (2012)

Koren, S., Schatz, M.C., et al.: Hybrid error correction and de novo assembly of single-molecule sequencing reads. Nature Biotechnology 30(7), 693–700 (2012)

Langmead, B., Salzberg, S.L.: Fast gapped-read alignment with Bowtie 2. Nature Methods 9(4), 357–359 (2012)

Medvedev, P., Brudno, M.: Maximum likelihood genome assembly. Journal of Computational Biology 16(8), 1101–1116 (2009)

Medvedev, P., Pham, S., Chaisson, M., Tesler, G., Pevzner, P.: Paired de Bruijn graphs: a novel approach for incorporating mate pair information into genome assemblers. Journal of Computational Biology 18(11), 1625–1634 (2011)

Myers, E.W.: The fragment assembly string graph. Bioinformatics 21(suppl 2), ii79–ii85 (2005)

Myers, E.W., Sutton, G.G., et al.: A whole-genome assembly of Drosophila. Science 287(5461), 2196–2204 (2000)

Pham, S.K., Antipov, D., Sirotkin, A., Tesler, G., Pevzner, P.A., Alekseyev, M.A.: Pathset graphs: a novel approach for comprehensive utilization of paired reads in genome assembly. Journal of Computational Biology 20(4), 359–371 (2013)

Quail, M.A., Smith, M., Coupland, P., Otto, T.D., Harris, S.R., Connor, T.R., Bertoni, A., Swerdlow, H.P., Gu, Y.: A tale of three next generation sequencing platforms: comparison of Ion Torrent, Pacific Biosciences and Illumina MiSeq sequencers. BMC Genomics 13(1), 341 (2012)

Rahman, A., Pachter, L.: CGAL: computing genome assembly likelihoods. Genome Biology 14(1), R8 (2013)

Salzberg, S.L., Phillippy, A.M., et al.: GAGE: a critical evaluation of genome assemblies and assembly algorithms. Genome Research 22(3), 557–567 (2012)

Simpson, J.T., Durbin, R.: Efficient construction of an assembly string graph using the FM-index. Bioinformatics 26(12), i367–i373 (2010)

Varma, A., Ranade, A., Aluru, S.: An improved maximum likelihood formulation for accurate genome assembly. In: Computational Advances in Bio and Medical Sciences (ICCABS 2011), pp. 165–170. IEEE (2011)

Zerbino, D.R., Birney, E.: Velvet: algorithms for de novo short read assembly using de Bruijn graphs. Genome Research 18(5), 821–829 (2008)

A Common Framework for Linear and Cyclic Multiple Sequence Alignment Problems

Sebastian Will[1] and Peter F. Stadler[1−7]

[1] Dept. Computer Science, and Interdisciplinary Center for Bioinformatics, Univ.
Leipzig, Härtelstr. 16-18, Leipzig, Germany
[2] MPI Mathematics in the Sciences, Inselstr. 22, Leipzig, Germany
[3] FHI Cell Therapy and Immunology, Perlickstr. 1, Leipzig, Germany
[4] Dept. Theoretical Chemistry, Univ. Vienna, Währingerstr. 17, Wien, Austria
[5] Bioinformatics and Computational Biology research group, University of Vienna,
A-1090 Währingerstraße 17, Vienna, Austria
[6] RTH, Univ. Copenhagen, Grønnegårdsvej 3, Frederiksberg C, Denmark
[7] Santa Fe Institute, 1399 Hyde Park Rd., Santa Fe, USA

Abstract. Circularized RNAs have received considerable attention is the last few years following the discovery that they are not only a rather common phenomenon in the transcriptomes of Eukarya and Archaea but also may have key regulatory functions. This calls for the adaptation of basic tools of sequence analysis to accommodate cyclic sequences. Here we discuss a common formal framework for linear and circular alignments as partitions that preserve (cyclic) order. We focus on the similarities and differences and describe a prototypical ILP formulation.

Keywords: cyclic sequence alignment, multiple sequence alignment, cyclic orders, integer linear programming, circular RNAs.

1 Introduction

While only recently considered a rare oddity, circular RNAs have been identified as a quite common phenomenon in Eukaryotic as well as Archaeal transcriptomes. In Mammalia, thousands of circular RNAs are reported together with evidence for regulation of miRNAs and transcription [1]; in Archaea, "expected" circRNAs like excised tRNA introns and intermediates of rRNA processing as well as many circular RNAs of unknown function have been revealed [2]. Most methods to comparatively analyze biological sequences require the computation of multiple alignments as a first step. While this task has received plenty of attention for linear sequences, comparably little is known for the corresponding circular problem. Although most bacterial and archaeal genomes are circular, this fact can be ignored for the purpose of constructing alignments, because genome-wide alignments are always anchored locally and then reduce to linear alignment problems. Even for mitochondrial genomes, with a typical size of 10-100kb, anchors are readily identified so that alignment algorithms for linear sequences are applicable. This situation is different, however, for short RNAs such as viroid

D. Brown and B. Morgenstern (Eds.): WABI 2014, LNBI 8701, pp. 135–147, 2014.

and other small satellite RNAs [3] or the abundant circular non-coding RNAs in Archaea [2, 4].

Cyclic pairwise alignment problems were considered e.g. in [5–8], often with applications to 2D shape recognition rather than to biological sequences. Cyclic alignments obviously can be computed by dynamic programming by linearizing the sequences at a given match, resulting in an quintic-time algorithm for general gap cost functions by solving $O(n^2)$ general linear alignment problems in cubic time [9]. For affine gap costs, the linear problem is solved in quadratic time by the Needleman-Wunsch algorithm, resulting in a quartic-time solution for the cyclic problem. For linear gap costs, a $O(n^2 \log n)$ cyclic alignment algorithm exists, capitalizing on the fact that alignment traces do not intersect in this case [7]. A variant was applied to identify cyclically permuted repeats [10]. This approach does not generalize to other types of cost functions, however. In [11], a cubic-time dynamic programming solution is described for affine gap cost. Multiple alignments of cyclic sequences have received very little attention. A progressive alignment algorithm was implemented and applied to viroid phylogeny [11].

Since the multiple alignment problem is NP-hard for all interesting scoring functions [12–14], we focus here on a versatile ILP formulation that can easily accommodate both linear and circular input strings. To this end, we start from the combinatorial characterization of multiple alignments of linear sequences [15, 16] and represent multiple alignments as order-preserving partitions, a framework that lends itself to a natural ILP formulation. This approach is not restricted to pure sequence alignments but also covers various models of RNA sequence-structure alignments, requiring modifications of the objective function only.

2   Cyclic Multiple Alignments

2.1   Formal Model of Linear Sequence Alignments

Morgenstern and collaborators [15, 16] introduced a combinatorial framework characterizing multiple alignments, which we recapitulate slightly paraphrasing. While multiple alignments are often viewed as matrices of sequence letters (and gaps) or –less frequently– as graphs, where sequence positions in the same *alignment column* are connected by *alignment edges*, this framework defines alignments in terms of a "multiple alignment partition". The connection between these concepts being that the sets of sequence positions in the same column (or mutually connected by alignment edges) form the classes of the partition.

We are given a set $\{s^{(1)}, \ldots, s^{(M)}\}$ of M sequence strings with lengths $n_a = |s^{(a)}|$ for $a = 1, \ldots, M$. For the moment, we need only the structure of the sequence strings as finite, totally ordered sets. The actual letters s_k^a impact only the scoring function, which we largely consider as given. For convenience, we refer to the *sequence index* $a \in \{1, \ldots, M\}$ of the sequence string $s^{(a)}$ simply as *sequence*. The tuple (a, k) denotes sequence position k in sequence a. We denote the set of sequence positions by $X = \{(a, k) | 1 \leq a \leq M, 1 \leq k \leq n_a\}$. Each sequence carries a natural order $<$ on its sequence position. The order $<$ on the

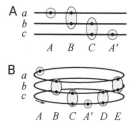

Fig. 1 Linear and cyclic conflicts. **A** Linear. The classes A, B, C, and A' are ordered by $\preceq$. $\prec$ is not transitive, since e.g. A and C are not comparable by $\prec$. From $A \prec B$, $B \prec C$ and $C \prec A'$ follows $(A \cup A') \preceq (A \cup A')$; this shows $\preceq$ is not irreflexive on $\{(A \cup A'), B, C\}$. **B** Cyclic. For example, $\blacktriangleleft EAB$ and $\blacktriangleleft EBC$ hold, but E, A, and C are not comparable by $\blacktriangleleft$; thus $\blacktriangleleft$ is not transitive. A valid cMSA cannot contain $(A \cup A')$, B, C, D, and E, since then $\overline{\blacktriangleleft}$ is not antisymmetric: $\blacktriangleleft EAB$ and $\blacktriangleleft EBC$ imply $\overline{\blacktriangleleft} E(A \cup A')C$, while $\blacktriangleleft CA'D$ and $\blacktriangleleft CDE$ lead to the contradiction $\overline{\blacktriangleleft} C(A \cup A')E$.

individual sequence positions then naturally extends to a relation $\prec$ on 2^X by setting $A \prec B$ if

(IR) $A \neq B$

(NC) $(a, i), (b, j) \in A$ and $(a, k) \in B$ then, for every $(b, l) \in B$ holds $i < k$ implies $j < l$ and $i > k$ implies $j > l$.

(C) There is $a \in \{1, \ldots, M\}$ such that $(a, i) \in A$, $(a, j) \in B$ and $i < j$.

By definition, $\prec$ is irreflexive (IR) and antisymmetric (i.e., $A \prec B$ implies not $B \prec A$). As the example in Fig. 2.1 shows, $\prec$ is not transitive. Note that $A \prec B$ implies $A \cap B = \emptyset$. We say that A and B are *non-crossing* if (NC) holds. We say that A and B are comparable if $A = B$, $A \prec B$, or $B \prec A$. The example in Fig. 2.1 shows that the transitive closure $\overline{\prec}$ of $\prec$ is not irreflexive (and consequently, not antisymmetric) in general.

A multiple sequence alignment $(X, \mathcal{A}, \prec)$ on X is a partition $\mathcal{A}$ of X such that, for all $A, B \in \mathcal{A}$, holds

(A1) If $(a, i) \in A$ and $(a, j) \in A$ implies $i = j$

(A2) A and B are non-crossing

(A3) The transitive closure $\overline{\prec}$ of $\prec$ is a partial order on $\mathcal{A}$.

The elements of $\mathcal{A}$ are the *alignment columns*. Two positions (a, i) and (b, j) are aligned if they appear in the same column. Note that $\mathcal{A}$ may also contain singletons, i.e., positions that are not aligned to any other position.

Condition (A1) ensures that each column contains at most one position from each sequence. This condition could be relaxed to obtain a block model of alignments, see e.g. [17]. This model relates to the usual picture of an alignment as a rectangular display of sequences with gaps in the following manner. A *display* $D(\mathcal{A})$ of $\mathcal{A}$ is a total ordering $\overline{\prec}'$ of $\mathcal{A}$. To see that this definition simply paraphrases the usual concept of MSAs we can argue as follows: Every partial order can be extended to a total order, thus $\overline{\prec}'$ exists. The restriction of $D(\mathcal{A})$ to a sequence a contains all sequence positions (a, i) of a because $\mathcal{A}$ is a partition. Condition (C) guarantees that the partial order $\overline{\prec}$, and hence also its completion $\overline{\prec}'$, preserves the original order $<$ on the sequences, i.e., any choice of $\overline{\prec}'$ equals $<$ when restricted to a single sequence.

2.2 Cyclic Orders

The above construction critically depends on the existence of the linear order $<$ on the input sequences. On circular sequences, however, such a linear order exists only locally. Instead there is a natural cyclic order. A ternary relation $\vartriangleleft i\,j\,k$ on a set V is a *cyclic order* [18] if for all $i, j, k \in V$ holds

(cO1) $\vartriangleleft i\,j\,k$ implies i, j, k pairwise distinct. (irreflexive)
(cO2) $\vartriangleleft i\,j\,k$ implies $\vartriangleleft k\,i\,j$. (cyclic)
(cO3) $\vartriangleleft i\,j\,k$ implies $\neg\vartriangleleft k\,j\,i$ (antisymmetric)
(cO4) $\vartriangleleft i\,j\,k$ and $\vartriangleleft i\,k\,l$ implies $\vartriangleleft i\,j\,l$. (transitive)
(cO5) If i, j, k are pairwise distinct then $\vartriangleleft i\,j\,k$ or $\vartriangleleft k\,j\,i$. (total)

If only (cO1) to (cO4) hold, CO is a partial cyclic order. A pair of points (p, q) is adjacent in a total cyclic order on V if there is no $h \in V$ such that $\vartriangleleft p\,h\,q$.

In contrast to recognizing partial (linear) orders, i.e., testing for acyclicity of (the transitive closure of) $(X, \prec)$, the corresponding problem for cyclic orders is NP-complete [19]. The conditions (CO1) through (CO4), however, are easy enough to translate to ILP constraints. Furthermore, the multiple sequence alignment problem is NP-complete already for linear sequences; thus, the extra complication arising from the cyclic ordering problem is irrelevant.

Cyclic orders can be linearized by cutting them at any point resulting in a linear order with the cut point as its minimal (or maximal) element [20]. A trivial variation on this construction is to insert an additional cut point 0 between adjacent points to obtain a linearized order that has one copy of the artificial cut point as its minimal and maximal element, respectively. Formally, let $\vartriangleleft$ be a total cyclic order on V; furthermore, let p and q be adjacent points in this order on V. Then, the relation $\overline{\vartriangleleft \cup \{p, 0, q\}}$ is a total cyclic order on the set $V \cup \{0\}$. The corresponding linearization is $(V, <_{p0q})$ with $i <_{p0q} j$ and $j <_{p0q} k$ iff $\vartriangleleft i\,j\,k$ for all $i, j, k \in V$. Of course, this can be extended by adding a (distinct) copy of 0 as both the minimal and maximal elements, i.e. $V \cup \{0_-, 0_+\}$ is also totally ordered, if we set $0_- <_{p0q} k$ and $k <_{p0q} 0_+$ for all $k \in V$.

2.3 Cyclic Multiple Alignments

Given a cyclic order $\vartriangleleft$ instead of a linear order $<$ on the sequence positions we can define a relation $\blacktriangleleft$ on 2^X such that, for $A, B, C \in 2^X$, we have $\blacktriangleleft ABC$ if

(IR) A, B, and C are pairwise distinct
(CNC) If $(a, i) \in A$, $(a, j) \in B$, $(a, k) \in C$, $(b, p) \in A$, $(b, q) \in B$, and $(b, r) \in C$ then $\vartriangleleft i\,j\,k$ implies $\vartriangleleft p\,q\,r$.
(CC) There exist $a \in \{1, \dots, M\}$ and $(a, i) \in A$, $(a, j) \in B$, $(a, k) \in C$ such that $\vartriangleleft i\,j\,k$.

We call three sets A, B, C *cyclically non-crossing* if (CNC) is satisfied. Three non-crossing sets are *cyclically comparable* if (CC) is true. Note that $\blacktriangleleft$ is irreflexive and antisymmetric but not transitive in general.

Definition 1. *A cyclic MSA (cMSA) $(X, \mathcal{A})$ is a partition $\mathcal{A}$ of X such that the following conditions are satisfied for all $A, B, C \in \mathcal{A}$*

(A1) $(a, i) \in A$ and $(a, j) \in A$ implies $i = j$
(cA2) A, B, C are cyclically non-crossing for all $A, B, C \in \mathcal{A}$
(cA3) The transitive closure $\overline{\blacktriangleleft}$ of $\blacktriangleleft$ is a partial cyclic order of $\mathcal{A}$.

The restriction of $(\mathcal{A}, \overline{\blacktriangleleft})$ to an individual sequence is just $\{1, \ldots, n_a\}$ with its cyclic order $\vartriangleleft$. Cyclic MSAs can be extended to totally ordered displays $D(\mathcal{A}, \overline{\blacktriangleleft})$.

Since the columns of cMSA can be cyclically ordered, we can find a cut (in any of its totally cyclically ordered extensions).

Definition 2. *A cMSA with cut $\varnothing = \{(a, 0_a) | 1 \leq a \leq M\}$ is a cMSA on $X^\varnothing := X \cup \{(a, 0_a) | 1 \leq a \leq M\}$ such that, for each a, 0^a is a cut point in sequence a, i.e., there is an adjacent pair $p_a, q_a \in V_a$ w.r.t. $\vartriangleleft$ such that $\vartriangleleft p_a \, 0_a \, q_a$.*

Note that $\varnothing$ by definition contains a cut point 0_a in every sequence $a = 1, \ldots, M$.

If there is an alignment column that touches every sequence, i.e., $|A| = M$, we can place a cut either to its left (cutting the incoming adjacency) or to the right (cutting the outgoing adjacency). Such a column need not exist, however. A subset $B \subseteq X$ is *contiguous* w.r.t. $\mathcal{A}$ if

(1) $(a, i), (a, j) \in B$ implies $(a, k) \in B$ for all k such that $\vartriangleleft i \, k \, j$, i.e. the projection of B onto sequence a is an interval.
(2) If $A \in \mathcal{A}$ and $A \cap B \neq \emptyset$ then $A \subseteq B$, i.e. B is a union of classes of $\mathcal{A}$.

A contiguous set of sequence positions is therefore a collection of consecutive alignment columns. We say that B is an *anchor* if, in addition,

(3) For every two sequences a and a' in $\{1, \ldots, M\}$, there is a *path* $a = b_1, \ldots, b_k = a'$ over $\{1, \ldots, M\}$, such that for all $i \in \{1, \ldots, k-1\}$ there exist A, j, j' such that $\{(a_i, j), (a_{i+1}, j')\} \subseteq A \in \mathcal{A}$.

Let P and Q be classes in $\mathcal{A}$, such that $P, Q \not\subseteq B$. Then, as an immediate consequence of the definition, for any $A \subseteq B$, $\blacktriangleleft PAQ$ implies $\blacktriangleleft PBQ$. Furthermore, let $(a, i) \in P$ and $(a, k) \in Q$. Then, either $\blacktriangleleft PBQ$ or $\blacktriangleleft QBP$, i.e. any pair of alignment columns that touch a common sequence is circularly comparable with every anchor. Of course, anchors can be defined for linear MSAs as well (with analogous properties.)

An cMSA is *irreducible* if it contains an anchor. Otherwise, there is a non-trivial partition of the set $\{1, \ldots, M\}$ of sequences such that the alignment can be split into alignments with anchors on these sub-sets. More visually, this simply means that subsets are not connected by any alignment edge.

For our purposes, the importance of anchors is that they define natural positions for cuts, namely as either the incoming edges or the outgoing edges of B. A simple consequence of this fact is

Lemma 1. *If $(X^\varnothing, \mathcal{A} \cup \{\varnothing\})$ is a cMSA with cut $\varnothing$ then $(X, \mathcal{A})$ is a cMSA. Conversely, for every cMSA $(X, \mathcal{A})$ there is a cut $\varnothing$ such that $(X^\varnothing, \mathcal{A} \cup \{\varnothing\})$ is a cMSA with cut.*

Proof. By construction, $\mathcal{A} \cup \{\varnothing\}$ is a partition of $X^{\varnothing}$ and (A1) holds, hence $\mathcal{A}$ is a partition of X. The properties of being non-crossing and cyclically partially ordered are inherited by subsets, hence $(X, \mathcal{A})$ is a cMSA. Conversely, if $(X, \mathcal{A})$ is irreducible, there is an anchor, which in turn provides us a with a cut to left and to right of it. If $(X, \mathcal{A})$ is reducible, we can find an anchor for each irreducible subset of sequences. Their disjoint union provides a cut in $(X, \mathcal{A})$.

Cyclic alignments thus can, as the intuition would tell us, equivalently, be characterized as linear alignments with cut. The virtue of the technicalities above is that we do not have to make the cut explicit by renumbering but instead by specifying its positions in terms of the circular orders on the input sequences.

3   Partition-Based ILP for MSA and cMSA

ILP approaches to (linear) MSAs so far were based on variables for individual alignment edges, i.e., $x_{ai,bj} = 1$ if position i of sequence a is aligned with position j in sequence b. In this picture, an alignment is viewed as a graph on X. The correspondence between partitions $\mathcal{A}$ of X and the graph $\Gamma(X, \mathcal{A}, <)$ is very simple: $\{(a, i), (b, j)\}$ is an alignment edge if and only if $a \neq b$ and there is $A \in \mathcal{A}$ such that $(a, i) \in A$, $(b, j) \in A$. It is customary to view $\Gamma(X, \mathcal{A}, <)$ by connecting consecutive positions of the same sequence by a directed arc [21, 22]. A cycle Z in $\Gamma(X, \mathcal{A}, <)$ is called *mixed* if it contains at least one directed arc and all arcs are oriented along the cycle. Z is *critical* if all vertices in the same sequence occur consecutively along Z.

Proposition 1 ([21, 22]). *A partition $(X, \mathcal{A}, \prec)$ satisfying (A1) is a MSA if and only if $\Gamma(X, \mathcal{A}, \prec)$ contains no critical mixed cycle.*

This observation forms the basis for the current ILP-based MSA implementations. Circular alignments, of course, have an analogous graph representation $\Gamma(X, \mathcal{A}, \lhd)$. The discussion of the previous section immediately implies

Proposition 2. *A partition $\mathcal{A}$ of X is a cyclic MSA if and only if there is a cut $\varnothing$ for $\mathcal{A}$ such that the graph $\Gamma(X^{\varnothing}, \mathcal{A} \bigcup \{\varnothing\})$ contains no critical mixed cycle that does not intersect $\varnothing$.*

The only extra complication for cMSAs is that an explicit representation of the cut is required and only mixed cycles that do not cross the cut are inconsistent with axioms (cA1), (cA2), and (cA3).

The main difficulty of using the mixed cycle condition in an ILP framework is that there are exponentially many potential critical mixed cycles. While, conceptually, the Maximum Weight Trace (MWT) ILP formulation includes all critical mixed cycle inequalities, it is strictly infeasible to feed all those constraints to an ILP solver and apply branch-and-bound. This dilemma was resolved in [21] by means of the branch and cut scheme: Starting without mixed cycle constraints, selected inequalities are iteratively added on demand during the branch-and-bound optimization. A polynomial separation algorithm works at the core of this approach: given a solution of the LP corresponding to the current ILP instance, it selects a critical mixed cycle inequality that removes this solution. This inequality is added to the current ILP and the process is iterated.

Instead of constructing the cyclic MSA ILP based on the graph formalization of MSA as MWT, we devise here novel ILP formulations that directly build on the partition formalization of MSAs. Remarkably, this model required only polynomially many variables and constraints, whereas the MWT formulation required an exponential number of constraints. Our formulation is based exclusively on Boolean variables; consequently, we omit the constraints of all variables to domains $\{0, 1\}$ for brevity.

Partition. A MSA (or cMSA) is represented by at most $N = \sum_{a=1}^{M} n_a = |X|$ classes α of positions $x \in X$. As before, we denote the set of classes by $\mathcal{A}$; we use Greek letters $\alpha, \beta, \gamma, \ldots$ to denote single classes (where we used A,B,C,... before.) The classes are modeled by membership variables $\mathbf{P}x\alpha = 1$ if $x \in \alpha$. The simple constraint $\sum_{\alpha} \mathbf{P}x\alpha = 1$ for all $x \in X$ ensures that this describes a partition of X; $\sum_{1 \leq i \leq n_a} \mathbf{P}(a, i)\alpha \leq 1$ for all $\alpha \in \mathcal{A}$ and sequences $1 \leq a \leq M$ guarantees that each partition contains at most one position per sequence.

Linear Order. Next, we model the partial ordering relation between classes; in the linear case, this is the transitive closure of relation $\prec$. For this purpose, we introduce ordering variables $\mathbf{O}\alpha\beta$ for $\alpha \neq \beta \in \mathcal{A}$, value 1 indicating $\alpha \preceq \beta$.

First, the ordering variables are related to the membership variables and relation $<$ on positions

$$\begin{pmatrix} \forall \alpha \neq \beta \in \mathcal{A}, \\ 1 \leq a \leq M, 1 < j \leq n_a \end{pmatrix} \quad \sum_{i<j} \mathbf{P}(a, i)\alpha + \sum_{i \geq j} \mathbf{P}(a, i)\beta \leq \mathbf{O}\alpha\beta + 1. \quad \text{(CI)}$$

This corresponds to condition (C). Furthermore, we constrain the ordering variables to describe a partial order, implementing (A3), by

$$\begin{array}{lrr} (\forall \alpha \in \mathcal{A}) & \mathbf{O}\alpha\alpha = 0 & \text{(OI1)} \\ (\forall \alpha \neq \beta \in \mathcal{A}) & \mathbf{O}\alpha\beta = 1 - \mathbf{O}\beta\alpha & \text{(OI2)} \\ (\forall \alpha, \beta, \gamma \in \mathcal{A}) & \mathbf{O}\alpha\beta + \mathbf{O}\beta\gamma \leq \mathbf{O}\alpha\gamma + 1. & \text{(OI3)} \end{array}$$

Together with (CI), the constraints (OI1)-(OI3), which model the properties antireflexive, antisymmetric, and transitive of the order relation, guarantee that the modeled classes are non-crossing (A2).

Cyclic Order. As immediate benefit of the partition-based formulation, we can move from the linear model to the cyclic model by redefining the order relation. The partial cyclic ordering relation $\blacktriangleleft$ on classes can be expressed analogously by variables $\mathbf{CO}\alpha\beta\gamma$, which are first related to the membership variables and relation $\lhd$ by

$$
\begin{pmatrix} \forall \alpha \neq \beta \neq \gamma \in \mathcal{A}, \\ 1 \leq a \leq M, \\ 1 < j < k \leq n_a \end{pmatrix} \quad \sum_{1 \leq i < j} \mathbf{P}(a,i)\alpha + \sum_{i \leq i < k} \mathbf{P}(a,j)\beta + \sum_{j \leq i \leq n_a} \mathbf{P}(a,k)\gamma
$$
$$
\leq \mathbf{CO}\alpha\beta\gamma + 2. \tag{CCI}
$$

Note that it is indeed sufficient to specify the above implication for $1 \leq i < j < k \leq n_a$ (instead of all i,j,k, where $\lhd i\,j\,k$), due to the cyclic implications in (cOI2) below. Secondly, we guarantee to describe a partial cyclic order by

$$
\begin{array}{lll}
(\forall \alpha, \beta, \gamma \in \mathcal{A}, |\{\alpha, \beta, \gamma\}| < 3) & \mathbf{CO}\alpha\beta\gamma = 0 & \text{(cOI1)} \\
(\forall \alpha \neq \beta \neq \gamma \in \mathcal{A}) & \mathbf{CO}\alpha\beta\gamma \neq \mathbf{CO}\gamma\beta\alpha & \text{(cOI2)} \\
(\forall \alpha \neq \beta \neq \gamma \in \mathcal{A}) & \mathbf{CO}\alpha\beta\gamma = \mathbf{CO}\gamma\alpha\beta & \text{(cOI3)} \\
(\forall \alpha \neq \beta \neq \gamma \neq \delta \in \mathcal{A}) & \mathbf{CO}\alpha\beta\gamma + \mathbf{CO}\alpha\gamma\delta \leq \mathbf{CO}\alpha\beta\delta + 1 & \text{(cOI4)}
\end{array}
$$

These inequalities guarantee that the described relation is irreflexive (cOI1), antisymmetric (cOI2), cyclic (cOI3), and transitive (cOI4), which amounts to (cA3). Analogously to the linear case, (cA2) follows together with (CCI).

Consequently, even in the ILP formulations we can simply change the type of ordering (linear *versus* circular) to switch between linear and circular alignments.

Objective Function. A sum-of-pairs alignment score sums up weights (i.e., match and mis-match scores) w_{xy} for positions $x, y \in X$ in the same partition. For this purpose, we introduce variables $\mathbf{E}xy$ and auxiliary variables $\mathbf{E}xy\alpha$ with inequalities

$$
\begin{array}{lll}
(\forall x \in X, y \in X, \alpha \in \mathcal{A}) & 2\mathbf{E}xy\alpha \leq \mathbf{P}x\alpha + \mathbf{P}y\alpha & \text{(EI1)} \\
(\forall x \in X, y \in X, \alpha \in \mathcal{A}) & \mathbf{P}x\alpha + \mathbf{P}y\alpha \leq \mathbf{E}xy\alpha + 1 & \text{(EI2)} \\
(\forall x \in X, y \in X) & \mathbf{E}xy = \sum_{\alpha \in \mathcal{A}} \mathbf{E}xy\alpha & \text{(EI3)}
\end{array}
$$

Note that without the linearity requirement, the same relation could be expressed by $\mathbf{E}xy = \sum_{\alpha \in \mathcal{A}} \mathbf{P}y\alpha\mathbf{P}y\alpha$. In the simplest case, the objective function is therefore given by the linear expression

$$
\sum_{x,y \in X} w_{xy}\mathbf{E}xy. \tag{OF}
$$

Linear gap costs. We model linear gap cost with cost g per gap by introducing *gap variables* $\mathbf{G}(a,i)b$ together with the equalities

$$
(\forall (a,i) \in X, 1 \leq b \leq M, b \neq a) \quad \mathbf{G}(a,i)b + \sum_{j:(b,j)\in X} \mathbf{E}(a,i)(b,j) = 1. \tag{GI1}
$$

Notably, we introduce gap variables for sequence a w.r.t. sequence b, so that we can model a sum-of-pairs score on gaps. To this end, we add the term $\sum_{(a,i)\in X, 1\leq b\leq M, b\neq a} g\mathbf{G}(a,i)b$ to the objective function. Note that GI1 serves the dual purpose of defining clique inequalities, which speed up the optimization (as known from branch-and-cut approaches, e.g. [21].)

Affine gap costs. Affine gap penalties of the form $h + kg$ for gaps of length k can be introduced by further variables $\mathbf{GO}(a,i)b$ together with

$$(\forall(a,i) \in X, 1 \leq b \leq M, b \neq a) \quad \mathbf{GO}(a,i)b \geq \mathbf{G}(a,i)b - \mathbf{G}(a,i+1)b$$
$$\mathbf{GO}(a,i)b \leq \mathbf{G}(a,i)b$$
$$\mathbf{GO}(a,i)b \leq 1 - \mathbf{G}(a,i+1)b \qquad \text{(GI2)}$$

By (GI2), $\mathbf{GO}(a,i)b$ equals one, if and only if there is a gap (w.r.t. b) at (a,i) and no gap at $(a,i+1)$, i.e. $\mathbf{GO}(a,i)b = 1$ signals gap opening (at the right end of each gap). For (GI2), we define $\mathbf{G}(a, n_a + 1)b := 0$ to penalize (right) end gaps, in the linear case, and $\mathbf{G}(a, n_a + 1)b := \mathbf{G}(a, 1)b$ to avoid double counting of gaps, in the circular case. Finally, we model the additional gap opening penalties by $\sum_{(a,i)\in X, 1\leq b\leq M, b\neq a} h\mathbf{GO}(a,i)b$.

Further preliminary tuning of the model. To allow more effective evaluation we extend our elementary model. First, we break permutation symmetries of partition classes by assigning the positions (a,i) of one of the sequences to the first n_a partition classes α by setting the corresponding $\mathbf{P}x\alpha$ to 1. Then, we add simple clique inequalities to allow at most one of several conflicting alignment edges. Moreover, the number of modeled partition classes can be reasonably limited. In the case of linear alignment it is furthermore common to model only a subset of the alignment edges, e.g. only edges where $|i - j| \leq \Delta$. Note that the same restriction usually does not make sense in the case of circular alignment.

4 Multiple RNA Secondary Structure Alignment

The extension of sequence alignments to alignment of general contact structures is straightforward in the partition-based ILP framework. It suffices to extend the objective function to reward matches of pairwise contacts.

Denote the set of base pairs to describe the secondary structure information of sequence a by $\mathcal{B}_a$. We define weights $w_{(a,i,i')(b,j,j')}$ for the match of secondary structure base pairs $(i, i') \in \mathcal{B}_a$ and $(j, j') \in \mathcal{B}_b$. For the purpose of simultaneous alignment and folding, informative weights can be derived from the probabilities of base pairs in the RNAs Boltzmann ensemble (c.f. [23–25]). The corresponding score contribution is modeled based on variables $\mathbf{B}aii'bjj'$, indicating the match of base pairs (i, i') of a with (j, j') of b and inequalities

$$\begin{pmatrix} \forall 1 \leq a < b \leq M, \\ (i, i') \in \mathcal{B}_a, (j, j') \in \mathcal{B}_b \end{pmatrix} \quad 2\mathbf{B}aii'bjj' \leq \mathbf{E}(a,i)(b,j) + \mathbf{E}(a,i')(b,j') \qquad \text{(SI)}$$

The structure contribution to the objective functions is therefore expressed by

$$\sum_{\substack{1 \leq a < b \leq M, \\ 1 \leq i < i' \leq n_a, \\ 1 \leq j < j' \leq n_b}} w_{(a,i,i')(b,j,j')} \mathbf{B}aii'bjj'. \tag{OFS}$$

Again straightforwardly, the above elementary structure alignment model can be extended to prevent the matching of crossing base pairs or more than one base pair per base, since those events are commonly considered conflicts in simultaneous folding and alignment. For this purpose, we introduce auxiliary variables $\mathbf{B}aii'$, indicating that a base pair (i, i') in sequence a is matched, are introduced together with constraints

$$\begin{pmatrix} \forall 1 \leq a \leq M, \\ (i, i') \in \mathcal{B}_a \end{pmatrix} \quad \mathbf{B}aii' \leq \sum_{\substack{1 \leq b \leq M, \\ (j,j') \in \mathcal{B}_b}} \mathbf{B}aii'bjj' \leq |\mathcal{B}_b| \, \mathbf{B}aii'$$

$$\tag{SI2}$$

$$\begin{pmatrix} \forall 1 \leq a \leq M, \\ (i, i') \in \mathcal{B}_a, (j, j') \in \mathcal{B}_a, \\ i < j < i' < j' \text{ or } |\{i, i', j, j'\}| < 4, \\ (i, i') \neq (j, j') \end{pmatrix} \quad 1 \geq \mathbf{B}aii' + \mathbf{B}ajj'. \tag{SINC}$$

By the inequalities (SI2), for each base pair (i,i') of a sequence a, $\mathbf{B}aii'=1$ if and only if there exists a base pair (j,j') second sequence b, s.t. $\mathbf{B}aii'bjj'=1$. Given the variables $\mathbf{B}aii'$, we can simply forbid all conflicts by pairwise constraints in (SINC). Only (SINC) is specific to linear alignment; in the circular case, the only necessary modification is to replace, in the all-quantor of (SINC), the linear ordering condition $i < j < i' < j'$ for the base pairs (i,i') and (j,j') by the corresponding expression for circular order $\lhd i' j j' \wedge \lhd j i' j'$.

5   Preliminary Computational Results

Our prototypical implementation generates models (in several variants) from multiple input sequences, writes them to CPLEX LP format, and runs them with the IBM CPLEX linear programming solver. For preliminary experiments, we manually designed several small input instances of multiple sequences, such that the sequences form typical pseudoknots that are hard to align by dynamic programming algorithms (e.g., [26]); furthermore, we introduce sequence mutations, such that pure sequence alignment programs fail to derive the (by design known) structurally correct alignment. We designed several instances of linear sequences, which let us evaluate the linear MSA model and compare to the cyclic model. Furthermore, we rotated the sequences to mimic the effect of decircularizing cyclic sequences at arbitrary (non-homologous) positions to obtain typical inputs for cyclic alignment.

We generated several models based on a simple made-up scoring scheme of invariant scores for base matches (4) and mismatches (−4), linear cost of gaps

Table 1. Preliminary results. See text and the electronic appendix[†] for details.

Instance				Model		Solving Time	
ID	#Seqs	Length	PK-type	Type	Delta	5% Tolerance	Optimal
1	3	10	2-knot	lin	3	0.6	0.6
1	3	10	2-knot	lin	-	1.4	1.4
1	3	10	2-knot	cyc	-	170	176
1R	3	10	2-knot	cyc	-	229	273
2	3	15	3-knot	lin	3	2.4	2.7
2	3	15	3-knot	lin	-	143	129
3	3	20	3-knot	lin	3	8.4	8.4
3	3	20	3-knot	lin	-	287	-
4	4	10	2-knot	lin	3	4.8	6.4
4	4	10	2-knot	lin	-	10	28

[†] http://www.bioinf.uni-leipzig.de/publications/supplements/14-006

(-3), and constant contributions per arc match (32). Subsequently, we applied the standalone CPLEX solver essentially out-of-the-box[1]. The features of our test instances and results are summarized in Table 1. The actual instances and alignments are reported in the Appendix. All tests were performed on a Lenovo T431s notebook; time-out was set to 10 minutes. Run-times are reported for solving to proven optimality and within 5% tolerance of the LP-relaxation bound (CPLEX configuration: tolerance mipgap 0.02). The found optimal alignments have been structurally correct. Linear alignments were performed with and without a "diagonal" restriction of alignment edges (i, j) to $|i - j| < \Delta$. The cyclic alignment failed for all instances but the one of three sequences of length 10. Whether the sequences are given in the correct rotation (ID 1) or rotated to each other (ID 1R) does not seem to make a large difference in this case.

6   Discussion

We have explored here a simple mathematical framework for multiple sequence alignments that highlights the similarities of circular and linear multiple sequence alignments. The key ingredient is, as in early work of Dress, Morgenstern, and collaborators [15, 16], the view of MSAs as set systems, more precisely as (cyclic) order preserving partitions. As it turns out, replacing the familiar linear order by a cyclic one suffices to obtain a full characterization of cyclic MSAs. Furthermore, the mathematical structure directly translates to a generic ILP formulations detailed in sect. 3. Remarkably, in contrast to previous ILP models, which define an exponential number of constraints, this model requires only polynomially many variables and constraints. A prototypical implementation indicates,

---

[1] We used the default solving strategy, but turned off the generation of Gomory cuts; this slightly sped up the calculations in our experiments.

however, that in this naïve form even highly efficient commercial solvers such as a CPLEX cannot accommodate instances large enough to be of practical interest. We therefore also investigated the cyclic analog of the "critical mixed cycle", which form the basis for branch and cut and Lagrangian relaxation approaches [21, 22]. Reassuringly, it can be phrased in terms of a cut linearizing the cyclic MSA and critical mixed cycles not interfering with the cut. Although our contribution does not *immediately* provide a production-grade software, it points out many promising directions for future work. Furthermore, it is – to our knowledge – the first systematic analysis of the cyclic *multiple* sequence alignment problem.

Acknowledgments. This work was funded, in part, by the *Deutsche Forschungsgemeinschaft* within the EUROCORES Programme EUROGIGA (project GReGAS) of the European Science Foundation.

References

1. Jeck, W.R., Sharpless, N.E.: Detecting and characterizing circular RNAs. Nat. Biotechnol. 32, 453–461 (2014)
2. Danan, M., Schwartz, S., Edelheit, S., Sorek, R.: Transcriptome-wide discovery of circular RNAs in Archaea. Nucleic Acids Res. 40, 3131–3142 (2012)
3. Ding, B.: Viroids: self-replicating, mobile, and fast-evolving noncoding regulatory RNAs. Wiley Interdiscip Rev. RNA 1, 362–375 (2010)
4. Doose, G., Alexis, M., Kirsch, R., Findeiß, S., Langenberger, D., Machné, R., Mörl, M., Hoffmann, S., Stadler, P.F.: Mapping the RNA-seq trash bin: Unusual transcripts in prokaryotic transcriptome sequencing data. RNA Biology 10, 1204–1210 (2013)
5. Bunke, H., Bühler, U.: Applications of approximate string matching to 2D shape recognition. Patt. Recogn. 26, 1797–1812 (1993)
6. Gregor, J., Thomason, M.G.: Dynamic programming alignment of sequences representing cyclic patterns. IEEE Trans. Patt. Anal. Mach. Intell. 15, 129–135 (1993)
7. Maes, M.: On a cyclic string-to-string correction problem. Inform. Process. Lett. 35, 73–78 (1990)
8. Mollineda, R.A., Vidal, E., Casacuberta, F.: Cyclic sequence alignments: approximate versus optimal techniques. Int. J. Pattern Rec. Artif. Intel. 16, 291–299 (2002)
9. Dewey, T.G.: A sequence alignment algorithm with an arbitrary gap penalty function. J. Comp. Biol. 8, 177–190 (2001)
10. Benson, G.: Tandem cyclic alignment. Discrete Appl. Math. 146, 124–133 (2005)
11. Mosig, A., Hofacker, I.L., Stadler, P.F.: Comparative analysis of cyclic sequences: Viroids and other small circular RNAs. In: Giegerich, R., Stoye, J. (eds.) Proceedings GCB 2006, vol. P-83. Lecture Notes in Informatics, pp. 93–102 (2006)
12. Wang, L., Jiang, T.: On the complexity of multiple sequence alignment. J. Comput. Biol. 1, 337–348 (1994)
13. Just, W.: Computational complexity of multiple sequence alignment with SP-score. J. Comput. Biol. 8, 615–623 (2001)
14. Elias, I.: Settling the intractability of multiple alignment. J. Comput. Biol. 13, 1323–1339 (2006)

15. Morgenstern, B., Frech, K., Dress, A., Werner, T.: DIALIGN: finding local similarities by multiple sequence alignment. Bioinformatics 14(3), 290–294 (1998)
16. Morgenstern, B., Stoye, J., Dress, A.W.M.: Consistent equivalence relations: a set-theoretical framework for multiple sequence alignments. Technical report, University of Bielefeld, FSPM (1999)
17. Otto, W., Stadler, P.F., Prohaska, S.J.: Phylogenetic footprinting and consistent sets of local aligments. In: Giancarlo, R., Manzini, G. (eds.) CPM 2011. LNCS, vol. 6661, pp. 118–131. Springer, Heidelberg (2011)
18. Meggido, N.: Partial and complete cyclic orders. Bull. Am. Math. Soc. 82, 274–276 (1976)
19. Galil, Z., Megiddo, N.: Cyclic ordering in NP-complete. Theor. Comp. Sci. 5, 179–182 (1977)
20. Novák, V.: Cuts in cyclically ordered sets. Czech. Math. J. 34, 322–333 (1984)
21. Reinert, K., Lenhof, H.P., Mutzel, P., Mehlhorn, K., Kececioglu, J.D.: A branch-and-cut algorithm for multiple sequence alignment. In: Proceedings of the First Annual International Conference on Research in Computational Molecular Biology (RECOMB), pp. 241–250. ACM (1997)
22. Lenhof, H.P., Morgenstern, B., Reinert, K.: An exact solution for the segment-to-segment multiple sequence alignment problem. Bioinformatics 15, 203–210 (1999)
23. Hofacker, I.L., Bernhart, S.H., Stadler, P.F.: Alignment of RNA base pairing probability matrices. Bioinformatics 20, 2222–2227 (2004)
24. Will, S., Reiche, K., Hofacker, I.L., Stadler, P.F., Backofen, R.: Inferring non-coding RNA families and classes by means of genome-scale structure-based clustering. PLoS Comput. Biol. 3, e65 (2007)
25. Bauer, M., Klau, G.W., Reinert, K.: Accurate multiple sequence-structure alignment of RNA sequences using combinatorial optimization. BMC Bioinformatics 8 (2007)
26. Möhl, M., Will, S., Backofen, R.: Lifting prediction to alignment of RNA pseudoknots. J. Comp. Biol. 17, 429–442 (2010)

Entropic Profiles, Maximal Motifs and the Discovery of Significant Repetitions in Genomic Sequences

Laxmi Parida[1], Cinzia Pizzi[2], and Simona E. Rombo[3]

[1] IBM T. J. Watson Research Center
[2] Department of Information Engineering, University of Padova
[3] Department of Mathematics and Computer Science, University of Palermo

Abstract. The degree of predictability of a sequence can be measured by its entropy and it is closely related to its repetitiveness and compressibility. Entropic profiles are useful tools to study the under- and over-representation of subsequences, providing also information about the scale of each conserved DNA region. On the other hand, compact classes of repetitive motifs, such as maximal motifs, have been proved to be useful for the identification of significant repetitions and for the compression of biological sequences. In this paper we show that there is a relationship between entropic profiles and maximal motifs, and in particular we prove that the former are a subset of the latter. As a further contribution we propose a novel linear time linear space algorithm to compute the function Entropic Profile introduced by Vinga and Almeida in [18], and we present some preliminary results on real data, showing the speed up of our approach with respect to other existing techniques.

1 Introduction

Sequence data is growing in volume with the availability of more and more precise, as well as accessible, assaying technologies. Patterns in biological sequences is central to making sense of this exploding data space, and its study continues to be a problem of vital interest. Natural notions of maximality and irredundancy have been introduced and studied in literature in order to limit the number of output patterns without losing information [3, 4, 6, 10, 11, 14–17]. Such notions are related to both the length and the occurrences of the patterns in the input sequence. Maximal patterns have been successfully applied to the identification of biologically significant repetitions, and compressibility of biological sequences, to list a few areas of use.

Different flavors of patterns, based either on combinatorics or statistics, can usually be shown to be a variation on this basic concept of maximal patterns. In particular, it is well known that the degree of predictability of a sequence can be measured by its entropy and, at the same time, it is also closely related with its repetitiveness and compressibility [9]. Entropic profile was introduced [7, 8, 18] to study the under- and over-representation of segments, and also the scale of each conserved DNA region.

D. Brown and B. Morgenstern (Eds.): WABI 2014, LNBI 8701, pp. 148–160, 2014.
© Springer-Verlag Berlin Heidelberg 2014

Due to the fundamental nature of maximality, a natural question arises about a possible relationship between maximal patterns and entropic profiles. We explore this question in the paper and show that entropic profiles are indeed a subset of maximal patterns. Based on this inshight, we improve the running time of the detection of entropic profiles by proposing an efficient algorithm to extract entropic profiles in $O(n)$ time and space. The algorithm exploits well known properties of the suffix tree to group together the subwords that are needed to compute the entropy for a specific position and for the input sequence as a whole.

Finally, we present an experimental validation of the proposed algorithm performed on the whole genome of *Haemophilus influenzae*, showing that our approach outperforms the other existing techniques in terms of time performance.

The manuscript is organized as follows. In the next section we recall some basic notions about DNA sequence entropic profiles and maximal motifs, while in the next section we show the relationship occurring between them. Section 4 presents our linear time linear space algorithms, and some preliminary experimental comparisons are discussed in Section 5. The paper ends with a summary of results and some further considerations.

2 Background

Let $x = x_1 \ldots x_n$ be a string defined over an alphabet Σ. We denote by $x_i \ldots x_j$ the subword of x starting at position i and ending at position $j > i$, and by $c([i, j])$ the number of occurrences of $x_i \ldots x_j$ in x.

2.1 Maximal Motifs

Among all the candidate over-represented subwords of an input string, those presenting special properties of maximal saturation have been proved to be a special compact class of motifs with high informative content, and they have been shown to be computable in linear time [12]. We next recall some basic definitions.

Definition 1. *(Left-maximal motif) The subword* $x' = x_i \ldots x_j$ *of* x *is a left-maximal motif if it does not extist any other subword* $x'' = x_{i-h} \ldots x_j$ $(0 < h \le i)$ *such that* $c([i, j]) = c([i - h, j])$.

Definition 2. *(Right-maximal motif) The subword* $x' = x_i \ldots x_j$ *of* x *is a right-maximal motif if there exist no subword* $x''' = x_i \ldots x_{j+k}$ $(0 < k < n - j)$ *such that* $c([i, j]) = c([i, j + k])$.

Definition 3. *(Maximal motif) The subword* x' *is a maximal subword if it is both left- and right- maximal.*

Maximal motifs are those subwords of the input string which cannot be *extended* at the left or at the right without loosing at least one of their occurrences.

2.2   Entropic Profiles

Entropic profiles may be estimated according to different entropy formulations. The definitions on entropic profiles recalled here are taken from the seminal papers [7, 8, 18], where the Rényi entropy of probability density estimation and the Parzen's window method applied to Chaos Game Representation/Universal Sequence Maps are exploited.

Let L be the chosen length resolution and ϕ be a smoothing parameter.

Definition 4. *(Main EP function) The main EP function is given by:*

$$\hat{f}_{L,\phi}(x_i) = \frac{1 + \frac{1}{n} \sum_{k=1}^{L} 4^k \phi^k \cdot c([i - k + 1, i])}{\sum_{k=0}^{L} \phi^k}$$

Definition 5. *(Normalized EP) Let $m_{L,\phi}$ be the mean and $S_{L,\phi}$ be the standard deviation using all positions $i = 1 \ldots n$. The normalized EP is:*

$$EP_{L,\phi}(x_i) = \frac{\hat{f}_{L,\phi}(x_i) - m_{L,\phi}}{S_{L,\phi}}$$

where:

$$m_{L,\phi} = \frac{1}{n} \sum_{i=1}^{n} \hat{f}_{L,\phi}(x_i) \ and \ S_{L,\phi} = \sqrt{\frac{1}{n - 1} \sum_{i=1}^{n} (\hat{f}_{L,\phi}(x_i) - m_{L,\phi})^2}$$

The main entropy function $\hat{f}$ is shown to be computable in linear time in [5]. In that work, however, the normalized entropy as defined in the original papers is not considered. A different normalization is defined instead:

$$FastEP_{L,\phi} = \frac{f_{L,\phi}(i)}{\max_{0 \le j < n}[f_{L,\phi}(j)]}$$

3   Entropic Profiles vs Maximal Motifs

We now discuss the relationship between entropic profiles and maximal motifs. The following theorem holds.

Theorem 1. *The entropic profiles scoring maximum values of the main EP function $\hat{f}$ are left-maximal motifs of the input string.*

Proof. Let i be a generic position of the input string and L' be the length of the subword x' starting at $i - L' + 1$ and ending at i, such that x' scores the maximum value of entropy at the position i. Then the following inequalities hold, with respect to the two subwords x'' and x''' of length $L' + 1$ and $L' - 1$, ending at i and starting at $i - L'$ and at $i - L' + 2$, respectively:

$$\begin{cases} \dfrac{n+\sum_{k=1}^{L'} 4^k \phi^k c([i-k+1,i])}{\sum_{k=0}^{L'} \phi^k} \geq \dfrac{n+\sum_{k=1}^{L'+1} 4^k \phi^k c([i-k+1,i])}{\sum_{k=0}^{L'+1} \phi^k} \\[4mm] \dfrac{n+\sum_{k=1}^{L'} 4^k \phi^k c([i-k+1,i])}{\sum_{k=0}^{L'} \phi^k} \geq \dfrac{n+\sum_{k=1}^{L'-1} 4^k \phi^k c([i-k+1,i])}{\sum_{k=0}^{L'-1} \phi^k} \end{cases}$$

As shown in the Appendix, the two inequalities above can be rewritten as:

$$\begin{cases} \dfrac{n+\sum_{k=1}^{L'} 4^k \phi^k c([i-k+1,i])}{\sum_{k=0}^{L'} \phi^k} \geq \dfrac{4^{L'+1} \phi^{L'+1} c([i-L',i])}{\phi^{L'+1}} \\[4mm] \dfrac{n+\sum_{k=1}^{L'-1} 4^k \phi^k c([i-k+1,i])}{\sum_{k=0}^{L'-1} \phi^k} \leq \dfrac{4^{L'} \phi^{L'} c([i-L'+1,i])}{\phi^{L'}} \end{cases}$$

leading to the following relation between the number of occurrences of x' and x'':

$$c([i - L' + 1]) \geq 4\, c([i - L', i]).$$

Let us now suppose that x' is not left-maximal. From Definition 3, it follows that all the occurrences of x' should be covered from another subword x'' extending x' at the left of at least one character. This would mean that $c([i - L' + 1])$ should be equal to $c([i - L', i])$, that is, a contraddiction.   □

Note that not necessarily a left-maximal motif corresponds to a peak of entropy, as shown by the following example.

Example 1. Let $\phi = 10$ and consider the following input string:

$$\begin{array}{ccccccccccccc} 0 & 1 & 2 & 3 & 4 & 5 & 6 & 7 & 8 & 9 & 10 & 11 \\ T & C & A & A & C & G & G & C & G & G & C & T \end{array}$$

We wonder if the maximal motif $CGGC$, ending at positions 7 and 10, corresponds to a peak of entropy at one of those positions. We have that $\hat{f}_{3,10}(7) = 9.85$, $\hat{f}_{4,10}(7) = 39.38$ and $\hat{f}_{4,10}(7) = 76.8$, therefore $CGGC$ has not a peak of $\hat{f}$ at that position. The same values occur for the position $i = 10$.

4   Methods

In this section the available algorithms to compute entropic profiles are discussed, and faster algorithms for entropic profiles computation and normalization are presented. We recall that we want to analyze an input string x of length n by means of entropic profiles of resolution L and for a fixed ϕ.

4.1 Existing Algorithms

There are two algorithms available in literature that compute entropic profiles.

The algorithm described in [8] is a faster version of the original algorithm proposed by Fernandes et al. [7]. It relies on a truncated suffix trie data structure, which is quadratic both in time and space occupation, enhanced with a list of side links that connect all the nodes at the same depth in the tree. This is needed to speed up the normalization because, in the formulas used to compute mean and standard deviation [7], the counting of subwords of the same length is a routine operation. With this approach the maximum value of L had to be set to 15.

The other method, presented in [5], uses a suffix tree on the reverse string to obtain linear time and space computation of the absolute values of entropy for some paramethers L and ϕ. These values are then normalized with respect to the maximum value of entropy among all the substrings of length L. To obtain the maximum value $\max_L$, in correspondence of a given L, all values $\max_l$, where $1 \leq l < L$, are needed. The algorithm has a worst case complexity $O(n^2)$, but being guided by a branch-and-cut technique in practice substantial savings are possible.

A key property of both suffix tries and suffix tree [12] is that, once the data structure is built on a text string x, the occurrences of a pattern $y = y_1 \ldots y_m$ in x can be found by following the path labelled with $y_1 \ldots y_m$ from the root of the tree. If such a path exists, the occurrences are given by the indexes of the leaves of the subtree rooted at the node in which the path ends. Moreover, being the suffix tree a compact version of a suffix trie, we have for it the further property that all the strings corresponding to paths that end in the "middle" of an arc between two nodes share the same set of occurrences. Figure 1 shows an example of trie and suffix tree.

4.2 Preprocessing

For the computation of the values needed to obtain both the absolute and the normalized values of entropy, we perform the same preprocessing procedure described in [5]. We recall here the main steps as we will need the annotated suffix tree for the subsequent description of the speed up to compute the mean and the standard deviation.

Consider the suffix tree T built on the reverse of the input string x. In such a tree, strings that are described from paths ending at the same locus share the same set of ending positions in the original input string. Hence, they are exactly the strings we need to consider when computing the values of entropy. Some care needs to be taken to map the actual positions during the computation, but this will not affect the time complexity. Therefore, in the following discussion we will just refer to the standard association between strings and positions in a suffix tree, keeping in mind they are actually reversed.

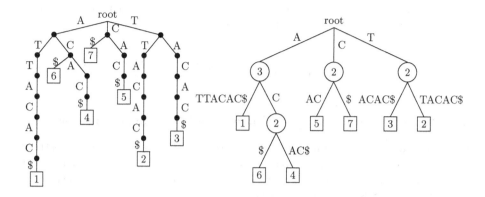

Fig. 1. A suffix trie (left) and a suffix tree (right) built on the same string $x = ATTACAC\$$. The leaves correspond to positions in x. The internal nodes in the suffix tree hold the number of occurrences of the strings that have the node as a locus.

The main observation in [5] is that in the reverse tree the absolute value of the EP function for $n - i$ is equal to:

$$f_{L,\phi}(x_i) = \frac{1 + \frac{1}{n}\sum_{k=1}^{L} 4^k \phi^k \cdot c([i, i + k - 1])}{\sum_{k=0}^{L} \phi^k}$$

In the suffix tree T each node v is annotated with a variable $count(v)$ which stores the number of occurrences of the subword $w(v)$, given by the concatenation of labels from the root to the node v. This can be done in linear time with a bottom-up traversal by setting up the value of the leaves to 1, and the value of the internal nodes to the sum of the values of their children.

Each node v is also annotated with the value of the main summation in the entropy formula. Let i be the position at which occurs the string $w(v)$:

$$main(v) = \sum_{k=1}^{L} 4^k \phi^k \cdot c([i, i + k - 1])$$

Note that once this value is available the absolute value of entropy for $w(v)$ can be computed in constant time:

$$\frac{1 + \frac{1}{n}main(v)(1 - \phi)}{1 - \phi^{L+1}}$$

Now let $h(v)$ be the length of $w(v)$ and $parent(v)$ be the parent node of v. The annotation takes linear time with a pre-order traversal of the tree that passes the contribution of shorter prefixes to the following nodes in the path:

$$main(v) = main(parent(v)) + \sum_{k=h(parent(v))+1}^{h(v)} (4\phi)^k count(v)$$

When *main(parent(v))* is known, the value of *main(v)* can be computed in constant time, since *count(v)* does not depend on *k*:

$$main(v) = main(parent(v)) + count(v) \frac{(4\phi)^{h(parent(v))+1} - (4\phi)^{h(v)+1}}{1 - 4\phi}$$

4.3   Efficient Computation of Entropy and Normalizing Factors

Once the annotation is complete, one can retrieve the entropy for a substring $x[i, i + L - 1]$ by following the path of length L from the root. If it ends at a node v the value of *main(v)* is retrieved, and the absolute value of entropy is computed, otherwise the additional factor:

$$\sum_{k=h(parent(v))+1}^{L} (4\phi)^k count(v) = count(v) \frac{(4\phi)^{h(parent(v))+1} - (4\phi)^{L+1}}{1 - 4\phi}$$

needs to be added to *main(parent(v))*.

In [5] each string of length L starting at each position one wants to analyze is searched for in the suffix tree, and the value of entropy is computed as described above (and normalized according to the maximum value of entropy for length L). In discovery frameworks, where no information about the motif position is known in advance, and there are potentially as many positions to analyze as the length of the input string, this might not be the fastest solution.

On the other hand, by exploiting well known properties of the suffix tree [12] it is possible to propose a different approach that is as simple as powerful, and allowed us to obtain linear time and space algorithms not only for the computation of the absolute value of entropy, but also for its normalization through mean and standard deviation.

Absolute Value of Entropy. We can collect the absolute value of entropy for all positions in the input string with a simple traversal of the tree at depth L (in terms of length of strings that labels the paths). The steps to follow when computing the entropy once we reach the last node of the path are the same we already described for computing the entropy of a given substring. Differently from before, when we reach the last node of a path we also store the value of entropy in an array of size n (or any other suitable data structure) at the positions corresponding to the leaves of the subtree rooted at the node, which are the occurrences of the string that labels the path.

Moreover, as a byside product of this traversal, we can also collect information to compute the mean and the standard deviation in linear time.

The Mean. Consider the mean first. We need to sum up the values of entropy over all possible substrings of length L in the input string. Indeed we can re-write the formula considering the contribution of all different subwords of length L. Let w be one of such subwords, $f_{L,\phi}(w)$ be the corresponding entropy, v_w be its locus and D_L be the set containing all the different subwords of length L in x. The mean can be rewritten as:

$$m_{L,\phi} = \frac{1}{n} \sum_{i=1}^{n} \hat{f}(x_i) = \frac{1}{n} \sum_{w \in D_L} count(v_w) \times \hat{f}_{L,\phi}(w)$$

Therefore, when traversing the tree, we also keep a variable in which we add the value of entropies found at length L, multiplied by the value of $count(\cdot)$ stored at their locus.

The Standard Deviation. The standard deviation can be rewritten as:

$$S_{L,\phi} = \sqrt{\frac{1}{n-1} \sum_{i=1}^{n} (\hat{f}_{L,\phi}(x_i) - m_{L,\phi})^2} = \sqrt{\frac{1}{n-1} \left(\sum_{i=1}^{n} (\hat{f}^2_{L,\phi}(x_i)) - n m^2_{L,\phi} \right)}$$

Again we aggregate the contribution coming from the same subwords, so that $\sum_{i=1}^{n} \hat{f}^2_{L,\phi}(x_i)$ becomes:

$$\sum_{w \in D_L} (count(v_w) \times \hat{f}^2_{L,\phi}(w))$$

To compute this sum, when traversing the tree we keep a variable in which we add the square of the entropies we compute at length L, multiplied by the value of $count(\cdot)$ stored at their locus.

Once the above summation and the mean have been computed with a single traversal at depth L of our tree, we have all the elements needed to compute the standard deviation in constant time.

The Maximum. As a side observation, one can also note as, in terms of asyntotic complexity, the maximum value of entropy can also be retrieved in linear time with a tree traversal without need to compute the value of $\max_l, 1 \leq l < L$.

4.4   Practical Considerations

We described our algorithms in terms of suffix tree, but we do not really need the entire tree. A truncated suffix tree [2, 13], with a truncation factor equal to the maximum L one is willing to investigate, would be sufficient. Alternatively, an enhanced suffix array [1] allowing a traversal of the virtual LCP tree could also be used.

Note also that if we keep track of the frontier at depth L, i.e., the last nodes of the paths we visit when traversing the tree to compute $f_{L,\phi}$, we can compute the entropic profiles for longer L without starting from the root. Indeed, even in the case we have to start from the root, the preprocessing step does not need to be repeated.

5   Experimental Analysis

In this section we present the results of the experimental analysis we performed on the whole genome of *Haemophilus influenzae*, which is one of the most extensively analyzed in this context. For all the considered methods, the time performance evaluations we show do not include the preprocessing step, i.e., the construction of the exploited data structures (suffix trees or suffix tries), whereas the time needed to annotate the tree is always included. All the tests were run on a laptop with a 3.06GHz Core 2 Duo and 8Gb of Ram.

Figure 2 shows a comparison among the original EP function computation by Vinga et al [18] (denoted by EP in the following), FASTEP by Comin and Antonello [5] and our approach, that is, LINEAREP. In particular, the running times in milliseconds are shown for $\phi = 10$, $L = 10$ and increasing values of n. As it is clear from the figure, LINEAREP outperfomes the other two methods, thus confirming the theoretic results.

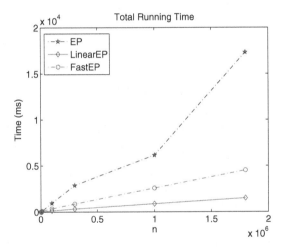

Fig. 2. Comparison among EP, FASTEP and LINEAREP ($\phi = 10$, $L = 10$) for increasing values of n. Total running time includes the computation of the normalizing factors, and the normalized EP values for the whole sequence.

We recall that both EP and LINEAREP compute the normalized EP function according to the same formulation, that is, with respect to the mean and standard deviation (for a fixed lenght L), whereas FASTEP computes a different normalization with respect to the maximum value of the main EP function. We then performed also a direct comparison between EP and LINEAREP for the computation of mean and standard deviation. Figure 3 shows that, except for $n = 1,000$, LINEAREP is faster than EP in such a computation (however, the total time is lower for LINEAREP also for $n = 1,000$, as shown in Figure 2).

We finally compared EP and LINEAREP on a window of lenght 1,000, for L varying between 3 and 15. Note that 15 is a technical limit imposed by the

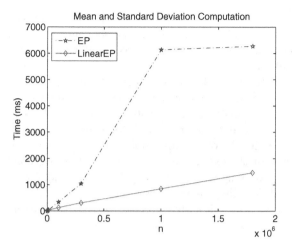

Fig. 3. Comparison between EP and LINEAREP for the computation of mean and standard deviation

software of EP. We do not have such limitation. Indeed we tested our algorithm till $L = 100$ obtaining results close to those for $L = 15$. Figure 4 shows the results for the time needed to compute the mean and the standard deviation. The first observation is that, in both cases, the performances of EP do not change significantly for increasing values of L, whereas the running times of LINEAREP sensibly increase for increasing values of L. This is due to the fact that n is fixed and L varies. To compute mean and standard deviation LINEAREP traverses a

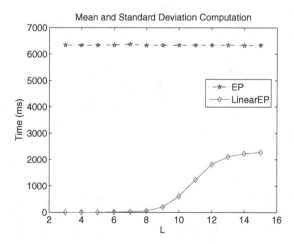

Fig. 4. Computation of mean and standard deviation for EP and LINEAREP ($L \in [3, 15]$)

portion of the tree that is dependent on L, while EP computation of mean and standard deviation is mainly dependent on the sequence length, that is fixed. The running times of LinearEP are from three magnitude order to three times faster than those of EP.

For sake of completeness of the presented results, we add some details about the preprocessing steps performed by each of the considered software tools. In particular, both our prototype and the software of [5] build a full suffix tree although they do not need it in principle. Moreover, the algorithm of [18] is implemented in C, while the others are implemented in Java. The efficiency of our linear algorithm for the extraction of entropic profiles overcomes the known gap between these two languages, but this does not hold for the suffix tree construction. Indeed, building the suffix trie needed to run EP took around 2.7 seconds, while building the full suffix tree, for both other software tools, took around 12 seconds.

6   Concluding Remarks

The research proposed here includes two main contributions. The first contribution is the study of possible relationships between two classes of motifs analyzed in the literature and both effective in singling out significant biological repetitions, that are, entropic profiles and maximal motifs. We proved that entropic profiles are a subset of maximal motifs, and, in particular, that they are left-maximal motifs of the input string. The second contribution of the present manuscript is the proposal of a novel linear time linear space algorithm for the extraction of entropic profiles, according to the original normalization reported in [7]. Experimental validations confirmed that the algorithm proposed here is faster than the others in the literature, including a recent approach where a different normalization was introduced [5].

From these contributions interesting considerations emerge. First of all, we observe that entropic profiles are related to a specific length which one can only guess when doing de novo discovery. So one could think of extracting maximal motifs first, and then investigate entropic profiles in the regions of the maximal motifs and for values of L around the maximal reported length. The process of discovery of the entropic profiles would be further improved then. Other improvements in the entropic profiles extraction could come from the exploitation of more efficient data structures such as enhanced suffix arrays [1]. In this regard, we note that the preprocessing step can also be speeded since, as already pointed out in the previous section, a full suffix tree is not necessary for the computation. Finally, open challenges still remain open about further issues concerning maximal motifs and entropic profiles. Notably among them, one may wonder if entropic profiles do not recover the complete information, so maximal motifs are more reliable when it comes to discovery problems or if, on the contrary, entropic profiles cover the complete information, i.e., they are a refinement of maximal motifs and should be preferred.

Acknowledgments. The research by C. Pizzi and S. E. Rombo was partially supported by the Project "Approcci composizionali per la caratterizzazione e il mining di dati omici" (PRIN 20122F87B2). C. Pizzi was also partially supported by the "Progetto di Ateneo" Univ. of Padova CPDA11023.

References

1. Abouelhoda, M.I., Kurtz, S., Ohlebusch, E.: Replacing suffix trees with enhanced suffix arrays. Journal of Discrete Algorithms 2, 53–86 (2004)
2. Allali, A., Sagot, M.-F.: The at most k deep factor tree. Technical Report (2004)
3. Apostolico, A., Parida, L.: Incremental paradigms of motif discovery. Journal of Computational Biology 11, 1 (2004)
4. Apostolico, A., Pizzi, C., Ukkonen, E.: Efficient algorithms for the discovery of gapped factors. Algorithms for Molecular Biology 6, 5 (2011)
5. Comin, M., Antonello, M.: Fast Computation of Entropic Profiles for the Detection of Conservation in Genomes. In: Ngom, A., Formenti, E., Hao, J.-K., Zhao, X.-M., van Laarhoven, T. (eds.) PRIB 2013. LNCS, vol. 7986, pp. 277–288. Springer, Heidelberg (2013)
6. Federico, M., Pisanti, N.: Suffix tree characterization of maximal motifs in biological sequences. Theoretical Computer Science 410(43), 4391–4401 (2009)
7. Fernandes, F., Freitas, A.T., Vinga, S.: Detection of conserved regions in genomes using entropic profiles. INESC-ID Tec. Rep. 33/2007
8. Fernandes, F., Freitas, A.T., Almeida, J.S., Vinga, S.: Entropic Profiler – detection of conservation in genomes using information theory. BMC Research Notes 2, 72 (2009)
9. Herzel, H., Ebeling, W., Schmitt, A.O.: Entropies of biosequences: The role of repeats. Physical Review E 50, 5061–5071 (1994)
10. Grossi, R., Pietracaprina, A., Pisanti, N., Pucci, G., Upfal, E., Vandin, F.: MADMX: a strategy for maximal dense motif extraction. Journal of Computational Biology 18(4), 535–545 (2011)
11. Grossi, R., Pisanti, N., Crochemore, M., Sagot, M.-F.: Bases of motifs for generating repeated patterns with wild cards. IEEE/ACM Transactions on Computational Biology and Bioinformatics 2(1), 40–50 (2005)
12. Gusfield, D.: Algorithms on Strings, Trees and Sequences: Computer Science and Computational Biology. Cambridge University Press (1997)
13. Na, J.C., Apostolico, A., Iliopulos, C.S., Park, K.: Truncated suffix trees and their application to data compression. Theoretical Computer Science 304(1-3), 87–101 (2003)
14. Parida, L.: Pattern Discovery in Bioinformatics: Theory & Algorithms. Chapman & Hall/CRC (2007)
15. Parida, L., Pizzi, C., Rombo, S.E.: Characterization and Extraction of Irredundant Tandem Motifs. In: Calderón-Benavides, L., González-Caro, C., Chávez, E., Ziviani, N. (eds.) SPIRE 2012. LNCS, vol. 7608, pp. 385–397. Springer, Heidelberg (2012)
16. Parida, L., Pizzi, C., Rombo, S.E.: Irredundant tandem motifs. Theoretical Computer Science 525, 89–102 (2014)
17. Rombo, S.E.: Extracting string motif bases for quorum higher than two. Theoretical Computer Science 460, 94–103 (2012)
18. Vinga, S., Almeida, J.S.: Local Renyi entropic profiles of DNA sequences. BMC Bioinformatics 8, 393 (2007)

Appendix

Let us start from the following two inequalities:

$$
\begin{cases}
\frac{N+\sum_{k=1}^{L'} 4^k \phi^k c([i-k+1,i])}{\sum_{k=0}^{L'} \phi^k} \geq \frac{N+\sum_{k=1}^{L'+1} 4^k \phi^k c([i-k+1,i])}{\sum_{k=0}^{L'+1} \phi^k} \\
\frac{N+\sum_{k=1}^{L'} 4^k \phi^k c([i-k+1,i])}{\sum_{k=0}^{L'} \phi^k} \geq \frac{N+\sum_{k=1}^{L'-1} 4^k \phi^k c([i-k+1,i])}{\sum_{k=0}^{L'-1} \phi^k}
\end{cases}
$$

from which:

$$
\begin{cases}
\frac{N+\sum_{k=1}^{L'} 4^k \phi^k c([i-k+1,i])}{\sum_{k=0}^{L'} \phi^k} \geq \frac{N+\sum_{k=1}^{L'} 4^k \phi^k c([i-k+1,i])+4^{L'+1}\phi^{L'+1}c([i-L',i])}{\sum_{k=0}^{L'} \phi^k+\phi^{L'+1}} \\
\frac{N+\sum_{k=1}^{L'-1} 4^k \phi^k c([i-k+1,i])+4^{L'}\phi^{L'} c([i-L'+1,i])}{\sum_{k=0}^{L'-1} \phi^k+\phi^{L'}} \geq \frac{N+\sum_{k=1}^{L'-1} 4^k \phi^k c([i-k+1,i])}{\sum_{k=0}^{L'-1} \phi^k}
\end{cases}
$$

Let us consider $A = N + \sum_{k=1}^{L'} 4^k \phi^k c([i - k + 1, i])$, $B = \sum_{k=0}^{L'} \phi^k$, $C = 4^{L'+1}\phi^{L'+1}c([i - L', i])$, $D = \phi^{L'+1}$; $A' = N + \sum_{k=1}^{L'-1} 4^k \phi^k c([i - k + 1, i])$, $B' = \sum_{k=0}^{L'-1} \phi^k$, $C' = 4^{L'}\phi^{L'} c([i - L' + 1, i])$ and $D' = \phi^{L'}$. Then:

$$
\begin{cases}
\frac{A}{B} \geq \frac{A+C}{B+D} \implies \frac{A}{B} \geq \frac{C}{D} \\
\frac{A'+C'}{B'+D'} \geq \frac{A'}{B'} \implies \frac{A'}{B'} \leq \frac{C'}{D'}
\end{cases}
$$

The two inequalities above can be then rewritten as:

$$
\begin{cases}
\frac{N+\sum_{k=1}^{L'} 4^k \phi^k c([i-k+1,i])}{\sum_{k=0}^{L'} \phi^k} \geq \frac{4^{L'+1}\phi^{L'+1}c([i-L',i])}{\phi^{L'+1}} \\
\frac{N+\sum_{k=1}^{L'-1} 4^k \phi^k c([i-k+1,i])}{\sum_{k=0}^{L'-1} \phi^k} \leq \frac{4^{L'}\phi^{L'} c([i-L'+1,i])}{\phi^{L'}}
\end{cases}
$$

$$
\begin{cases}
N + \sum_{k=1}^{L'-1} 4^k \phi^k c([i - k + 1, i]) \geq \frac{\sum_{k=0}^{L'} \phi^k}{\phi^{L'+1}} 4^{L'+1}\phi^{L'+1} c([i - L', i]) - 4^{L'}\phi^{L'} c([i - L' + 1, i]) \\
N + \sum_{k=1}^{L'-1} 4^k \phi^k c([i - k + 1, i]) \leq \frac{\sum_{k=0}^{L'-1} \phi^k}{\phi^{L'}} 4^{L'}\phi^{L'} c([i - L' + 1, i])
\end{cases}
$$

Then:

$$
\frac{\sum_{k=0}^{L'} \phi^k}{\phi^{L'+1}} 4^{L'+1}\phi^{L'+1} c([i - L', i]) - 4^{L'}\phi^{L'} c([i - L' + 1, i]) \leq \frac{\sum_{k=0}^{L'-1} \phi^k}{\phi^{L'}} 4^{L'}\phi^{L'} c([i - L' + 1, i])
$$

$$
\sum_{k=0}^{L'} \phi^k 4 c([i - L', i]) - \phi^{L'} c([i - L' + 1, i]) - \sum_{k=0}^{L'-1} \phi^k c([i - L' + 1, i]) \leq 0
$$

$$
(\phi^{L'} + \sum_{k=0}^{L'-1} \phi^k) c([i - L' + 1, i]) \geq 4 \sum_{k=0}^{L'} \phi^k c([i - L', i])
$$

$$
c([i - L' + 1, i]) \geq 4 c([i - L', i])
$$

Estimating Evolutionary Distances
from Spaced-Word Matches

Burkhard Morgenstern[1,2], Binyao Zhu[3], Sebastian Horwege[1],
and Chris-André Leimeister[1]

[1] University of Göttingen, Institute of Microbiology and Genetics,
Department of Bioinformatics, Goldschmidtstr. 1, 37077 Göttingen, Germany
`bmorgen@gwdg.de`
[2] Université d'Evry Val d'Essonne, Laboratoire Statistique et Génome, UMR CNRS
8071, USC INRA, 23 Boulevard de France, 91037 Evry, France
[3] University of Göttingen, Institute of Microbiology and Genetics,
Department of General Microbiology, Grisebachstr. 8, 37077 Göttingen, Germany

Abstract. Alignment-free methods are increasingly used to estimate
distances between DNA and protein sequences and to reconstruct phylogenetic trees. Most distance functions used by these methods, however,
are heuristic measures of dissimilarity, not based on any explicit model
of evolution. Herein, we propose a simple estimator of the evolutionary
distance between two DNA sequences calculated from the number of
(spaced) word matches between them. We show that this distance function estimates the evolutionary distance between DNA sequences more
accurately than other distance measures used by alignment-free methods. In addition, we calculate the variance of the number of (spaced)
word matches depending on sequence length and mismatch probability.

1 Introduction

Alignment-free methods are increasingly used for DNA and protein sequence
comparison since they are much faster than traditional alignment-based approaches [1]. Most alignment-free algorithms compare the *word* or *k-mer composition* of the input sequences [2]. They use standard metrics such as the *Euclidean*
or the *Jensen-Shannon (JS)* distance [3] on the relative word frequency vectors
of the input sequences to estimate their distances.

Recently, we proposed an alternative approach to alignment-free sequence
comparison. Instead of considering *contiguous* subwords of the input sequences,
our approach considers *spaced words, i.e.* words containing *wildcard* or *don't
care* characters at positions defined by a pre-defined *pattern P*, similar as the
spaced seeds that are used in database searching [4]. As in existing alignment-
free methods, the (relative) frequencies of these spaced words are compared using
standard distance measures [5]. In [6], we extended this approach by using whole
sets $\mathcal{P} = \{P_1, \ldots, P_m\}$ of patterns and calculating the spaced-word frequencies
with respect to *all* patterns in $\mathcal{P}$. In this *multiple-pattern* approach, the distance
between two sequences is defined as the *average* of the distances between by the

D. Brown and B. Morgenstern (Eds.): WABI 2014, LNBI 8701, pp. 161–173, 2014.
© Springer-Verlag Berlin Heidelberg 2014

spaced-word frequency vectors with respect to the individual patterns $P_i \in \mathcal{P}$, see also [7].

Phylogeny reconstruction is one of the main applications of alignment-free sequence comparison. Consequently, most alignment-free methods were benchmarked by applying them to phylogeny problems. The distance metrics used by these methods, however, are only rough measures of dissimilarity, not derived from any explicit model of molecular evolution. This may be one reason why the distances calculated by alignment-free algorithms are usually not directly evaluated, but they are used as input for distance-based phylogeny methods such as *Neighbour-Joining* [8]. The resulting tree topologies are than compared to trusted reference topologies.

Obviously, this is only a very rough way of evaluating these methods since the resulting trees do not only depend on the calculated distance values but also on the tree-reconstruction method that is used. Also, comparing topologies ignores branch lengths, so the results of these benchmark studies depend only indirectly on the distance values calculated by the alignment-free methods that are to be evaluated. A remarkable exception is the paper by Haubold *et al.* [9]. The program K_r developed by these authors estimates evolutionary distances based on a probabilistic model of evolution, and the authors compare the estimated distances directly to the known distances of simulated sequences. To our knowledge, K_r is the only alignment-free method that aims to estimate distances in a rigorous way. The authors of K_r have shown that this program can correctly estimate evolutionary distances between DNA sequences up to a distance of around 0.5 mutations per site.

In previous papers, we have shown, that our *spaced-word* approach is useful for phylogeny reconstruction. Tree topologies calculated with *Neighbour-Joining* based on *spaced-word* frequency vectors are usually superior to topologies calculated from the *contiguous* word frequency vectors that are used by traditional alignment-free methods [6]. Moreover, the 'multiple-pattern approach' led to much better results than the 'single-pattern approach'. We also showed experimentally that the distance values and tree topologies produced by *spaced words* are statistically more stable than distances and trees produced using *contiguous* subwords of the sequences. In fact, the main difference between our *spaced words* and the *contiguous words* used by established methods is that spaced word matches at neighbouring positions are statistically less dependent on each other.

In these previous papers, we investigated the difference between spaced word frequencies and on contiguous word frequencies. Therefore, we applied the same distance metrics to our spaced-word frequencies that are applied by standard methods to k-mer frequencies, namely *Jensen-Shannon* and the *Euclidean* distance. In the present paper, we propose a new distance measure on DNA sequences that estimates their distances from the number N of space-word matches based on a probabilistic model. We show that this distance measure is more accurate and works for more distantly related sequences than existing alignment-free distance measures. Secondly, we calculate the *variance* of N for *contiguous k-mers* that are used in standard approaches, as well as to *spaced words* with the single and multiple pattern approach. We show that the variance of N is lower for spaced words than for contiguous words and that the variance is further reduced in our *multiple* pattern approach.

2   Motifs and Spaced Words

As usual, for an alphabet Σ and $\ell \in \mathbf{N}$, Σ^ℓ denotes the set of all sequences of length ℓ over Σ. For a sequence $S \in \Sigma^\ell$ and $0 < i \leq \ell$, $S[i]$ denotes the i-th character of S. A *pattern* of length ℓ is a word $P \in \{0, 1\}^\ell$, *i.e.* a sequence over $\{0, 1\}$ of length ℓ. A position i with $P[i] = 1$ is called a *match position* while a position i with $P[i] = 0$ is called a *don't care position*. The number of all *match positions* in a patterns P is called the *weight* of P. For a pattern P of weight k, $\hat{P} = \{\hat{P}_1, \ldots \hat{P}_k\}$, $\hat{P}_i < \hat{P}_{i+1}$, denotes the set of all match positions.

A *spaced word* w of weight k over an alphabet Σ is a pair (P, w') such that P is a pattern of weight k and w' is a word of length k over Σ. We say that a spaced word (P, w') occurs at position i in a sequence S over Σ, if $S[i + \hat{P}_r - 1] = w'[r - 1]$ for all $1 \leq r \leq k$. For example, for

$$\Sigma = \{A, T, C, G\}, \quad P = 1101, \quad w' = ACT,$$

we have $\hat{P} = \{1, 2, 4\}$, and the spaced word $w = (P, w')$ occurs at position 2 in sequence $S = CACGTCA$ since

$$S[2]S[3]S[5] = ACT = w'.$$

A pattern is called *contiguous* if it consists of match positions only, a spaced word is called contiguous if the underlying pattern is contiguous.

For a pattern P of weight k and two sequences S_1 and S_2 over an alphabet Σ, we say that there is a *spaced-word match* with respect to P at (i, j) if

$$S_1[i + \hat{P}_r - 1] = S_2[j + \hat{P}_r - 1]$$

holds for all $1 \leq r \leq k$. For example, for sequences $S_1 = ACTCTAA$ and $S_2 = TATAGG$ and P as above, there is a spaced-word match at $(3, 1)$ since one has $S_1[3] = S_2[1], S_1[4] = S_2[2]$ and $S_1[6] = S_2[4]$.

3   The Number N of Spaced-Word Matches for a Pair of Sequences with Respect to a Set $\mathcal{P}$ of Patterns

We consider sequences S_1 and S_2 as above and a fixed set $\mathcal{P} = \{P_1, \ldots, P_m\}$ of patterns. For simplicity, we assume that all patterns in $\mathcal{P}$ have the same length ℓ and the same weight k. For now, we use a simplified model of sequence evolution without insertions and deletions, with a constant mutation rate and with different sequence positions evolving independently of each other. Moreover, we assume that we have the same substitution rates for all substitutions $a \rightarrow$

$b, a \neq b$. We therefore consider two sequences S_1 and S_2 of the same length L with match probabilities

$$P(S_1[i] = S_2[j]) = \begin{cases} p & \text{for } i = j \\ q & \text{for } i \neq j \end{cases}$$

where $q = \sum_{a \in \mathcal{A}} q_a^2$ is the background match probability with q_a denoting the relative frequency of a single character $a \in \mathcal{A}$ and $p \geq q$ is the match probability for a pair of 'homologous' positions.

We want to study the number $N = N(S_1, S_2, \mathcal{P})$ of $\mathcal{P}$-matches between S_1 and S_2, i.e. the number of spaced-word matches with respect to patterns $P \in \mathcal{P}$. N can be seen as the *inner product* of the count vectors for spaced words with respect to the set of patterns $\mathcal{P}$. In the special case where $\mathcal{P}$ consists of a *single contiguous* pattern, N is also called the D_2 score [10]. The statistical behaviour of the D_2 score has been studied under the *null model* that S_1 and S_2 are *unrelated* [11,12]. By contrast, we want to investigage the number N of spaced-word matches for *evolutionarily related* sequence pairs under a model as specified above. To this end, we define $X_{i,j}^P$ to be the Bernoulli random variable that is 1 if there is a P-match between S_1 and S_2 at (i, j), $P \in \mathcal{P}$, and 0 otherwise, so N can be written as

$$N = \sum_{\substack{P \in \mathcal{P} \\ i,j}} X_{i,j}^P$$

If we want to calculate the expectation value and variance of N, we have to distinguish between 'homologue' spaced-word matches, that is matches that are due do 'common ancestry' and 'background matches' due to chance. In the simplest case where we do not consider insertions and deletions in our model of evolution, a P-match at (i, j) is homologue if and only if $i = j$ holds. So in this special case, we can define

$$\mathcal{X}^{Hom} = \left\{ X_{i,i}^P \mid 1 \leq i \leq L - \ell + 1, P \in \mathcal{P} \right\},$$

$$\mathcal{X}^{BG} = \left\{ X_{i,j}^P \mid 1 \leq i, j \leq L - \ell + 1, i \neq j, P \in \mathcal{P} \right\}.$$

Note that if sequences do not contain insertions and deletions, *every* spaced-word match is either entirely a *homologous* match or entirely a *background* match. If indels are considered, a spaced-word match can cover both, homologous and background regions, and the above definitions need to be adapted. The set $\mathcal{X}$ of all random variables $X_{i,j}^P$ can then be written as $\mathcal{X} = \mathcal{X}^{Hom} \cup \mathcal{X}^{BG}$, the total sum N of spaced-word matches with respect to the set $\mathcal{P}$ of patterns is

$$N = \sum_{X \in \mathcal{X}} X$$

and the expected number of spaced-word matches is

$$E(N) = E\left(\sum_{X \in \mathcal{X}} X \right) = \sum_{X \in \mathcal{X}^{Hom}} E(X) + \sum_{X \in \mathcal{X}^{BG}} E(X),$$

where the expectation value of a single random variable $X \in \mathcal{X}$ is

$$E(X) = \begin{cases} p^k & \text{if } X \in \mathcal{X}^{Hom} \\ q^k & \text{if } X \in \mathcal{X}^{BG} \end{cases} \tag{1}$$

There are $L - \ell + 1$ positions (i, i) and $(L - \ell) \cdot (L - \ell + 1)$ positions $(i, j), i \neq j$ where spaced-word matches can occur, so we obtain

$$E(N) = m \cdot \left[(L - \ell + 1) \cdot p^k + (L - \ell) \cdot (L - \ell + 1) \cdot q^k \right] \tag{2}$$

4   Estimating Evolutionary Distances from the Number N of Spaced-Word Matches

If the *weight* of the patterns – *i.e.* the number of match positions – in the *spaced-words* approach is sufficiently large, random space-word matches can be ignored. In this case, the *Jensen-Shannon* distance between two DNA sequences approximates the number of (spaced) words that occur in one of the compared sequences, but not in the other one. Thus, if two sequences of length L are compared and N is the number of (spaced) words that two sequences have in common, their *Jenson-Shannon* distance can be approximated by $L - N$. (Similarly, the *Euclidean* distances between two sequences can be approximated by the square root of this value if the distance is small and k is large enough.) For *small* evolutionary distances, the *Jensen-Shannon* distance grows therefore roughly linearly with the distance between two sequences, and this explains why it is possible to produce reasonable phylogenies based on this metric. It is clear, however, that the *Jensen-Shannon* distance is far from linear to the real distance for larger distances. We therefore propose an alternative estimator of the evolutionary distance between two sequences in terms of the number N of *spaced-word* matches between them. Again, we first consider sequences without insertions and deletions.

From the expected number $E(N)$ of spaced words shared by sequences S_1 and S_2 with respect to a set of patterns $\mathcal{P}$ as given in equation (2), we obtain

$$\hat{p} = \sqrt[k]{\frac{N}{m \cdot (L - \ell + 1)} - (L - \ell) \cdot q^k} \tag{3}$$

as an estimator for the match probability p for sequences without indels, and with Jukes-Cantor [13] we obtain

$$\hat{d} = -\frac{3}{4} \cdot \ln \left[\frac{4}{3} \sqrt[k]{\frac{N}{m \cdot (L - \ell + 1)} - (L - \ell) \cdot q^k} - \frac{1}{3} \right] \tag{4}$$

as an estimator for the distance d between the sequences S_1 and S_2.

Equation (2) for the expected number N of spaced-word matches between two sequences S_1 and S_2 can be easily generalized to the case where S_1 and S_2 have different lengths and contain insertions and deletions. Let L_1 and L_2 be the lengths of S_1 and S_2, respectively and $L^{Hom} \leq \min\{L_1, L_2\}$ the length of the 'homologous' part of the sequences consisting u un-gapped pairs of segments. Ignoring spaced-word matches that cover both homologous and random regions, we can estimate the expectation value of the number of spaced-word matches as

$$E(N) \approx m \cdot \left[\left(L^{Hom} - u(l+1) \right) \cdot p^k + \left((L - \ell)^2 - L^{Hom} \right) \cdot q^k \right]$$

and equations (3) and (4) can be adapted accordingly.

5   The Variance of N

To calculate the *variance* of N, we adapt results on the occurrence of words in a sequence as outlined in [14]. First, we calculate the *joint* probability of *overlapping* spaced-word matches for (different or equal) patterns from $\mathcal{P}$ at different sequence positions. Note that an overlap between a P-match at i, j and a P'-match at (i', j') can occur only if $i' - i = j' - j$ (and for non-overlapping P-matches, their joint probability is, of course, the product of their individual probabilities). We therefore consider a P-match at (i, j) and a P' match at $(i + s, j + s)$ for some $s \geq 0$.

For patterns P, P' and $s \in \mathbf{N}$ we define $n(P, P', s)$ to be the number of integers that are match positions of P or match positions of P' shifted by s positions to the right (or both). Formally, if

$$\hat{P}_s = \{\hat{P}_1 + s, \ldots, \hat{P}_k + s\}$$

denotes the set of match positions of a pattern P shifted by s positions to the right, we define

$$n(P, P', s) = |\hat{P} \cup \hat{P}'_s| = |\hat{P}| + |\hat{P}'_s| - |\hat{P} \cap \hat{P}'_s|$$

For example, for $P = 101011, P' = 111001$ and $s = 2$, there are 6 positions that are match positions of P or of P' shifted by 2 positions to the right, namely positions 1, 3, 4, 5, 6, 8:

$$\begin{aligned} P: &\ 1\,0\,1\,0\,1\,1 \\ P': &\quad\ \ 1\,1\,1\,0\,0\,1 \end{aligned}$$

so one has $n(P, P', s) = 6$. In particular, one has $n(P, P, 0) = k$ for all patterns P of weight k, and

$$n(P, P, s) = k + \max\{s, k\}$$

for all *contiguous* patterns P of weight (or length) k. With this notation, we can write

$$E\left(X_{i,j}^P \cdot X_{i+s,j+s}^{P'} \right) = \begin{cases} p^{n(P,P',s)} & \text{if } i = j \\ q^{n(P,P',s)} & \text{else} \end{cases} \tag{5}$$

for all $X_{i,j}^P, X_{i+s,j+s}^{P'}$

To calculate the covariance of two random variables from $\mathcal{X}$, we distinguish again between homologue and random matches. Note that the covariance of two *non-overlapping* random variables X and X' is always zero (in particular, the covariance $Cov(X, X')$ is zero for a 'homologue' $X \in \mathcal{X}^{Hom}$ and a 'random' $X' \in \mathcal{X}^{BG}$). We first consider 'homologue' pairs $X_{i,i}^{P}, X_{i+s,i+s}^{P'} \in \mathcal{X}^{Hom}$. Here, we obtain with (5)

$$Cov\left(X_{i,i}^{P}, X_{i+s,i+s}^{P'}\right) = p^{n(P,P',s)} - p^{2k} \qquad (6)$$

Similarly, for a pair of 'background' variables $X_{i,j}^{P}, X_{i+s,j+s}^{P'} \in \mathcal{X}^{BG}$, one obtains

$$Cov\left(X_{i,j}^{P}, X_{i+s,j+s}^{P'}\right) = q^{n(P,P',s)} - q^{2k}. \qquad (7)$$

Since 'homologue' and 'background' variables are uncorrelated, the variance of N can be written as

$$Var(N) = Var\left(\sum_{X \in \mathcal{X}} X\right) = Var\left(\sum_{X \in \mathcal{X}^{Hom}}\right) + Var\left(\sum_{X \in \mathcal{X}^{BG}}\right)$$

We express the variance of these sums of random variable as the sum of all of their covariances, so for the 'homologue' random variables we can write

$$Var\left(\sum_{X \in \mathcal{X}^{Hom}} X\right) = \sum_{P,P' \in \mathcal{P}} \sum_{i,i'=1}^{L-l+1} Cov\left(X_{i,i}^{P}, X_{i',i'}^{P'}\right)$$

Since the covariance for non-correlated random variables vanishes, we can ignore the covariances of all pairs $\left(X_{i,i}^{P}, X_{i',i'}^{P'}\right)$ with $|i - i'| \geq l$ so, ignoring side effects, we can write the above sum as

$$Var\left(\sum_{X \in \mathcal{X}^{Hom}} X\right) \approx \sum_{i=1}^{L-\ell+1} \sum_{P,P' \in \mathcal{P}} \sum_{s=-\ell+1}^{\ell-1} Cov\left(X_{i,i}^{P}, X_{i+s,i+s}^{P'}\right)$$

and since the above covariances depend only on s but not on i, we can use (5) and (7) and obtain

$$Var\left(\sum_{X \in \mathcal{X}^{Hom}} X\right) \approx (L - \ell + 1) \cdot \sum_{P,P' \in \mathcal{P}} \sum_{s=-\ell+1}^{\ell-1} \left(p^{n(P,P',s)} - p^{2k}\right)$$

and similarly

$$Var\left(\sum_{X \in \mathcal{X}^{BG}} X\right) \approx (L - \ell + 1) \cdot (L - \ell) \cdot \sum_{P,P' \in \mathcal{P}} \sum_{s=-\ell+1}^{\ell-1} \left(q^{n(P,P',s)} - q^{2k}\right)$$

Together, we get

$$
\begin{aligned}
Var(N) \approx \quad & (L - \ell + 1) \cdot \sum_{P,P' \in \mathcal{P}} \sum_{s=-\ell+1}^{\ell-1} \left(p^{n(P,P',s)} - p^{2k} \right) \\
& + (L - \ell + 1) \cdot (L - \ell) \cdot \sum_{P,P' \in \mathcal{P}} \sum_{s=-\ell+1}^{\ell-1} \left(q^{n(P,P',s)} - q^{2k} \right)
\end{aligned}
\tag{8}
$$

6   Test Results

To evaluate the distance function defined by equation (4), we simulated pairs of DNA sequences with an (average) length of 100,000 and with an average of d substitutions per sequence position. We varied d between 0 and 1 and compared the distances estimated by our distance measure and by various other alignment-free programs to the 'real' distance d. We performed these experiments for sequence pairs without insertions and deletions and for sequence pairs where we included insertions and deletions with a probability of 1% at every position. The length of indels was randomly chosen between 1 and 50 with uniform probability.

Figure 1 shows the results of these experiments. Our new distance measure applied to spaced-word frequencies is well in accordance with the real distances d for values of $d \leq 0.8$ on sequence pairs without insertions and deletions if the *single-pattern* version of our program is used. For the *multiple-pattern* version, our distance function estimates the real distances correctly for all values of $d \leq 1$. If indels are added as specified above, our distance functions slightly overestimates the real distance d. By contrast, the *Jensen-Shannon* distance applied to the same spaced-word frequencies increased non-linearly with d and flattened for values of around $d \geq 0.4$.

As mentioned, the only other alignment-free method that estimates evolutionary distances on the basis of a probabilistic model of evolution is K_r [15]. In our study, K_r correctly estimated the true distance d for values of around $d \leq 0.6$, this precisely corresponds to the results reported by the authors of the program. For larger distances, K_r grossly overestimates the distance d, though. The distance values calculated by the program k *mismatch average common substring (kmacs)* that we previously developed [16] are roughly linear to the real distances d for values of up to around $d = 0.3$. From around $d = 0.5$ on, the curve becomes flat. With $k = 30$ mismatches, the performance of *kmacs* was better than with $k = 0$, in which case *kmacs* corresponds to the *Average Common Substring (ACS)* approach [17].

Next, we applied various distance measures to a set of 27 mitochondrial genomes from primates that were previously used by [15] to evaluate alignment-free approaches. We used our *multiple spaced-words* approach with the parameters that we used in [6], that is with a pattern weight (number of match positions) of $k = 9$ and with pattern lengths ℓ between 9 and 30, *i.e.* with up to 30 *don't care* positions in the patterns. For each value of ℓ, we randomly generated sets $\mathcal{P}$

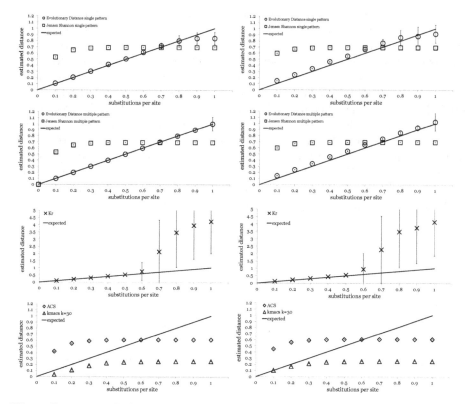

Fig. 1. Distances calculated by different alignment-free methods for pairs of simulated DNA sequences plotted against their 'real' distances d measured in substitutions per site. Plots on the left-hand side are for sequence pairs without insertions and deletions, on the right-hand side the corresponding results are shown for sequences with an indel probability of 1% for each site and an average indel length of 25. From top to bottom, the applied methods were: (1) spaced words with the *single-pattern* approach and the *Jensen-Shannon* distance and the distance defined in this paper, (2) the *multiple-pattern* version of spaced words using sets $\mathcal{P}$ of $m = 100$ patterns with the same distance functions, (3) K_r and (4) *kmacs* and *ACS*.

of $m = 100$ patterns. In addition, we used the competing approaches *FFP* [18], CVTree [19], K_r [9], *kmacs* [16] and *ACS* [17].

For each method, we calculated a distance matrix for the input sequences, and we compared the obtained distance matrices to a reference distance matrix that we calculated with the program *Dnadist* from the *PHYLIP* package [20] based on a reference multiple alignment. For comparison with the reference matrix, we used a software program based on the *Mantel* test [21] that was also used in [22]. Figure 2 shows the results of this comparison. As can be seen, our new distance measure, applied to multiple spaced-word frequencies, produced distance matrices close to the reference matrix and outperformed the *Jenson-Shannon* distance for all pattern lengths ℓ that we tested. Our new distance also outperformed some of the existing alignment-free methods, with the exception of K_r and *kmacs*.

As a third sequence set, we used a set of 112 HIV-1 genomes from the HIV-1/SIVcpz database at *Los Alamos National Laboratory* [23]. Again, we compared the distance matrices produced by various alignment-free methods to a reference matrix calculated with *Dnadist* from a trusted reference alignment from the HIV database. Here, the *Jenson-Shannon* distance applied to *multiple spaced-word* frequencies was slightly superior to our new distance function if the length ℓ (and therefore the number of *don't care positions* of the underlying patterns) was small. Only for larger values of ℓ, our new distance was superior to *Jensen-Shannon*. As in the previous example, *kmacs* was among the best performing methods. On the HIV sequences, we could also apply a multiple-alignment program. We used *CLUSTAL Ω* [24] and applied *Dnadist* to the resulting alignment. Not surprisingly, this slow but accurate method of sequence comparison performed better than all alignment-free approaches that we tested.

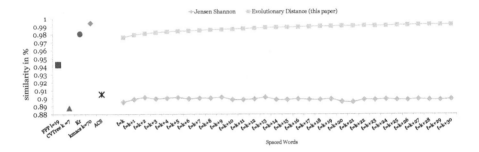

Fig. 2. Evaluation of distance matrices calculated with various alignment-free methods for a set of 27 primate mitochondrial genomes. Each distance matrix was compared to a trusted *reference* distance matrix based on the *Mantel test*. The similarity between the calculated matrices and the reference matrix is plotted. We evaluated our new distance measure defined by equation (4) using sets $\mathcal{P}$ of 100 randomly calculated patterns with $k = 9$ *match positions* and varying length (yellow), as well as the *Jensen-Shannon* distance applied to the same spaced-word frequency vectors (green). In addition, distance matrices calculated by various other alignment-free methods were evaluated.

Fig. 1 shows not only striking differences in the shape of the distance functions used by various alignment-free programs. There are also remarkable differences in the *variance* of the distances calculated with the new distance measure that we defined in equation (4). This distance is defined in terms of the number N of *(spaced) word* matches between two sequences. As mentioned above, the established *Jensen-Shannon* and *Euclidean* distances on (spaced) word frequency vectors also depend on N, as they can be approximated by $L - N$ and $\sqrt{L - N}$, respectively. Thus, the variances of these three distance measures directly depend on the variance of N. As can be seen in Fig. 1, the variance of the distances calculated with our new distance function increases with the frequency of substitutions. Also, the variance is higher for the single-pattern approach than for the multiple-pattern approach. To explain this observation, we calculated the *variance* of the number N of spaced-

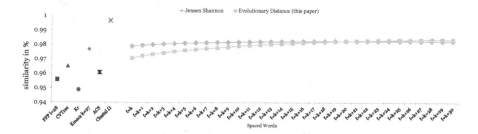

Fig. 3. Evaluation of distance matrices calculated with various alignment-free methods for a set of 112 HIV-1 genomes. Similarities between calculated matrices and a reference matrix were calculated as in Figure2. In addition, we evaluated a distance matrix based on a multiple alignment calculated by *CLUSTAL Ω*.

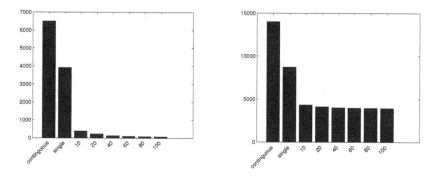

Fig. 4. Variance of the normalized number $\frac{N}{m}$ of spaced-word matches where $m = |\mathcal{P}|$ is the number of patterns in the multiple-pattern approach. Formula (8) was applied to contiguous and to single and multiple spaced words for un-gapped sequence pairs of length 16,000 *nt* with a mismatch frequency of 0.7 (left) and 0.25 (right)

word matches using equation 8. Fig. 4 summarizes the results for a sequence length of $L = 16.000$ and mismatch frequencies of 0.7 and 0.25, respectively.

7   Discussion

In this paper, we proposed a new estimator for the evolutionary distance between two DNA sequences based on the number N of *spaced-word* matches between them. While most alignment-free methods use ad-hoc distance measures, the distance function that we defined is based on a probabilistic model of evolution and seems to be a good estimator for the number of substitutions per site that have occurred since two sequences have evolved separately. For simplicity, we used a model of evolution without insertions and deletions. Nevertheless, our test results show that our distance function is still a reasonable estimator if the input sequences contain a moderate number of insertions and deletions. Obviously, our distance function would drastically overestimate the distances between sequence pairs that share only local homologies. This seems to be a major limitation of our approach. However, as indicated in section 4, our distance measure can be adapted to the case of local homologies if the length of these homologies and the number of gaps in the homologous regions can be estimated. In principle, it should therefore be possible to apply our method to locally related sequences by first estimating the extent of their shared homologies and then adapting our distance measure accordingly.

The distance introduced in this paper and other distance measures that we previously used for our *spaced words* approach depend on the number N of space-word matches between two sequences with respect to a set $\mathcal{P}$ of patterns of 'match' and 'don't care' positions. This is similar for more traditional, k-mer based distance measures where $\mathcal{P}$ consists of one single contiguous pattern $P = 1 \ldots 1$. Obviously, the expected number of (spaced) word matches is essentially the same for contiguous and for spaced words of the corresponding *weight*. Herein, we showed how the *variance* of N can be calculated and demonstrated that this variance is considerably lower for our spaced-words approach than for the standard approach that is based on contiguous words, and that our *multiple-pattern* approach further reduces the variance of N/m where m is the number of patterns in $\mathcal{P}$. This seems to be the main reason why our *multiple spaced words* approach outperforms the *single-pattern* approach that we previously introduced as well as the classical k-mer approach when used for phylogeny reconstruction.

As we have shown, the variance of N depends on the number of overlapping 'match' positions if patterns from $\mathcal{P}$ are shifted against each other. Consequently, in our single-pattern approach, the variance of N is higher for periodic patterns than for non-periodic patterns, and if a pattern like $101010\ldots$ is used, the variance is equal to the variance of the contiguous pattern with the same weight. In our benchmark studies, we could experimentally confirm that on phylogeny benchmark data, *spaced words* performs worse with periodic patterns than with non-periodic patterns. Therefore, the theoretical results of this study may be useful to find patterns or sets of patterns that minimize the variance of N and thereby improve our spaced-words approach.

Acknowledgements. We would like to thank Marcus Boden, Sebastian Lindner, Alec Guyomard and Claudine Devauchelle for help with the program evaluation and Gilles Didier for help with the software to compare distance matrices.

References

1. Vinga, S.: Editorial: Alignment-free methods in computational biology. Briefings in Bioinformatics 15, 341–342 (2014)
2. Blaisdell, B.E.: A measure of the similarity of sets of sequences not requiring sequence alignment. Proceedings of the National Academy of Sciences of the United States of America 83, 5155–5159 (1986)
3. Lin, J.: Divergence measures based on the shannon entropy. IEEE Transactions on Information theory 37, 145–151 (1991)
4. Ma, B., Tromp, J., Li, M.: PatternHunter: faster and more sensitive homology search. Bioinformatics 18, 440–445 (2002)

5. Boden, M., Schöneich, M., Horwege, S., Lindner, S., Leimeister, C.-A., Morgenstern, B.: Alignment-free sequence comparison with spaced k-mers. In: German Conference on Bioinformatics 2013. OpenAccess Series in Informatics (OASIcs), vol. 34, pp. 24–34 (2013)
6. Leimeister, C.-A., Boden, M., Horwege, S., Lindner, S., Morgenstern, B.: Fast alignment-free sequence comparison using spaced-word frequencies. Bioinformatics 30, 2000–2008 (2014)
7. Horwege, S., Sebastian, L., Boden, M., Hatje, K., Kollmar, M., Leimeister, C.-A., Morgenstern, B.: *Spaced words* and *kmacs*: fast alignment-free sequence comparison based on inexact word matches. Nucleic Acids Research 42, W7–W11 (2014)
8. Saitou, N., Nei, M.: The neighbor-joining method: a new method for reconstructing phylogenetic trees. Molecular Biology and Evolution 4, 406–425 (1987)
9. Haubold, B., Pierstorff, N., Möller, F., Wiehe, T.: Genome comparison without alignment using shortest unique substrings. BMC Bioinformatics 6, 123 (2005)
10. Lippert, R.A., Huang, H., Waterman, M.S.: Distributional regimes for the number of k-word matches between two random sequences. Proceedings of the National Academy of Sciences 99, 13980–13989 (2002)
11. Kantorovitz, M., Robinson, G., Sinha, S.: A statistical method for alignment-free comparison of regulatory sequences. Bioinformatics 23, 249–255 (2007)
12. Reinert, G., Chew, D., Sun, F., Waterman, M.S.: Alignment-free sequence comparison (i): Statistics and power. Journal of Computational Biology 16, 1615–1634 (2009)
13. Jukes, T.H., Cantor, C.R.: Evolution of Protein Molecules. Academy Press (1969)
14. Robin, S., Rodolphe, F., Schbath, S.: DNA, Words and Models: Statistics of Exceptional Words. Cambridge University Press, Cambridge (2005)
15. Haubold, B., Pfaffelhuber, P., Domazet-Loso, M., Wiehe, T.: Estimating mutation distances from unaligned genomes. Journal of Computational Biology 16, 1487–1500 (2009)
16. Leimeister, C.-A., Morgenstern, B.: kmacs: the k-mismatch average common substring approach to alignment-free sequence comparison. Bioinformatics 30, 1991–1999 (2014)
17. Ulitsky, I., Burstein, D., Tuller, T., Chor, B.: The average common substring approach to phylogenomic reconstruction. Journal of Computational Biology 13, 336–350 (2006)
18. Sims, G.E., Jun, S.-R., Wu, G.A., Kim, S.-H.: Alignment-free genome comparison with feature frequency profiles (FFP) and optimal resolutions. Proceedings of the National Academy of Sciences 106, 2677–2682 (2009)
19. Qi, J., Luo, H., Hao, B.: CVTree: a phylogenetic tree reconstruction tool based on whole genomes. Nucleic Acids Research 32(suppl 2), W45–W47 (2004)
20. Felsenstein, J.: PHYLIP - Phylogeny Inference Package (Version 3.2). Cladistics 5, 164–166 (1989)
21. Bonnet, E., de Peer, Y.V.: zt: A sofware tool for simple and partial mantel tests. Journal of Statistical Software 7, 1–12 (2002)
22. Didier, G., Laprevotte, I., Pupin, M., Hénaut, A.: Local decoding of sequences and alignment-free comparison. J. Computational Biology 13, 1465–1476 (2006)
23. Kuiken, C., Leitner, T., Foley, B., Hahn, B., Marx, P., McCutchan, F., Wolinsky, S., Korber, B.T. (eds.): HIV Sequence Compendium 2009. Theoretical Biology and Biophysics Group, Los Alamos National Laboratory, Los Alamos, New Mexico (2009)
24. Sievers, F., Wilm, A., Dineen, D., Gibson, T.J., Karplus, K., Li, W., Lopez, R., McWilliam, H., Remmert, M., Söding, J., Thompson, J.D., Higgins, D.G.: Fast, scalable generation of high-quality protein multiple sequence alignments using Clustal Omega. Molecular Systems Biology 7, 539 (2011)

On the Family-Free DCJ Distance

Fábio V. Martinez[1,2], Pedro Feijão[2],
Marília D.V. Braga[3], and Jens Stoye[2]

[1] Faculdade de Computação, Universidade Federal de Mato Grosso do Sul, Brazil
[2] Technische Fakultät and CeBiTec, Universität Bielefeld, Germany
[3] Inmetro – Instituto Nacional de Metrologia, Qualidade e Tecnologia, Brazil

Abstract. Structural variation in genomes can be revealed by many (dis)similarity measures. Rearrangement operations, such as the so called double-cut-and-join (DCJ), are large-scale mutations that can create complex changes and produce such variations in genomes. A basic task in comparative genomics is to find the rearrangement distance between two given genomes, i.e., the minimum number of rearragement operations that transform one given genome into another one. In a family-based setting, genes are grouped into gene families and efficient algorithms were already proposed to compute the DCJ distance between two given genomes. In this work we propose the problem of computing the DCJ distance of two given genomes without prior gene family assignment, directly using the pairwise similarity between genes. We propose a new family-free DCJ distance, prove that the family-free DCJ distance problem is APX-hard, and provide an integer linear program to its solution.

1 Introduction

Genomes are subject to mutations or rearrangements in the course of evolution. Typical large-scale rearrangements change the number of chromosomes and/or the positions and orientations of genes. Examples of such rearrangements are inversions, translocations, fusions and fissions. A classical problem in comparative genomics is to compute the rearrangement distance, that is, the minimum number of rearrangements required to transform a given genome into another given genome [14].

In order to study this problem, one usually adopts a high-level view of genomes, in which only "relevant" fragments of the DNA (e.g., genes) are taken into consideration. Furthermore, a pre-processing of the data is required, so that we can compare the content of the genomes.

One popular method, adopted for more than 20 years, is to group the genes in both genomes into *gene families*, so that two genes in the same family are said to be equivalent. This setting is said to be *family-based*. Without gene duplications, that is, with the additional restriction that each family occurs exactly once in each genome, many polynomial models have been proposed to compute the genomic distance [3,4,12,17]. However, when gene duplications are allowed, the problem is more intricate and all approaches proposed so far are NP-hard, see for instance [1,7,8,15,16].

D. Brown and B. Morgenstern (Eds.): WABI 2014, LNBI 8701, pp. 174–186, 2014.

It is not always possible to classify each gene unambiguously into a single gene family. Due to this fact, an alternative to the family-based setting was proposed recently and consists in studying the rearrangement distance without prior family assignment. Instead of families, the pairwise similarity between genes is directly used [5, 10]. This approach is said to be *family-free*. Although the family-free setting seems to be at least as difficult as the family-based setting with duplications, its complexity is still unknown for various distance models.

In this work we are interested in the problem of computing the distance of two given genomes in a family-free setting, using the *double cut and join* (DCJ) model [17]. The DCJ operation, that consists of cutting a genome in two distinct positions and joining the four resultant open ends in a different way, represents most of large-scale rearrangements that modify genomes. After preliminaries and a formal definition of the family-free DCJ distance, we present in Section 4 a hardness result, before giving a linear programming solution and showing its feasibility for practical problem instances in Section 5. Section 6 concludes.

2 Preliminaries

Let A and B be two distinct genomes and let $\mathcal{A}$ be the set of genes in genome A and $\mathcal{B}$ be the set of genes in genome B.

Each gene g in a genome is an oriented DNA fragment that can be represented by the symbol g itself, if it has direct orientation, or by the symbol $-g$, if it has reverse orientation. Furthermore, each one of the two extremities of a linear chromosome is called a *telomere*, represented by the symbol ∘. Each chromosome in a genome can be represented by a string that can be circular, if the chromosome is circular, or linear and flanked by the symbols ∘ if the chromosome is linear. For the sake of clarity, each chromosome is also flanked by parentheses. As an example, consider the genome $A = \{(\circ\ 3\ -1\ 4\ 2\ \circ), (\circ\ 5\ -6\ -7\ \circ)\}$ that is composed of two linear chromosomes.

Since a gene g has an orientation, we can distinguish its two ends, also called its *extremities*, and denote them by g^t (*tail*) and g^h (*head*). An *adjacency* in a genome is either the extremity of a gene that is adjacent to one of its telomeres, or a pair of consecutive gene extremities in one of its chromosomes. If we consider again the genome A above, the adjacencies in its first chromosome are 3^t, 3^h1^h, 1^t4^t, 4^h2^t and 2^h.

2.1 Adjacency Graph and Family-Based DCJ Distance

In the family-based setting we are given two genomes A and B with the same content, that is, $\mathcal{A} = \mathcal{B}$. When there are no duplications, that is, when each family is represented by exactly one gene in each genome, the DCJ distance can be easily computed with the help of the *adjacency graph* $AG(A, B)$, a bipartite multigraph such that each partition corresponds to the set of adjacencies of one of the two input genomes and an edge connects the same extremities of genes in both genomes. In other words, there is a one-to-one correspondence between the set of edges in $AG(A, B)$ and the set of gene extremities. Vertices have degree one or two and thus an adjacency graph is a collection of paths and cycles. An example of an adjacency graph is given in Figure 1.

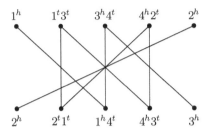

Fig. 1. The adjacency graph for the two unichromosomal and linear genomes $A = \{(\circ\ -1\ 3\ 4\ 2\ \circ)\}$ and $B = \{(\circ\ -2\ 1\ 4\ 3\ \circ)\}$

The family-based DCJ distance d_{DCJ} between two genomes A and B without duplications can be computed in linear time and is closely related to the number of components in the adjacency graph $AG(A, B)$ [4]:

$$d_{DCJ}(A, B) = n - c - i/2 \,,$$

where $n = |\mathcal{A}| = |\mathcal{B}|$ is the number of genes in both genomes, c is the number of cycles and i is the number of odd paths in $AG(A, B)$.

Observe that, in Figure 1, the number of genes is $n = 4$ and $AG(A, B)$ has one cycle and two odd paths. Consequently the DCJ distance is $d_{DCJ}(A, B) = 4 - 1 - 2/2 = 2$.

The formula for $d_{DCJ}(A, B)$ can also be derived using the following approach. Given a component C in $AG(A, B)$, let $|C|$ denote the length, or number of edges, of C. From [6,11] we know that each component in $AG(A, B)$ contributes independently to the DCJ distance, depending uniquely on its length. Formally, the contribution $d(C)$ of a component C in the total distance is given by:

$$d(C) = \begin{cases} \frac{|C|}{2} - 1 \,, & \text{if } C \text{ is a cycle}\,, \\ \frac{|C|-1}{2} \,, & \text{if } C \text{ is an odd path}\,, \\ \frac{|C|}{2} \,, & \text{if } C \text{ is an even path}\,. \end{cases}$$

The sum of the lengths of all components in the adjacency graph is equal to $2n$. Let $\mathcal{C}$, $\mathcal{I}$, and $\mathcal{P}$ represent the sets of components in $AG(A, B)$ that are cycles, odd paths and even paths, respectively. Then, the DCJ distance can be calculated as the sum of the contributions of each component:

$$
\begin{aligned}
d_{\text{DCJ}}(A, B) &= \sum_{C \in AG(A,B)} d(C) \\
&= \sum_{C \in \mathcal{C}} \left(\frac{|C|}{2} - 1 \right) + \sum_{C \in \mathcal{I}} \left(\frac{|C| - 1}{2} \right) + \sum_{C \in \mathcal{P}} \left(\frac{|C|}{2} \right) \\
&= \frac{1}{2} \left(\sum_{C \in AG(A,B)} |C| \right) - \sum_{C \in \mathcal{C}} 1 - \sum_{C \in \mathcal{I}} \frac{1}{2} \\
&= n - c - i/2 \, .
\end{aligned}
$$

2.2 Gene Similarity Graph for the Family-Free Model

In the family-free setting, each gene in each genome is represented by a distinct symbol, thus $\mathcal{A} \cap \mathcal{B} = \emptyset$ and the cardinalities $|\mathcal{A}|$ and $|\mathcal{B}|$ may be distinct. Let a be a gene in A and b be a gene in B, then their *normalized similarity* is given by the value $\sigma(a, b)$ that ranges in the interval $[0, 1]$.

We can represent the similarities between the genes of genome A and the genes of genome B with respect to σ in the so called *gene similarity graph* [5], denoted by $GS_\sigma(A, B)$. This is a weighted bipartite graph whose partitions $\mathcal{A}$ and $\mathcal{B}$ are the sets of genes in genomes A and B, respectively. Furthermore, for each pair of genes (a, b), such that $a \in \mathcal{A}$ and $b \in \mathcal{B}$, if $\sigma(a, b) > 0$ there is an edge e connecting a and b in $GS_\sigma(A, B)$ whose weight is $\sigma(e) := \sigma(a, b)$. An example of a gene similarity graph is given in Figure 2.

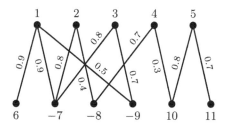

Fig. 2. A possible gene similarity graph for the two unichromosomal linear genomes $A = \{(\circ\ 1\ 2\ 3\ 4\ 5\ \circ)\}$ and $B = \{(\circ\ 6\ -7\ -8\ -9\ 10\ 11\ \circ)\}$

3   Reduced Genomes and Family-Free DCJ Distance

Let A and B be two genomes and let $GS_\sigma(A, B)$ be their gene similarity graph. Now let $M = \{e_1, e_2, \ldots, e_n\}$ be a matching in $GS_\sigma(A, B)$ and denote by $w(M) = \sum_{e_i \in M} \sigma(e_i)$ the weight of M, that is the sum of its edge weights. Since the endpoints of each edge $e_i = (a, b)$ in M are not saturated by any other edge of M, we can unambiguously define the function $s(a, M) = s(b, M) = i$. The *reduced genome* A^M is obtained by deleting from A all genes that are not saturated by M, and renaming each saturated gene a to $s(a, M)$, preserving its orientation. Similarly, the reduced genome B^M is obtained by deleting from B all genes that are not saturated by M, and renaming each saturated gene b to $s(b, M)$, preserving its orientation. Observe that the set of genes in A^M and in B^M is $\mathcal{G}(M) = \{s(g, M) \colon g \text{ is saturated by the matching } M\} = \{1, 2, \ldots, n\}$.

3.1   The Weighted Adjacency Graph of Reduced Genomes

Let A^M and B^M be the reduced genomes for a given matching M of $GS_\sigma(A, B)$. The *weighted adjacency graph* of A^M and B^M, denoted by $AG_\sigma(A^M, B^M)$, is obtained by constructing the adjacency graph of A^M and B^M and adding weights to the edges as follows. For each gene i in $\mathcal{G}(M)$, both edges $i^t i^t$ and $i^h i^h$ inherit the weight of edge e_i in M, that is, $\sigma(i^t i^t) = \sigma(i^h i^h) = \sigma(e_i)$. Observe that, for each edge $e \in M$, we have two edges of weight $\sigma(e)$ in $AG_\sigma(A^M, B^M)$, thus $w(AG_\sigma(A^M, B^M)) = 2\,w(M)$ (the weight of $AG_\sigma(A^M, B^M)$ is twice the weight of M). Examples of weighted adjacency graphs are shown in Figure 3.

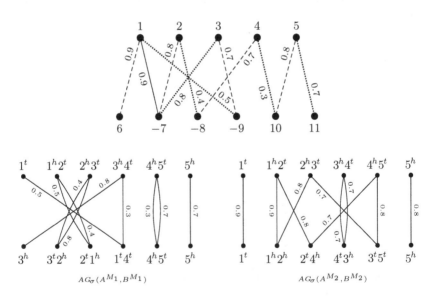

Fig. 3.   Considering the same genomes $A = \{(\circ\ 1\ 2\ 3\ 4\ 5\ \circ)\}$ and $B = \{(\circ\ 6\ -7\ -8\ -9\ 10\ 11\ \circ)\}$ as in Figure 2, let M_1 (dotted edges) and M_2 (dashed edges) be two distinct matchings in $GS_\sigma(A, B)$, shown in the upper part. The two resulting weighted adjacency graphs $AG_\sigma(A^{M_1}, B^{M_1})$, that has two odd paths and three cycles, and $AG_\sigma(A^{M_2}, B^{M_2})$, that has two odd paths and two cycles, are shown in the lower part.

3.2   The Weighted DCJ Distance of Reduced Genomes

Based on the weighted adjacency graph, in [5] a family-free DCJ similarity measure has been proposed. To be more consistent with the comparative genomics literature, where distance measures are more common than similarities, here we propose a general family-free DCJ distance. Moreover, edge weights are treated in a way that, when all weights are equal to 1, the definition falls back to the (unweighted) family-based DCJ distance.

To define the distance measure, we consider the components of the graph $AG_\sigma(A^M, B^M)$ separately, similarly to the approach described in Section 2.1 for the family-based model. Now, the contribution of each component C is denoted by $d_\sigma(C)$ and must include not only the length $|C|$ of the component, but also information about the weights of the edges in C. Basically, we need a function $f(C)$ to use instead of $|C|$ in the contribution function $d_\sigma(C)$, such that: (i) when all edges in C have weight 1, $f(C) = |C|$, that is, the contribution of C is the same as in the family-based version; (ii) when the weights decrease, f should increase, because smaller weights mean less similarity, or increased distance between the genomes.

The simplest linear function f that satisfies both conditions is $f(C) = 2|C| - w(C)$, where $w(C) = \sum_{e \in C} \sigma(e)$ is the sum of the weights of all the edges in C. Then, the *weighted contribution* $d_\sigma(C)$ of the different types of components is:

$$d_\sigma(C) = \begin{cases} \frac{2|C|-w(C)}{2} - 1\,, & \text{if } C \text{ is a cycle}\,, \\ \frac{2|C|-w(C)-1}{2}\,, & \text{if } C \text{ is an odd path}\,, \\ \frac{2|C|-w(C)}{2}\,, & \text{if } C \text{ is an even path}\,. \end{cases}$$

Let $\mathcal{C}$, $\mathcal{I}$, and $\mathcal{P}$ represent the sets of components in $AG_\sigma(A^M, B^M)$ that are cycles, odd paths and even paths, respectively. Summing the contributions of all the components, the resulting distance for a certain matching M is computed as follows:

$$\begin{aligned} d_\sigma(A^M, B^M) &= \sum_{C \in AG_\sigma(A^M,B^M)} d_\sigma(C) \\ &= \sum_{C \in \mathcal{C}} \left(\frac{2|C|-w(C)}{2} - 1\right) + \sum_{C \in \mathcal{I}} \left(\frac{2|C|-w(C)-1}{2}\right) + \sum_{C \in \mathcal{P}} \left(\frac{2|C|-w(C)}{2}\right) \\ &= \sum_{C \in AG_\sigma(A^M,B^M)} |C| - \frac{1}{2}\left(\sum_{C \in AG_\sigma(A^M,B^M)} w(C)\right) - \sum_{C \in \mathcal{C}} 1 - \sum_{C \in \mathcal{I}} \frac{1}{2} \\ &= 2|M| - w(AG_\sigma(A^M, B^M))/2 - c - i/2 \\ &= d_{\text{DCJ}}(A^M, B^M) + |M| - w(M)\,, \end{aligned}$$

since the number of genes in $\mathcal{G}(M)$ is equal to the size of M.

In Figure 3, matching M_1 gives the weighted adjacency graph with more components, but whose distance $d_\sigma(A^{M_1}, B^{M_1}) = 1 + 5 - 2.7 = 3.3$ is larger. On the other hand, M_2 gives the weighted adjacency graph with less components, but whose distance $d_\sigma(A^{M_2}, B^{M_2}) = 2 + 5 - 3.9 = 3.1$ is smaller.

3.3   The Family-Free DCJ Distance

Our goal in the remainder of this paper is to study the problem of computing the family-free DCJ distance, i.e., to find a matching in $GS_\sigma(A, B)$ that minimizes d_σ. First of all, it is important to observe that the behaviour of this function does not correlate with the size of the matching. Often smaller matchings, that possibly discard gene assignments, lead to smaller distances. genomes with any gene similarity graph, a trivial empty matching leads to the minimum distance, equal to zero.

Due to this fact we restrict the distance to *maximal matchings* only. This ensures that no pairs of genes with positive similarity score are simply discarded, even though they might increase the overall distance. Hence we have the following optimization problem:

> **Problem** FFDCJ-DISTANCE(A, B): Given genomes A and B and their gene similarities σ, calculate their family-free DCJ distance
>
> $$d_{\text{FFDCJ}}(A, B) = \min_{M \in \mathbb{M}} \{d_\sigma(A^M, B^M)\},$$
>
> where $\mathbb{M}$ is the set of all maximal matchings in $GS_\sigma(A, B)$.

4   Complexity of the Family-Free DCJ Distance

In order to assess the complexity of FFDCJ-DISTANCE, we use a restricted version of the family-based *exemplar DCJ distance problem* [8, 15]:

> **Problem.** (s, t)-EXDCJ-DISTANCE(A, B): Given genomes A and B, where each family occurs at most s times in A and at most t times in B, obtain *exemplar* genomes A' and B' by removing all but one copy of each family in each genome, so that the DCJ distance $d_{\text{DCJ}}(A', B')$ is minimized.

We establish the computational complexity of the FFDCJ-DISTANCE problem by means of a polynomial time and approximation preserving (AP-) reduction from the problem $(1, 2)$-EXDCJ-DISTANCE, which is NP-hard [8]. Note that the authors of [8] only consider unichromosomal genomes, but the reduction can be extended to multichromosomal genomes, since an algorithm that solves the multichromosomal case also solves the unichromosomal case.

Theorem 1. *Problem* FFDCJ-DISTANCE(A, B) *is APX-hard, even if the maximum degrees in the two partitions of* $GS_\sigma(A, B)$ *are respectively one and two.*

Proof. Using notation from [2] (Chapter 8), we give an AP-reduction (f, g, β) from $(1, 2)$-EXDCJ-DISTANCE to FFDCJ-DISTANCE as follows:

Algorithm f receives as input a positive rational number δ and an instance (A, B) of $(1, 2)$-EXDCJ-DISTANCE where A and B are genomes from a set of genes $\mathcal{G}$ and each gene in $\mathcal{G}$ occurs at most once in A and at most twice in B, and constructs an instance $(A', B') = f(\delta, (A, B))$ of FFDCJ-DISTANCE as follows.

Let the genes of A be denoted $a_1, a_2, \ldots, a_{|A|}$ and the genes of B be denoted $b_1, b_2, \ldots, b_{|B|}$. Then A' and B' are copies of A and B, respectively, except that symbol a_i in A' is relabeled by i, keeping its orientation, and b_j in B' is relabeled by $j + |A|$, also keeping its orientation. Furthermore, the similarity σ for genes in A' and B' is defined as $\sigma(i, k) = 1$ for i in A' and k in B', such that a_i is in A, b_j is in B, a_i and b_j are in the same gene family, and $k = j + |A|$. Otherwise, $\sigma(i, k) = 0$. Figure 4 gives an example of a $GS_\sigma(A', B')$ for this construction.

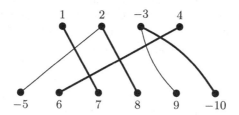

Fig. 4. Gene similarity graph $GS_\sigma(A', B')$ constructed from the input genomes $A = \{(\circ\ a\ c\ -b\ d\ \circ)\}$ and $B = \{(\circ\ -c\ \ d\ a\ c\ b\ -b\ \circ)\}$ of $(1, 2)$-EXDCJ-DISTANCE, where all edge weights are 1. Highlighted edges represent a maximal matching in $GS_\sigma(A', B')$.

Algorithm g receives as input a positive rational number δ, an instance (A, B) of $(1, 2)$-EXDCJ-DISTANCE and a solution M' of FFDCJ-DISTANCE, and transforms M' into a solution (A_X, B_X) of $(1, 2)$-EXDCJ-DISTANCE. This is a simple construction: for each edge (i, k) in M' we add symbols a_i to A_X and b_j to B_X, where $j = k - |A|$. For the example of Figure 4, a matching $M' = \{(1, 7), (2, 8), (-3, -10), (4, 6)\}$, which is a solution to FFDCJ-DISTANCE(A', B'), is transformed by g into the genomes $A_X = \{(\circ\ a_1\ a_2\ a_3\ a_4\ \circ)\} = \{(\circ\ a\ c\ -b\ d\ \circ)\}$ and $B_X = \{(\circ\ b_2\ b_3\ b_4\ b_6\ \circ)\} = \{(\circ\ d\ a\ c\ -b\ \circ)\}$, which is a solution to $(1, 2)$-EXDCJ-DISTANCE(A, B).

Notice that for any positive rational number δ, functions f and g are polynomial time algorithms on the size of their respective instances. Let $A_X := A$ and let B_X be an exemplar genome of B, such that $(A_X, B_X) = g(\delta, (A, B), M')$. Denote by c_{AG} and i_{AG} the number of cycles and odd paths in $AG(A_X, B_X)$, and by c_{AG_σ} and i_{AG_σ} the number of cycles and odd paths in $AG_\sigma(A^{M'}, B^{M'})$. Observe that we have $|A_X| = |B_X| = |M'|$, $c_{AG} = c_{AG_\sigma}$, $i_{AG} = i_{AG_\sigma}$, and thus

$$
\begin{aligned}
d_\sigma(A', B') &= 2|M'| - w(M') - c_{AG_\sigma} - i_{AG_\sigma}/2 \\
&= |A_X| - c_{AG} - i_{AG}/2 \\
&= d(A_X, B_X),
\end{aligned}
$$

that is, $\mathrm{opt}(\text{FFDCJ-DISTANCE}(A', B')) = \mathrm{opt}((1, 2)\text{-EXDCJ-DISTANCE}(A, B))$.

Therefore, $d_\sigma(A', B') \leq (1+\delta)\,\mathrm{opt}((1,2)\text{-EXDCJ-DISTANCE}(A, B))$ for any positive δ, and the last condition for the AP-reduction holds by setting $\beta := 1$:

$$
d(A_X, B_X) \leq (1 + \beta\delta)\,\mathrm{opt}((1, 2)\text{-EXDCJ-DISTANCE}(A, B)).
$$

□

Corollary 2. *There exists no polynomial-time algorithm for* FFDCJ-DISTANCE *with approximation factor better than 1237/1236, unless P = NP.*

Proof. As shown in [8], $(1,2)$-EXDCJ-DISTANCE is NP-hard to approximate within a factor of $1237/1236 - \varepsilon$ for any $\varepsilon > 0$. Therefore, the result follows immediately from [8] and from the AP-reduction in the proof of Theorem 1. □

Since the weight plays an important role in d_σ, a matching with maximum weight, that is obviously maximal, could be a candidate for the design of an approximation algorithm for FFDCJ-DISTANCE. However, we can demonstrate that it is not possible to obtain such an approximation, with the following example.

Consider an integer $k \geq 1$ and let $A = \{(\circ\ 1\ -2\ \cdots\ (2k{-}1)\ -2k\ \circ)\}$ and $B = \{(\circ\ -(2k{+}1)\ (2k{+}2)\ \cdots\ -(2k{+}2k{-}1)\ (2k{+}2k)\ \circ)\}$ be two unichromosomal linear genomes. Observe that A and B have an even number of genes with alternating orientation. While A starts with a gene in direct orientation, B starts with a gene in reverse orientation. Now let σ be the normalized similarity measure between the genes of A and B, defined as follows:

$$\sigma(i,j) = \begin{cases} 1, & \text{for each } i \in \{1,2,\ldots,2k\} \text{ and } j = 2k{+}i\,; \\ 1-\varepsilon, & \text{for each } i \in \{1,3,\ldots,2k{-}1\} \text{ and } j = 2k{+}i{+}1, \text{with } \varepsilon \in [0,1); \\ 0, & \text{otherwise.} \end{cases}$$

Figure 5 shows $GS_\sigma(A, B)$ for $k = 3$ and σ as defined above.

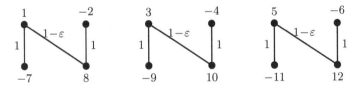

Fig. 5. Gene similarity graph $GS_\sigma(A, B)$ for $k = 3$

There are several matchings in $GS_\sigma(A, B)$. We are interested in two particular maximal matchings:

- M^* is composed of all edges that have weight $1 - \varepsilon$. It has weight $w(M^*) = (1{-}\varepsilon)|M^*|$. Its corresponding weighted adjacency graph $AG_\sigma(A^{M^*}, B^{M^*})$ has $|M^*| - 1$ cycles and two odd paths, thus $d_{\text{DCJ}}(A^{M^*}, B^{M^*}) = 0$. Consequently, we have $d_\sigma(A^{M^*}, B^{M^*}) = |M^*| - (1 - \varepsilon)|M^*| = \varepsilon|M^*|$.
- M is composed of all edges that have weight 1. It is the only matching with the maximum weight $w(M) = |M|$. Its corresponding weighted adjacency graph $AG_\sigma(A^M, B^M)$ has two even paths, but no cycles or odd paths, giving $d_{\text{DCJ}}(A^M, B^M) = |M|$. Hence, $d_\sigma(A^M, B^M) = 2|M| - |M| = |M|$.

Notice that $d_{\mathrm{FFDCJ}}(A, B) \leq d_\sigma(A^{M^*}, B^{M^*})$. Furthermore, since $|M| = 2|M^*|$,

$$\frac{d_\sigma(A^M, B^M)}{d_\sigma(A^{M^*}, B^{M^*})} = \frac{|M|}{\varepsilon|M^*|} = \frac{2}{\varepsilon}$$

and $2/\varepsilon \to +\infty$ when $\varepsilon \to 0$.

This shows that, for any genomes A and B, a matching of maximum weight in $GS_\sigma(A, B)$ can have d_σ arbitrarily far from the optimal solution and cannot give an approximation for FFDCJ-DISTANCE(A, B).

5   ILP to Compute the Family-Free DCJ Distance

We propose an integer linear program (ILP) formulation to compute the family-free DCJ distance between two given genomes. This formulation is a slightly different version of the ILP for the maximum cycle decomposition problem given by Shao et al. [16] to compute the DCJ distance between two given genomes with duplicate genes. Besides the cycle decomposition in a graph, as was made in [16], we also have to take into account maximal matchings in the gene similarity graph and their weights.

Let A and B be two genomes with extremity sets X_A and X_B, respectively, and let $G = GS_\sigma(A, B)$ be their gene similarity graph. The weight $w(e)$ of an edge e in G is also denoted by w_e. Let M be a maximal matching in G. For the ILP formulation, a weighted adjacency graph $H = AG_\sigma(A^M, B^M)$ is such that $V(H) = X_A \cup X_B$ and $E(H)$ has three types of edges: (i) matching edges that connect two extremities in different extremity sets, one in X_A and the other in X_B, if there exists one edge in M connecting these genes in G; the set of matching edges is denoted by E_m; (ii) adjacency edges that connect two extremities in the same extremity set if they are an adjacency; the set of adjacency edges is denoted by E_a; and (iii) self edges that connect two extremities of the same gene in an extremity set; the set of self edges is denoted by E_s. All edges in H are in $E_m \cup E_a \cup E_s = E(H)$. Matching edges have weights defined by the normalized similarity σ, all adjacency edges have weight 1, and all self edges have weight 0. Notice that any edge in G corresponds to two matching edges in H.

Now we describe the ILP. For each edge e in H, we create the binary variable x_e to indicate whether e will be in the final solution. We require first that each adjacency edge be chosen:

$$x_e = 1, \qquad \forall\, e \in E_a\,.$$

We require then that, for each vertex in H, exactly one incident edge to it be chosen:

$$\sum_{uv \in E_m \cup E_s} x_{uv} = 1, \forall\, u \in X_A\,, \quad \text{and} \quad \sum_{uv \in E_m \cup E_s} x_{uv} = 1, \forall\, v \in X_B\,.$$

Then, we require that the final solution be consistent, meaning that if one extremity of a gene in A is assigned to an extremity of a gene in B, then the other extremities of these two genes have to be assigned as well:

$$x_{a^h b^h} = x_{a^t b^t} , \qquad \forall\; ab \in E(G) .$$

We also require that the matching be maximal. It can be easily ensured if we garantee that at least one of the vertices connected by an edge in the gene similarity graph be chosen, which is equivalent to not allowing both of the corresponding self edges in the weighted adjacency graph be chosen:

$$x_{a^h a^t} + x_{b^h b^t} \leq 1 , \qquad \forall\; ab \in E(G) .$$

To count the number of cycles, we use the same strategy as described in [16]. We first give an arbitrary index for each vertex in H such that $V(H) = \{v_1, v_2, \ldots, v_k\}$ with $k = |V(H)|$. For each vertex v_i, we define a variable y_i that labels v_i such that

$$0 \leq y_i \leq i , \qquad 1 \leq i \leq k .$$

We also require that all vertices in the same cycle in the solution have the same label:

$$y_i \leq y_j + i \cdot (1 - x_e) , \; \forall\; e = v_i v_j \in E(H) ,$$
$$y_j \leq y_i + j \cdot (1 - x_e) , \; \forall\; e = v_i v_j \in E(H) .$$

And we create a binary variable z_i, for each vertex v_i, to verify whether y_i is equal to its upper bound i:

$$i \cdot z_i \leq y_i , \qquad 1 \leq i \leq k .$$

Notice that the way as variables z_i were defined, they count the number of cycles in H [16].

Finally, we set the objective function as follows:

$$\text{minimize} \quad 2 \sum_{e \in E_m} x_e - \sum_{e \in E_m} w_e x_e - \sum_{1 \leq i \leq k} z_i ,$$

which is exactly the family-free DCJ distance $d_{\text{FFDCJ}}(A, B)$ as defined in Section 3.

We performed some initial simulated experiments of our integer linear program formulation. We produced some datasets using the Artificial Life Simulator (ALF) [9]. Genome sizes varied from 1000 to 3000 genes, where the gene lengths were generated according to a gamma distribution with shape parameter $k = 3$ and scale parameter $\theta = 133$. A birth-death tree with 10 leaves was generated, with PAM distance of 100 from the root to the deepest leaf. For the amino acid evolution, the WAG substitution model with default parameters was used, with Zipfian indels at a rate of 0.000005. For structural evolution, gene duplications and gene losses were applied with a 0.001 rate, with a 0.0025 rate for reversals and translocations. To test different ration of rearrangement events, we also simulated datasets where the structural evolution ratios had a 2- and 5-fold increase.

To solve the ILPs, we ran the CPLEX Optimizer[1] on the 45 pairwise comparisons of each simulated dataset. All simulations were run in parallel on a cluster consisting of machines with an Intel(R) Xeon(R) E7540 CPU, with 48 cores and as many as 2 TB of memory, but for each individual CPLEX run only 4 cores and 2 GB of memory were allocated. The results are summarized on Table 1.

Table 1. ILP results for datasets with different genome sizes and evolutionary rates. Each dataset has 10 genomes, totalling 45 pairwise comparisons. Maximum running time was set to 20 minutes. For each dataset, it is shown the number of runs that found an optimal solution in time and their average running time. For the runs that did not finish, the last row shows the gap between the upper bound and the current solution. Rate $r = 1$ means the default rate for ALF evolution, and $r = 2$ and $r = 5$ mean 2-fold and 5-fold increase for the gene duplication, gene deletion and rearrangement rates.

	1000 genes			2000 genes			3000 genes		
	$r = 1$	$r = 2$	$r = 5$	$r = 1$	$r = 2$	$r = 5$	$r = 1$	$r = 2$	$r = 5$
Finished	45/45	22/45	6/45	45/45	9/45	1/45	45/45	7/45	3/45
Avg. Time (s)	0.66	11.09	24.26	1.29	2.76	16.97	2.24	16.36	36.01
Avg. Gap (%)	0	1.08	3.9	0	1.93	12.4	0	3.9	6.03

6 Conclusion

In this paper, we have defined a new distance measure for two genomes that is motivated by the double cut and join model, while not relying on gene annotations in form of gene families. In case gene families are known and each family has exactly one member in each of the two genomes, the distance equals the family-based DCJ distance and thus can be computed in linear time. In the general case, however, it is NP-hard and even hard to approximate. Nevertheless, we could give an integer linear program for the exact computation of the distance that is fast enough to be applied to realistic problem instances.

The family-free model has many potentials when gene family assignments are not available or ambiguous, in fact it can even be used to improve family assignments [13]. The work presented in this paper is another step in this direction.

Acknowledgments. We would like to thank Tomáš Vinař who suggested that the NP-hardness of FFDCJ-DISTANCE could be proven via a reduction from the exemplar distance problem. FVM and MDVB are funded from the Brazilian research agency CNPq grants Ciência sem Fronteiras Postdoctoral Scholarship 245267/2012-3 and PROMETRO 563087/2010-2, respectively.

---

[1] http://www-01.ibm.com/software/commerce/optimization/cplex-optimizer/

References

1. Angibaud, S., Fertin, G., Rusu, I., Thévenin, A., Vialette, S.: On the approximability of comparing genomes with duplicates. J. Graph Algorithms Appl. 13(1), 19–53 (2009)

2. Ausiello, G., Protasi, M., Marchetti-Spaccamela, A., Gambosi, G., Crescenzi, P., Kann, V.: Complexity and Approximation: Combinatorial Optimization Problems and Their Approximability Properties. Springer (1999)

3. Bafna, V., Pevzner, P.: Genome rearrangements and sorting by reversals. In: Proc. of FOCS 1993, pp. 148–157 (1993)

4. Bergeron, A., Mixtacki, J., Stoye, J.: A unifying view of genome rearrangements. In: Bücher, P., Moret, B.M.E. (eds.) WABI 2006. LNCS (LNBI), vol. 4175, pp. 163–173. Springer, Heidelberg (2006)

5. Braga, M.D.V., Chauve, C., Dörr, D., Jahn, K., Stoye, J., Thévenin, A., Wittler, R.: The potential of family-free genome comparison. In: Chauve, C., El-Mabrouk, N., Tannier, E. (eds.) Models and Algorithms for Genome Evolution, ch. 13, pp. 287–307. Springer (2013)

6. Braga, M.D.V., Stoye, J.: The solution space of sorting by DCJ. J. Comp. Biol. 17(9), 1145–1165 (2010)

7. Bryant, D.: The complexity of calculating exemplar distances. In: Sankoff, D., Nadeau, J.H. (eds.) Comparative Genomics, pp. 207–211. Springer, Netherlands (2000)

8. Bulteau, L., Jiang, M.: Inapproximability of (1,2)-exemplar distance. IEEE/ACM Trans. Comput. Biol. Bioinf. 10(6), 1384–1390 (2013)

9. Dalquen, D.A., Anisimova, M., Gonnet, G.H., Dessimoz, C.: ALF–a simulation framework for genome evolution. Mol. Biol. Evol. 29(4), 1115–1123 (2012)

10. Dörr, D., Thévenin, A., Stoye, J.: Gene family assignment-free comparative genomics. BMC Bioinformatics 13(Suppl 19), S3 (2012)

11. Feijão, P., Meidanis, J.: SCJ: A breakpoint-like distance that simplifies several rearrangement problems. IEEE/ACM Trans. Comput. Biol. Bioinf. 8(5), 1318–1329 (2011)

12. Hannenhalli, S., Pevzner, P.: Transforming men into mice (polynomial algorithm for genomic distance problem). In: Proc. of FOCS 1995, pp. 581–592 (1995)

13. Lechner, M., Hernandez-Rosales, M., Doerr, D., Wieseke, N., Thévenin, A., Stoye, J., Hartmann, R.K., Prohaska, S.J., Stadler, P.F.: Orthology detection combining clustering and synteny for very large datasets (unpublished manuscript)

14. Sankoff, D.: Edit distance for genome comparison based on non-local operations. In: Apostolico, A., Galil, Z., Manber, U., Crochemore, M. (eds.) CPM 1992. LNCS, vol. 644, pp. 121–135. Springer, Heidelberg (1992)

15. Sankoff, D.: Genome rearrangement with gene families. Bioinformatics 15(11), 909–917 (1999)

16. Shao, M., Lin, Y., Moret, B.: An exact algorithm to compute the DCJ distance for genomes with duplicate genes. In: Sharan, R. (ed.) RECOMB 2014. LNCS, vol. 8394, pp. 280–292. Springer, Heidelberg (2014)

17. Yancopoulos, S., Attie, O., Friedberg, R.: Efficient sorting of genomic permutations by translocation, inversion and block interchanges. Bioinformatics 21(16), 3340–3346 (2005)

New Algorithms for Computing Phylogenetic Biodiversity

Constantinos Tsirogiannis[1], Brody Sandel[1], and Adrija Kalvisa[2]

[1] MADALGO* and Department of Bioscience
Aarhus University, Denmark
[2] Faculty of Biology
University of Latvia, Latvia

Abstract. A common problem that appears in many case studies in ecology is the following: given a rooted phylogenetic tree $\mathcal{T}$ and a subset R of its leaf nodes, we want to compute the distance between the elements in R. A very popular distance measure that can be used for this reason is the *Phylogenetic Diversity* (PD), which is defined as the cost of the minimum weight Steiner tree in $\mathcal{T}$ that spans the nodes in R. To analyse the value of the PD for a given set R it is important also to calculate the variance of this measure. However, the best algorithm known so far for computing the variance of the PD is inefficient; for any input tree $\mathcal{T}$ that consists of n nodes, this algorithm has $\Theta(n^2)$ running time. Moreover, computing efficiently the variance and higher order statistical moments is a major open problem for several other phylogenetic measures. We provide the following results:

- We describe a new algorithm that computes efficiently in practice the variance of the PD. This algorithm has $O(\mathrm{SI}(\mathcal{T})+\mathrm{DSSI}^2(\mathcal{T}))$ running time; here $\mathrm{SI}(\mathcal{T})$ denotes the Sackin's Index of $\mathcal{T}$, and $\mathrm{DSSI}(\mathcal{T})$ is a new index whose value depends on how balanced $\mathcal{T}$ is.
- We provide for the first time exact formulas for computing the mean and the variance of another popular biodiversity measure, the Mean Nearest Taxon Distance (MNTD). These formulas apply specifically to ultrametric trees. For an ultrametric tree $\mathcal{T}$ of n nodes, we show how we can compute the mean of the MNTD in $O(n)$ time, and its variance in $O(\mathrm{SI}(\mathcal{T}) + \mathrm{DSSI}^2(\mathcal{T}))$ time.
- We introduce a new measure which we call the *Core Ancestor Cost* (CAC). A major advantage of this measure is that for any integer $k > 0$ we can compute all first k statistical moments of the CAC in $O(\mathrm{SI}(\mathcal{T}) + nk + k^2)$ time in total, using $O(n + k)$ space.

We have implemented the new algorithms for computing the variance of the PD and of the MNTD, and the statistical moments of the CAC. We conducted experiments on large phylogenetic datasets and we show that our algorithms perform efficiently in practice.

* Center for Massive Data Algorithmics, a Center of the Danish National Research Foundation.

D. Brown and B. Morgenstern (Eds.): WABI 2014, LNBI 8701, pp. 187–203, 2014.
© Springer-Verlag Berlin Heidelberg 2014

1   Introduction

Researchers in the field of ecology, but also from other disciplines in biology, are frequently confronted with the following problem: given a set of species, they want to measure if these species are close evolutionary relatives. The most common way to measure this is to use a phylogenetic tree $\mathcal{T}$, where each leaf of the tree corresponds to a species, and the weights of the tree edges represent some concept of distance e.g. time since the last speciation event. From $\mathcal{T}$ we select a subset of leaves R which correspond to the species that we want to examine. The next step is then to choose a method for computing the distance between the leaves in R based on the structure of $\mathcal{T}$. In the related literature, such methods are refered to as *phylogenetic biodiversity measures*. Two measures of this kind that are widely used are the *Phylogenetic Diversity* (PD) and the *Mean Nearest Taxon Distance* (MNTD). For a given tree $\mathcal{T}$ and a subset R of its leaves, the value of the PD is equal to the cost of the minimum-weight Steiner tree in $\mathcal{T}$ that spans the nodes in R. The value of the MNTD is the average path cost in $\mathcal{T}$ between any node $v \in R$ and its closest neighbour in $R \setminus \{v\}$.

Whichever method we choose for computing the distance between the elements in R, we need to know if the returned distance value is relatively small or large compared to other sets of leaves in $\mathcal{T}$. More specifically, we need to compare the distance value that we got for R with the distance values of all possible subsets of leaves in $\mathcal{T}$ that have exactly the same number of elements. In several case studies in biology this is done by computing the mean and the variance of the distance values among all those subsets of species [10,4,9,8]. We can then use these to calculate a standardized index; from the distance value that we got for R we subtract the mean and divide by the standard deviation. Depending on the distance measure that we choose, we can use this method to produce several indices. Some of the most widely used indices of this kind are the Net Relatedness Index (NRI), the Nearest Taxon Index (NTI, based on the MNTD) and the Phylogenetic Diversity Index (PDI, based on the PD) [15,17].

In a previous paper we introduced algorithms that compute the values for the mean and the variance the PD [15]. For a tree $\mathcal{T}$ that consists of n nodes in total, and for a non-negative integer r, we introduced an algorithm that computes in $O(n)$ time the mean value of the PD among all possible subsets that consist of r leaves. We also introduced an algorithm that computes the variance of the PD in $\Theta(n^2)$ time. The latter algorithm is quite inefficient since it takes $\Theta(n^2)$ time to execute, not only in the worst case but for every input tree. This makes the use of this algorithm limited in practice, since in some applications it is required to calculate the variance of some measure for a large number of different trees (for example, constructed algorithmically by slightly changing the structure of a given reference tree).

On the other hand, there are no known algorithms for computing the exact value of the mean and the variance of the MNTD. So far, researchers try to estimate these values using a random sampling technique; for a given subset size r, a few subsets of exactly r leaves in $\mathcal{T}$ are selected at random. Then, the mean and the variance of the MNTD is calculated using the values of this measure only for the selected subsets. The number of the sampled subsets is usually around a thousand. For sufficiently large values of r and n, this is a very small number of samples compared to the number of all possible subsets of r leaves in $\mathcal{T}$. This implies that the sampling approach is inexact, and may yield estimated values for the mean and the variance that are very different from the original ones. Hence, there is need to introduce exact and efficient algorithms for computing these statistics for the MNTD, which are required to derive the commonly used NTI [11].

Furthermore, in some studies it is required to compute not only the mean and the variance, but also the higher order moments of a given measure [3]. Unfortunately, for the most popular phylogenetic biodiversity measures computing the higher order statistics appears to be a difficult task. For the PD and the MNTD, any preliminary attempts that we made to compute the higher order moments lead to algorithms with running time that scales exponentially as the order of the moment increases. Yet, to this point we have not proven that designing more efficient algorithms is impossible; this is a conjecture. On the other hand, the skewness of another popular measure, the *Mean Pairwise Distance* (MPD), can be computed in $O(n)$ time [14]. However, the analytical expression that yields the value of the MPD skewness is particularly involved. Worse than that, it appears that deriving an expression for the higher order moments of the MPD may be overwhelmingly complicated. Therefore, there is the need for a non-trivial biodiversity measure for which we can efficiently compute its higher order moments.

Our Results. In this paper we present several results that have to do with the efficient computation of the statistical moments of certain phylogenetic biodiversity measures. Given a phylogenetic tree $\mathcal{T}$ and a positive integer r, we describe an algorithm that computes the variance of the PD among all subsets of r leaves in $O(\mathrm{SI}(\mathcal{T}) + \mathrm{DSSI}^2(\mathcal{T}))$ time, using $O(n)$ space. Here, we use $\mathrm{SI}(\mathcal{T})$ to denote the Sackin's Index of $\mathcal{T}$ which is equal to the sum of the numbers of leaves that appear at the subtree of each node in $\mathcal{T}$ [2]. We use $\mathrm{DSSI}(\mathcal{T})$ to denote a new index that we introduce, which we call the Distinct Subtree Sizes Index. We provide a formal definition of this new index later in this paper. The values of both the $\mathrm{SI}(\mathcal{T})$ and the $\mathrm{DSSI}(\mathcal{T})$ depend on the structure of the tree $\mathcal{T}$. When $\mathcal{T}$ is relatively balanced, the new algorithm has a very good performance, and is much more efficient in practice than the already known $\Theta(n^2)$ algorithm. It is only in the worst case, when $\mathcal{T}$ has $\Omega(n)$ height, that the new algorithm runs in $\Theta(n^2)$ time. Moreover, we present for the first time algorithms for computing the exact value of the mean and the variance of the MNTD for ultrametric trees; a tree is called ultrametric if any simple path from its root to a leaf node has the same cost. Given an ultrametric tree $\mathcal{T}$ of n nodes and a positive integer r, we provide an algorithm that runs in $O(n)$ time, and computes the mean of the MNTD among all subsets of r leaves in $\mathcal{T}$. We also present an algorithm that computes the variance of the MNTD in $O(\mathrm{SI}(\mathcal{T}) + \mathrm{DSSI}^2(\mathcal{T}))$ time, using $O(n)$ space. This algorithm is based on the on the same method as our new algorithm that computes the variance of the PD.

Furthermore, we present a new phylogenetic biodiversity measure which we call the *Core Ancestor Cost* (CAC). For a phylogenetic tree $\mathcal{T}$, a subset R of r leaves in $\mathcal{T}$, and a real $\chi \in (0.5, 1]$, the CAC of R is equal to the cost of the simple path that connects the root of $\mathcal{T}$ with the deepest common ancestor node of at least χr of the nodes in R. Among the many existing measures for phylogenetic diversity (for a review of such measures, see the work of Vellend et al. [16]) the CAC has the following advantage; we can compute efficiently in practice any of its statistical moments. In particular, we prove that for any integer $k > 0$ we can compute all of the first k moments of the CAC in $O(\text{SI}(\mathcal{T}) + nk + k^2)$ time in total, using $O(n + k)$ space. At the same time, the CAC is conceptually related to existing measures such as the Net Relatedness Index (NRI), which seeks to assess the degree to which species in a community are aggregated in particular sections of the tree.

We have implemented all the algorithms that we introduce in this paper, and we have measured their efficiency using large phylogenetic tree datasets that are publicly available. We show that all of the new algorithms have a very good performance in practice; the new algorithm that computes the variance of the PD appears to clearly outperform its predecessor that runs in $\Theta(n^2)$ time.

Related literature. The definition of the PD that we provide in this paper (that is the cost of the min-weight Steiner tree of a subset of leaves) is known in the related literature as the *unrooted* version of the PD. Steel was the first to provide a formula for the exact computation of the mean of the PD over all subsets of r leaves of a tree $\mathcal{T}$ [13]. This formula describes the value of the mean for the *rooted* variant of the PD; in this variant, for a given subset of leaves $R \in \mathcal{T}$ the value of the PD is equal to the value of the unrooted PD, plus the cost of the path that connects the root of $\mathcal{T}$ with the deepest common ancestor of all elements in R. In a previous paper we introduced exact expressions for computing the mean and the variance of the unrooted PD, and we examined issues related to their efficient computation [15]. Nipperess and Matsen [12] yield a related result for a more general version of the problem. They derive formulas for the mean and the variance of the PD for subsets of nodes in $\mathcal{T}$ that may also include internal nodes. They provide such formulas both for the rooted and the unrooted version of the PD. Faller et al. [6] and O'Dwyer et al. [5] consider several probability distributions for sampling subsets of leaves from a tree. In the version of the problem that they examine, formulas for the mean and the variance of the PD are derived among subsets of leaves that do not have the same number of elements.

To our knowledge, except our previous work, none of the above papers is concerned with analysing the running time of an algorithm that evaluates the derived formulas. Unlike these works, in the current paper we do not provide a new formula for the variance of the PD. Instead, among other results, we describe a novel non-trivial method for speeding up significantly the evaluation of the existing formula.

2 Computing Efficiently the Variance of Known Biodiversity Measures

Preliminaries. For a phylogenetic tree $\mathcal{T}$ we denote the set of the edges of $\mathcal{T}$ by E. For any edge $e \in E$ we use $w(e)$ to represent the weight of e. We consider that $w(e) > 0$ for every $e \in E$. We use V to denote the set of nodes of $\mathcal{T}$, and we use S to denote the set of leaf nodes of $\mathcal{T}$. We use n to indicate the total number of nodes in $\mathcal{T}$, and we use s to indicate the number of leaves in $\mathcal{T}$. For any node $v \in V$ we use $\mathrm{Ch}(v)$ to indicate the set of the child nodes of v. In this paper we consider only phylogenetic trees that are rooted. We denote the root node of $\mathcal{T}$ by $\mathrm{root}(\mathcal{T})$. Hence, in the rest of this work, whenever we use the term "phylogenetic tree" we mean a rooted tree with edges that have positive weights. We use $h(\mathcal{T})$ to denote the height of the tree, that is the maximum number of edges that appear on a simple path between the root of $\mathcal{T}$ and a leaf. Since $\mathcal{T}$ is a rooted tree, for any edge $e \in E$ we can distinguish the two nodes adjacent to e into a *parent* node and a *child* node. Here, the child node of e is the one for which the simple path between this node and the root contains e.

Let v be a node in $\mathcal{T}$ and let e be the edge whose child node is v. We use interchangeably $S(e)$ and $S(v)$ to denote the set of leaves that appear in the subtree of v. We denote the number of these leaves by $s(e)$ and $s(v)$. We call this number the *subtree size* of v. For a tree edge $e \in E$, we denote the set of the edges that appear in the subtree of e by $\mathrm{Off}(e)$. For any tree edge e we denote the set of the edges that appear on the simple path between $\mathrm{root}(\mathcal{T})$ and the child node of e by $\mathrm{Anc}(e)$. From this definition we get that $e \in \mathrm{Anc}(e)$. We also use $\mathrm{Ind}(e)$ to denote the set $E \setminus (\mathrm{Off}(e) \cup \mathrm{Anc}(e))$. For a given node $v \in V$, we use $\mathrm{Anc}(v)$ to represent the set $\mathrm{Anc}(e)$ where e is the edge whose child node is v.

We use $\mathrm{Sub}(S, r)$ to denote the set whose elements are all the subsets of S that have cardinality exactly r. For an edge $e \in E$ and a subset R of the leaves of $\mathcal{T}$, we use $S_R(e)$ to denote the elements of $S(e)$ that are also elements of R, that is $S_R(e) = S(e) \cap R$. We indicate the number of these leaves by $sr(e)$. Let $u, v \in S$ be two leaves in $\mathcal{T}$ and let p be the simple path that connects these leaves. We refer to the sum of the weights of the edges in p as the *cost* of this path. We represent this cost as $\mathrm{cost}(u, v)$.

A tree $\mathcal{T}$ is *ultrametric* if all simple paths between the root and the leaves have the same cost. This means also that for every internal node $x \in \mathcal{T}$ any simple path that connects x with a leaf in $S(x)$ has the same cost.

For a given tree $\mathcal{T}$ the *Sackin's index* of $\mathcal{T}$ is defined as the sum of the number of leaves that appear at the subtree of each node in $\mathcal{T}$. More formally, the Sackin's index of $\mathcal{T}$ is defined as:

$$\mathrm{SI}(\mathcal{T}) = \sum_{v \in V} s(v).$$

Alternatively, in the related literature the Sackin's index is described as the sum of the depths of all leaf nodes in $\mathcal{T}$. Both definitions are equivalent since they lead to exactly the same value. The Sackin's index is mainly used in the literature as a function for measuring how balanced a phylogenetic tree is [2].

Let $\mathcal{T}$ be a phylogenetic tree, and let R be a subset of r leaves in $\mathcal{T}$. Let $f(\mathcal{T}, R)$ be a function that maps the pair $\mathcal{T}, R$ to a non-negative real. Let r be a positive integer such that $r \leq s$. The expected value of f over all subsets that consist of exactly r leaves is equal to:

$$\mu(\mathcal{T}, r) = \mathrm{E}_{R \in \mathrm{Sub}(S,r)} \left[f(\mathcal{T}, R) \right] \ .$$

The variance of f over all subsets of r leaves is equal to:

$$var(\mathcal{T}, r) = \mathrm{E}_{R \in \mathrm{Sub}(S,r)} \left[(f(\mathcal{T}, R) - \mu(\mathcal{T}, r))^2 \right] \ .$$

We call the expected value and the variance of f the *lower order moments* of f. Let γ be a positive integer such that $\gamma \geq 3$. We define the γ *order moment* of f to be the normalised γ-th central moment of f, which is equal to the following quantity:

$$\frac{\mathrm{E}_{R \in \mathrm{Sub}(S,r)} \left[(f(\mathcal{T}, R) - \mu(\mathcal{T}, r))^\gamma \right]}{var^{\gamma/2}(\mathcal{T}, r)} \ .$$

We call the moments that are described by the last expression the *higher order moments* of f. In the present work, whenever we refer to calculating a statistical moment of some measure for a leaf subset size r, we consider a uniform probability distribution for selecting any subset of exactly r leaves in $\mathcal{T}$. In other words, all subsets of exactly r leaves in $\mathcal{T}$ are considered with the same probability when computing a statistical moment of a given measure.

2.1 A New Algorithm for Calculating the Variance of the PD

In a previous paper, we provided a formal expression for the exact value of the standard deviation of the PD [15]. Based on that expression, for a tree $\mathcal{T}$ and a sample size of r leaves, the variance of the PD is equal to:

$$var_{\mathrm{PD}}(\mathcal{T}, r) = \sum_{e \in E} \sum_{l \in E} w(e) \cdot w(l) \cdot (1 - \mathcal{F}(S, e, l, r)) - \mu_{\mathrm{PD}}^2(\mathcal{T}, r), \qquad (1)$$

where:

$$\mathcal{F}(S, e, l, r) = \begin{cases} \mathcal{F}_{\mathrm{Off}}(S, e, l, r) = \dfrac{\binom{s(e)}{r} + \binom{s-s(l)}{r} - \binom{s(e)-s(l)}{r}}{\binom{s}{r}} & \text{if } l \in \mathrm{Off}(e). \\[3mm] \mathcal{F}_{\mathrm{Off}}(S, l, e, r) = \dfrac{\binom{s(l)}{r} + \binom{s-s(e)}{r} - \binom{s(l)-s(e)}{r}}{\binom{s}{r}} & \text{if } e \in \mathrm{Off}(l). \\[3mm] \mathcal{F}_{\mathrm{Ind}}(S, e, l, r) = \dfrac{\binom{s-s(e)}{r} + \binom{s-s(l)}{r} - \binom{s-s(e)-s(l)}{r}}{\binom{s}{r}} & \text{otherwise.} \end{cases}$$

and where $\mu_{\mathrm{PD}}(\mathcal{T}, r)$ is the mean value of the PD over all possible subsets of exactly r leaves of $\mathcal{T}$. In our previous paper we showed how we can compute this mean value for a given r in $O(n)$ time. Hence, the bottleneck for calculating the variance of this metric is the computation of the following quantity:

$$\sum_{e \in E} \sum_{l \in E} w(e) \cdot w(l) \cdot (1 - \mathcal{F}(S, e, l, r)) \tag{2}$$

Given that we can evaluate function $\mathcal{F}$ in constant time[1], the expression in (1) leads to a trivial algorithm that runs in $O(n^2)$ time; for every pair of edges in $e, l \in E$ we calculate explicitly the value of $\mathcal{F}(S, e, l, r)$. However, as we mentioned earlier, the large size of recent phylogenetic datasets makes the use of this algorithm infeasible. Next we show how we can design an algorithm that can be much more efficient in practice, depending on how balanced the input tree $\mathcal{T}$ is. To describe this better, first we introduce a new concept that has to do with the structure of a rooted tree. In particular, let $\mathcal{D}(\mathcal{T})$ denote the set of all subtree sizes that are observed in the tree $\mathcal{T}$, that is $\mathcal{D}(\mathcal{T}) = \{s(e) : e \in E\}$.

We call this set the *distinct subtree sizes set* of $\mathcal{T}$. We represent the size of this set by $\mathrm{DSSI}(\mathcal{T})$, that means $\mathrm{DSSI}(\mathcal{T}) = |\mathcal{D}(\mathcal{T})|$. We call this value the *Distinct Subtree Sizes Index* of the tree $\mathcal{T}$. Based on this definition, we provide the following theorem.

Theorem 1. *Let $\mathcal{T}$ be a phylogenetic tree that consists of n nodes, and let r be a positive integer such that $r \leq n$. The variance of the Phylogenetic Diversity over all subsets of r leaves in $\mathcal{T}$ can be computed in $O(\mathrm{SI}(\mathcal{T}) + \mathrm{DSSI}(\mathcal{T})^2)$ time, using $O(n)$ memory.*

Proof. Based on the description that we provided earlier in this section, to prove the time bound it suffices to describe how we can evaluate efficiently the expression in (2). We can rewrite this expression as follows:

$$\sum_{e \in E} \sum_{l \in E} w(e) \cdot w(l) \cdot (1 - \mathcal{F}(S, e, l, r)) \tag{3}$$

$$= \left(\sum_{e \in E} w(e) \right)^2 - \sum_{e \in E} w(e)^2 \cdot \mathcal{F}_{\mathrm{Off}}(S, e, e, r) - 2 \sum_{e \in E} \sum_{l \in \mathrm{Off}(e)} w(e) \cdot w(l) \cdot \mathcal{F}_{\mathrm{Off}}(S, e, l, r) \tag{4}$$

$$- 2 \sum_{e \in E} \sum_{l \in \mathrm{Ind}(e)} w(e) \cdot w(l) \cdot \mathcal{F}_{\mathrm{Ind}}(S, e, l, r) . \tag{5}$$

It is easy to show that the first and the second sum in (4) consist of $\Theta(n)$ terms, and therefore they can be computed in $O(n)$ time. The third sum in (4)

---

[1] In the definition of $\mathcal{F}$, all the required values that involve binomial coefficients can be precomputed in $O(n)$ time in total in the RAM model. Each of the precomputed values can then be accessed in constant time each time we have to evaluate this expression.

consists of $\mathrm{SI}(\mathcal{T})$ terms since for every edge $e \in E$ there exist $s(e)$ terms in this sum. Since we can evaluate each of these terms in constant time, the expression in (4) can be evaluated in $O(\mathrm{SI}(\mathcal{T}))$ time in total.

The two nested sums of the quantity in (5) can be analysed as follows:

$$\sum_{e \in E} \sum_{l \in \mathrm{Ind}(e)} w(e) \cdot w(l) \cdot \mathcal{F}_{\mathrm{Ind}}(S, e, l, r) = \sum_{e \in E} \sum_{l \in E} w(e) \cdot w(l) \cdot \mathcal{F}_{\mathrm{Ind}}(S, e, l, r)$$

$$- 2 \sum_{e \in E} \sum_{l \in \mathrm{Off}(e)} w(e) \cdot w(l) \cdot \mathcal{F}_{\mathrm{Ind}}(S, e, l, r) - \sum_{e \in E} w(e)^2 \cdot \mathcal{F}_{\mathrm{Ind}}(S, e, e, r) . \quad (6)$$

Based on the same arguments as for the expression in (4), the two last sums in (6) can be evaluated in $O(\mathrm{SI}(\mathcal{T}))$ time in total. Let α be a positive integer such that $\alpha \in \mathcal{D}(\mathcal{T})$. Recall that $\mathcal{D}(\mathcal{T})$ is the set of all values $s(e)$ that we can observe among the edges of $\mathcal{T}$. Let $\zeta(\alpha)$ denote the sum of the weights of all the edges $e \in E$ for which it holds $s(e) = \alpha$, that means:

$$\zeta(\alpha) = \sum_{\substack{e \in E \\ s(e) = \alpha}} w(e)$$

Using this notation, the first sum in (6) can be written as:

$$\sum_{e \in E} \sum_{l \in E} w(e) \cdot w(l) \cdot \mathcal{F}_{\mathrm{Ind}}(S, e, l, r) = \sum_{\alpha \in \mathcal{D}(\mathcal{T})} \sum_{\beta \in \mathcal{D}(\mathcal{T})} \zeta(\alpha) \cdot \zeta(\beta) \cdot \mathcal{F}_{\mathrm{Ind}}(S, \alpha, \beta, r) .$$

$$(7)$$

In the last expression, we abuse slightly the notation for function $\mathcal{F}_{\mathrm{Ind}}$; for two integers $\alpha, \beta \in \mathcal{D}$ we imply that $\mathcal{F}_{\mathrm{Ind}}(S, \alpha, \beta, r) = \mathcal{F}_{\mathrm{Ind}}(S, e, l, r)$, where $s(e) = \alpha$ and $s(l) = \beta$. The sum in (7) consists of $\Theta(\mathrm{DSSI}^2(\mathcal{T}))$ terms. Each of these terms can be evaluated in constant time given that we have precomputed the values $\zeta(\alpha)$, $\forall \alpha \in \mathcal{D}(\mathcal{T})$. The values $\zeta(\alpha)$ can be precomputed trivially in $\Theta(n)$ time altogether, hence the expression in (7) can be evaluated in $\Theta(\mathrm{DSSI}^2(\mathcal{T}))$ time in total. Given the description that we provided for evaluating the expressions from (4) to (7), we conclude that the variance of the PD can be computed in $O(\mathrm{SI}(\mathcal{T}) + \mathrm{DSSI}(\mathcal{T})^2)$ time overall. To do this, we need to store the values of the functions $\mathcal{F}_{\mathrm{Off}}$, and $\mathcal{F}_{\mathrm{Ind}}$, and the values $\zeta(\alpha)$ for every $\alpha \in \mathcal{D}(\mathcal{T})$. These require $O(n)$ memory in total, and the theorem follows. $\qquad \square$

According to Theorem 1, we can compute the variance of the PD using an algorithm whose performance depends on the parameters $\mathrm{SI}(\mathcal{T})$ and $\mathrm{DSSI}(\mathcal{T})$. For every tree $\mathcal{T}$ it holds that $\mathrm{DSSI}(\mathcal{T}) \geq h(\mathcal{T})$ and $\mathrm{DSSI}(\mathcal{T}) \geq \mathrm{SI}(\mathcal{T})/n$. In the best case, when the input tree is balanced and has height $\Theta(\log n)$, the new algorithm runs in $\Theta(n \log n)$ time. But when it comes to the worst case performance, the new approach is not better than the trivial algorithm that was previously known; if $\mathrm{SI}(\mathcal{T}) = \Theta(n^2)$ or $\mathrm{DSSI}(\mathcal{T}) = \Theta(n)$ then the computation of the variance takes $O(n^2)$ time. In Section 4 we present experimental results that indicate that the new approach is much more efficient in practice. For different tree data sets that we use there, the values of SI($\mathcal{T}$) and DSSI($\mathcal{T}$) are much smaller than in the worst case scenario. In fact, we can prove a non-trivial tight worst case bound for DSSI($\mathcal{T}$); this bound depends on the number of nodes and the height of $\mathcal{T}$. The bound that we provide applies to trees that have a height that is at least logarithmic to the number of tree nodes (for example, trees where the nodes have constant maximum degree). The proof of the following lemma appears in the full version of this paper.

Lemma 1. *Let $\mathcal{T}$ be a phylogenetic tree that consists of n nodes and has height $h(\mathcal{T})$. In the worst case, the value of $\mathrm{DSSI}(T)$ can be as large as $\Theta(\sqrt{n \cdot h(\mathcal{T})})$.*

2.2   Computing the Mean Nearest Taxon Distance

Next we show how we can use the main result of the previous section in order to efficiently compute the variance of another popular phylogenetic measure. Let $\mathcal{T}$ be a phylogenetic tree, and let R be a subset of its leaves that consists of $|R| = r$ elements. The Mean Nearest Taxon Distance(MNTD) of the leaves in R is equal to the average distance between an element in R and its closest neighbour in R [17]. More formally, the MNTD is defined as:

$$\mathrm{MNTD}(\mathcal{T}, R) = \frac{1}{r} \sum_{v \in R} \min_{u \in R/\{v\}} cost(u, v) . \tag{8}$$

Like with other phylogenetic measures, in order to analyse the value of the MNTD for a set of leaves R it is important to compute the mean and the variance of this measure for all possible subsets of $|R|$ leaves in $\mathcal{T}$. Next we provide for the first time formal expressions that lead to the efficient computation of the exact value of the mean and the variance of the MNTD. The expressions that we provide hold only for ultrametric phylogenetic trees; recall that a tree $\mathcal{T}$ is ultrametric if all simple paths between the root and the leaves of $\mathcal{T}$ have the same cost. Ultrametric tree datasets are very common in phylogenetic research; for instance, ultrametric trees are produced for a given set of taxa when the weights of the tree edges represent specific notions of distance, such as time between speciation events. In the next lemma we show how we can simplify the expression in (8) when we specifically consider ultrametric trees.

Lemma 2. *Let $\mathcal{T}$ be an ultrametric phylogenetic tree and let $R \subseteq S$ be a subset of r leaves. The value of the MNTD for this subset is equal to:*

$$\mathrm{MNTD}(\mathcal{T}, R) = \frac{2}{r} \sum_{\substack{e \in E \\ sr(e)=1}} w(e) . \tag{9}$$

Proof. Let v be a leaf in R, and let u be the closest leaf to v in $R/\{v\}$. That means $cost(u, v) = \min_{x \in R/\{v\}} cost(v, x)$. Let $p(u, v)$ be the simple path that connects u and v in $\mathcal{T}$. We can partition $p(u,v)$ into two subpaths $p(u,a)$ and $p(v,a)$, where a is the deepest node in $\mathcal{T}$ that is a common ancestor of u and v. Since $\mathcal{T}$ is ultrametric, for every internal node $x \in \mathcal{T}$ any simple path that connects x with a leaf in $S(x)$ has the same cost. Therefore, it holds that $cost(u,a) = cost(v,a) = cost(u,v)/2$. Also, for any edge e that appears in the path $p(v,a)$ we have that $sr(e) = 1$. If that was not the case then there would exist an edge e in $p(v,a)$ and a leaf u' in $S(e)$ such that $u' \notin \{u,v\}$ and $cost(u',v) < cost(u,v)$, which contradicts the assumption that u is the closest leaf to v in R. From the above, we conclude that:

$$\mathrm{MNTD}(\mathcal{T},R) = \frac{1}{r} \sum_{v \in R} \min_{u \in R/\{v\}} cost(u,v) = \frac{2}{r} \sum_{v \in R} \sum_{\substack{e \in \mathrm{Anc}(v) \\ sr(e)=1}} w(e) = \sum_{\substack{e \in E \\ sr(e)=1}} w(e) \ .$$

□

Next we use the expression in (9) to obtain expressions for efficiently computing the mean and the variance of the MNTD for ultrametric trees.

Theorem 2. *Let $\mathcal{T}$ be an ultrametric phylogenetic tree that has s leaves and consists of n nodes in total. Let r be a non-negative integer with $r \leq s$. The expected value of the MNTD for a subset of exactly r leaves in $\mathcal{T}$ is equal to:*

$$\mu_{\mathrm{MNTD}}(\mathcal{T},r) = \frac{2}{r} \sum_{e \in E} \frac{w(e) \cdot s(e) \cdot \binom{s-s(e)}{r-1}}{\binom{s}{r}}, \tag{10}$$

and can be computed in $\Theta(n)$ time in the RAM model.

Proof. Let R be a subset of r leaves in $\mathcal{T}$, and let e be any edge in $\mathcal{T}$. We use $SP(e,R)$ to denote the function that has value 1 when $sr(e) = 1$, otherwise it has value zero. Based on Lemma 2, the expectation of the MNTD for a subset of r leaves in $\mathcal{T}$ is equal to:

$$\mathrm{E}_{\mathrm{MNTD}}(\mathcal{T},r) = \mathrm{E}_{R \in \mathrm{Sub}(S,r)} \left[\frac{2}{r} \sum_{\substack{e \in E \\ sr(e)=1}} w(e) \right] = \mathrm{E}_{R \in \mathrm{Sub}(S,r)} \left[\frac{2}{r} \sum_{e \in E} w(e) \cdot SP(e,R) \right] \tag{11}$$

$$= \frac{2}{r} \sum_{e \in E} w(e) \cdot \mathrm{E}_{R \in \mathrm{Sub}(S,r)} \left[SP(e,R) \right] \ . \tag{12}$$

Considering that every subset R of exactly r leaves is picked with the same probability, the expected value of the function $SP(e,R)$ is equal to:

$$\mathrm{E}_{R \in \mathrm{Sub}(S,r)} \left[SP(e,R) \right] = \frac{s(e) \cdot \binom{s-s(e)}{r-1}}{\binom{s}{r}} \ , \tag{13}$$

which leads to the expression in (10).

To compute the value of this expression, we first precompute values $\binom{x}{r-1}/\binom{s}{r}$ for every integer $x \in [r-1, s]$. This can be done alltogether in $O(n)$ time in the RAM model. Given these values, the rest of the expression (10) can be straightforwardly evaluated in $O(n)$ time.   □

The proofs of the next theorem appears in the full version of this paper.

Theorem 3. *Let $\mathcal{T}$ be an ultrametric phylogenetic tree that has s leaves and consists of n nodes in total. Let r be a natural number with $r \leq s$. The variance of the MNTD for a sample of exactly r leaves in $\mathcal{T}$ is equal to:*

$$\mathrm{var}_{\mathrm{MNTD}}(\mathcal{T}, r) = \frac{4}{r^2} \sum_{e \in E} \sum_{l \in E} w(e) \cdot w(l) \cdot \mathcal{G}(S, e, l, r) - \mu^2_{\mathrm{MNTD}}(\mathcal{T}, r), \quad (14)$$

where:

$$\mathcal{G}(S, e, l, r) = \begin{cases} \mathcal{G}_{\mathrm{Off}}(S, e, l, r) = \dfrac{s(l) \cdot \binom{s - s(e)}{r-1}}{\binom{s}{r}} & \text{if } l \in \mathrm{Off}(e). \\[12pt] \mathcal{G}_{\mathrm{Off}}(S, l, e, r) = \dfrac{s(e) \cdot \binom{s - s(l)}{r-1}}{\binom{s}{r}} & \text{if } e \in \mathrm{Off}(l). \\[12pt] \mathcal{G}_{\mathrm{Ind}}(S, e, l, r) = \dfrac{s(e) \cdot s(l) \cdot \binom{s - s(e) - s(l)}{r-2}}{\binom{s}{r}} & \text{otherwise.} \end{cases}$$

The variance of the MNTD can be computed in $O(\mathrm{SI}(\mathcal{T}) + \mathrm{DSSI}(\mathcal{T})^2)$ time, using $O(n)$ memory.

3   A New Biodiversity Measure

Earlier in this paper, we indicated that in several case studies there is the need to compute the higher order moments of a phylogenetic biodiversity measure. Yet, we argued that for a few popular measures this appears to be infeasible. Next we introduce a new non-trivial measure, for which we prove that we can calculate any of its statistical moments efficiently in practice.

Let $\mathcal{T}$ be a phylogenetic tree and let R be a subset of its leaves. Let χ be a any real in the interval $(0.5, 1]$. We use $v_{\mathrm{anc}}(R, \chi)$ to denote the deepest node in the tree that has at least χr elements of R in its subtree. We call this node the *core ancestor of R given χ*. We call the cost of the simple path that connects $v_{\mathrm{anc}}(R, \chi)$ with the root of $\mathcal{T}$ the *Core Ancestor Cost of R given χ* (CAC), and we denote this cost by $\mathrm{CAC}(\mathcal{T}, R, \chi)$.

We consider that the CAC can be a useful tool for phylogenetic analyses; the CAC can be used to measure whether a sample of leaves R consists mostly of a single group of closely related species, or R is made of several small unrelated groups. For example, if $\mathrm{CAC}(\mathcal{T}, R, 0.8)$ is relatively large and comparable to the average path cost between the root and any leaf in $\mathcal{T}$ then about 80% of the species in R have a common ancestor which is deep in the tree, and they are closely related. On the other hand, if $\mathrm{CAC}(\mathcal{T}, R, 0.51)$ is zero then R consists of at least two main unrelated groups of species. Early experiments that we conducted have demonstrated that the CAC is strongly positively related to the NRI and weakly negatively related to the PD, relationships which we intend to explore further in a future publication. In the present paper we focus on the computational aspects of this measure; we examine how we can compute efficiently the CAC and the values of its statistical moments.

For a given sample of leaves R and an integer $\chi \in (0.5, 1]$, value $\mathrm{CAC}(\mathcal{T}, R, \chi)$ can be computed in $O(n)$ time in the following way; first, we compute bottom-up the values $sr(e)$ for every $e \in E$. Then, we start from the root of $\mathcal{T}$ and we compute $\mathrm{CAC}(\mathcal{T}, R, \chi)$ by constructing incrementally the path that connects the root with $v_{\mathrm{anc}}(R, \chi)$.

The major advantage of using the CAC in phylogenetic analysis is that, for a given χ and size of R, we can efficiently compute in practice the value of any statistical moment of this measure. To describe how can do this, we define the following quantity:

$$\mathcal{C}_\chi(\mathcal{T}, r, k) = \mathrm{E}_{R \in \mathrm{Sub}(S,r)} \left[\mathrm{CAC}^k(\mathcal{T}, R, \chi) \right] .$$

We can compute any of the moments of CAC by using the values $\mathcal{C}_\chi(\mathcal{T}, r, k)$. In particular, The expectation of CAC for r leaves is equal to $\mathcal{C}_\chi(\mathcal{T}, r, 1)$, and the variance is equal to $\mathcal{C}_\chi(\mathcal{T}, r, 2) - \mathcal{C}_\chi^2(\mathcal{T}, r, 1)$. Using a standard formula from the mathematical literature, for any integer $k > 3$ the k-th order moment of CAC for r leaves can be expressed as:

$$\frac{\sum_{i=0}^{k} \binom{k}{i} (-\mathcal{C}_\chi(\mathcal{T}, r, 1))^i \, \mathcal{C}_\chi^{k-i}(\mathcal{T}, r, i)}{(\mathcal{C}_\chi(\mathcal{T}, r, 2) - \mathcal{C}_\chi(\mathcal{T}, r, 1))^{k/2}} . \tag{15}$$

Therefore, computing the k-th order moment of CAC boils down to calculating values $\mathcal{C}_\chi(\mathcal{T}, r, i)$ for every $i = 1, 2, \ldots, k$. In the next lemma we show that this can be done efficiently in practice. The proof of this lemma is provided in the full version of this paper.

Lemma 3. *Let $\mathcal{T}$ be a phylogenetic tree that has s leaves and consists of n nodes in total. Let $r \leq s$ be a positive integer and let χ be real number such that $\chi \in (0.5, 1]$. For any positive integer k it holds that:*

$$\mathcal{C}_\chi(\mathcal{T}, r, k) = \mathrm{E}_{R \in \mathrm{Sub}(S,r)} \left[\mathrm{CAC}^k(\mathcal{T}, R, \chi) \right]$$

$$= \sum_{v \in V} \mathrm{cost}(v, \mathrm{root}(\mathcal{T})) \cdot \frac{\sum_{i=\lceil r\chi \rceil}^{s(v)} \binom{s(v)}{i} \binom{s-s(v)}{r-i} - \sum_{u \in \mathrm{Ch}(v)} \sum_{j=\lceil r\chi \rceil}^{s(u)} \binom{s(u)}{j} \binom{s-s(u)}{r-j}}{\binom{s}{r}} . \tag{16}$$

We can compute the values $\mathcal{C}_\chi(\mathcal{T}, r, t)$ for all $t = 1, 2, \ldots, k$ in $O(\mathrm{SI}(\mathcal{T}) + kn)$ time, using $O(n + k)$ space.

The following theorem follows directly from combining Lemma 3 with Equation (15).

Theorem 4. *Let $\mathcal{T}$ be a phylogenetic tree that consists of n nodes and s leaves. Let r, k be two non-negative integers such that $r \leq s$, and let χ be a real such that $\chi \in (0.5, 1]$. We can compute the k first statistical moments of the Core Ancestor Cost among all possible subsets of exactly r leaf nodes of $\mathcal{T}$ given χ in $O(\mathrm{SI}(\mathcal{T}) + kn + k^2)$ time, using $O(n + k)$ space.*

4   Experiments and Benchmarks

We have implemented all of the algorithms that we introduced in the previous sections, and we have conducted experiments in order to measure their performance. In these experiments we also used an implementation of the old approach for computing the variance of the PD; this is the algorithm that always takes quadratic time to execute with respect to the size of the input tree. We use this implementation as a point of reference for our new algorithm that computes the variance of the PD. All of the implementations were developed in $C++$. The experiments were executed on an Intel i7-3770 eight-core CPU where each core is a 3.40 GHz processor. The main memory of this computer is 16 Gigabytes. The operating system that we used on this computer is Microsoft Windows 7.

In all the experiments that we conducted, we observed that the algorithm that computes the variance of the MNTD had an almost identical performance with the new algorithm that computes the variance of the PD. Therefore, for the sake of brevity, we chose not to illustrate the running times of the MNTD algorithm in this version of the paper.

We performed two sets of experiments; in the first set of experiments we used phylogenetic trees that were produced based on real-world biological data, representing the phylogenetic relations between existing species. We used two datasets of this kind; one dataset is a phylogenetic tree that represents the phylogeny of all mammal species [1]. This tree has 4510 leaf nodes and 6618 nodes in total. We refer to this tree as the `mammals` dataset. The other real-world dataset that we used is a tree that was constructed by Goloboff et. al [7]. This is the largest evolutionary tree of eukaryotic organisms that has been so far constructed from molecular and morphological data. It consists of 71181 leaves and 83751 nodes in total. This tree is unrooted; for the needs of our experiments we picked arbitrarily an internal node and used this as the root. We call this dataset the `eukaryotes` dataset. In the first set of experiments we ran our three new algorithms plus the old algorithm that computes the PD variance using as input the `mammals` and the `eukaryotes` datasets. We executed each algorithm several times on each dataset and we measured the total running time of the algorithm for all these executions. We did this because for the three algorithms that we introduce in this paper the time taken for a single execution was quite short, and comparable to the time spent by our software to read the input dataset. Hence, we executed each of the algorithms on each of the datasets ninety-nine times, each time using a different value of r, ranging from two to one hundred. Preliminary measurements showed that the value of r does not affect in practice the performance of any of the examined algorithms. This is also the case with the value of the χ parameter and the performance of the CAC algorithm. In the experiments we ran this algorithm with parameter values $\chi = 0.6$ and $k = 3$. We also calculated the values of the SI and the DSSI for each dataset. These results are presented in Table 1.

Table 1. The results of the experiments that involve trees which represent relations between species in the real-world. The running time of each algorithm is measured over ninety-nine consecutive executions on the same dataset (PD Old = the old approach for computing the PD variance, PD New = the new algorithm for computing the PD variance, CAC = the algorithm that computes the k first moments of the CAC for $k = 3$ and $\chi = 0.6$). Running times are presented in seconds.

Dataset	n	PD Old	PD New	CAC	SI	DSSI^2
eukaryotes	83751	> 3 hours	38.9	14.8	998850	109561
mammals	6618	1672	3.6	1.0	79984	26569

According to the results of these experiments, it becomes evident that the new algorithm that computes the variance of the PD outperforms clearly the old approach. For the two datasets that we considered, the new algorithm appears to be hundreds of times faster than the old one. Given that the running times are measured over ninety-nine executions, it appears that the new algorithm for the PD can process a tree of more than 80, 000 nodes in less than half a second. The algorithm that computes the first three moments of the CAC appears to be even faster than that. As it comes to the values of the SI and the DSSI, we see that the Sackin's Index is larger than the square of the DSSI. This may be an indication that, in practice, the SI is the dominating quantity in the analysis of the running time of the new algorithm. For both datasets, the SI appears to be equal to roughly twelve times the size of the input.

In the second set of experiments we used trees of various sizes that we generated algorithmically. These trees were created in the following manner; first we generated twenty trees using a randomised pure birth process. In this process, a tree is grown in a series of steps from a single root node; at each step we choose a leaf node v, and we add two child nodes to v. Node v is chosen uniformly at random among all the leaves of the current tree. Using this process we generated twenty binary trees, each having exactly 4, 000 leaves. From each of these trees we extracted sixteen subtrees; these subtrees have $250k$ leaves with k ranging from one to sixteen. The subtrees were produced by successively pruning chunks of 250 leaves from the original tree of 4, 000 leaves. In this way we produced 320 trees in total. We denote the set of these trees by $\mathcal{U}$.

We ran each of the implemented algorithms using as input the trees in $\mathcal{U}$. As we did in the previous set of experiments, we executed each algorithm ninety-nine times for each input tree, and we measured the total time taken for these executions. Figure 1 illustrates the running times of the old and the new algorithm that compute the variance of the PD, and the running times of the algorithm that computes the first k moments of the CAC for $\chi = 0.6$ and $k = 3$. Also, for each $\mathcal{T} \in \mathcal{U}$ we measured the values of the SI and the DSSI. Furthermore, we measured the running time of the algorithm that computes the moments of the CAC for a fixed tree of 4,000 leaves and for different values of k–see Figure 2.

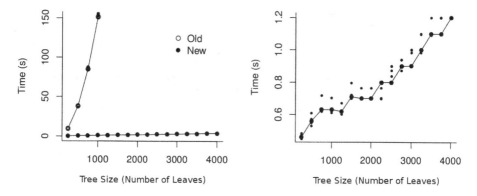

Fig. 1. The running times of three of the implemented algorithms using as input randomly generated trees. For each algorithm, the continuous line segments connect the median values of the measured running times for input trees that have the same number of leaves. Left: The running times of the old and the new algorithms that compute the variance of the PD. For each algorithm, the running times for input trees of the same number of leaves have very small difference in value, and hence they are almost indistinguishable. Right: The running time of the algorithm that computes the first k moments of the CAC for $k = 3$ and $\chi = 0.6$.

Again, as can be seen in Figure 1, the new algorithm for the PD variance has a much better performance than the old one. We see also that the algorithm that computes the moments of the CAC runs very fast, processing almost a hundred trees of a few thousand nodes in less than 1.5 seconds. In Figure 2 we see that the *SI* is evidently larger the DSSI for the randomly generated trees. Still, the value of the SI is not much larger the size of the input trees; given that the total number of nodes of a binary tree is roughly at most twice the number of its leaves, the SI in this set of experiments is not larger than ten times the size of the input. This possibly explains the very good performance of all the new algorithms that we introduce in this paper. Also, as expected, in Figure 2 we can see that the running time of the algorithm that calculates the moments CAC scales almost linearly as the value of k increases.

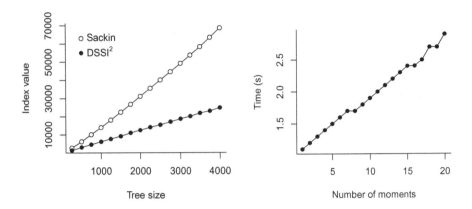

Fig. 2. Left: The values of the SI and of the square of the DSSI for the trees that we generated using a pure birth process. For each number of leaves, we illustrate only the median of these values. The rest of the values are quite close to this median, having at most an absolute difference of roughly two thousand units. Right: The running time of the algorithm that computes the first k moments of the CAC for a single tree of $4,000$ leaves and for k ranging from one to twenty.

References

1. Bininda-Emonds, O.R.P., Cardillo, M., Jones, K.E., MacPhee, R.D.E., Beck, R.M.D., Grenyer, R., Price, S.A., Vos, R.A., Gittleman, J.L., Purvis, A.: The Delayed Rise of Present-Day Mammals. Nature 446, 507–512 (2007)
2. Blum, M.G.B., François, O.: On Statistical Tests of Phylogenetic Tree Imbalance: The Sackin and Other Indices Revisited. Mathematical Biosciences 195, 14–153 (2005)
3. Cadotte, M., Albert, C.H., Walker, S.C.: The Ecology of Differences: Assessing Community Assembly with Trait and Evolutionary Distances. Ecology Letters 16, 1234–1244 (2013)
4. Cooper, N., Rodriguez, J., Purvis, A.: A Common Tendency for Phylogenetic Overdispersion in Mammalian Assemblages. Proceedings of the Royal Society B 275, 2031–2037 (2008)
5. O'Dwyer, J.P., Kembel, S.W., Green, J.L.: Phylogenetic Diversity Theory Sheds Light on the Structure of Microbial Communities. PLoS Computational Biology 8(12), e1002832(2012)
6. Faller, B., Pardi, F., Steel, M.: Distribution of Phylogenetic Diversity Under Random Extinction. Journal of Theoretical Biology 251, 286–296 (2008)
7. Goloboff, P.A., Catalano, S.A., Mirandeb, J.M., Szumika, C.A., Ariasa, J.S., Kallersjoc, M., Farris, J.S.: Phylogenetic Analysis of 73 060 Taxa Corroborates Major Eukaryotic Groups. Cladistics 25, 211–230 (2009)
8. Graham, C.H., Parra, J.L., Rahbek, C., McGuire, J.A.: Phylogenetic Structure in Tropical Hummingbird Communities. Proceedings of the National Academy of Sciences USA 106, 19673–19678 (2009)
9. Kembel, S.W., Hubbell, S.P.: The Phylogenetic Structure of a Neotropical Forest Tree Community. Ecology 87, S86–S99 (2006)

10. Kissling, W.D., Eiserhardt, W.L., Baker, W.J., Borchsenius, F., Couvreur, T.L.P., Balslev, H., Svenning, J.-C.: Cenozoic Imprints on the Phylogenetic Structure of Palm Species Assemblages Worldwide. Proceedings of the National Academy of Sciences USA 109, 7379–7384 (2012)
11. Kraft, N.J.B., Cornwell, W.K., Webb, C.O., Ackerly, D.D.: Trait Evolution, Community Assembly, and the Phylogenetic Structure of Ecological Communities. The American Naturalist 170, 271–283 (2007)
12. Nipperess, D.A., Matsen IV., F.A.: The Mean and Variance of Phylogenetic Diversity Under Rarefaction. Methods in Ecology and Evolution 4, 566–572 (2013)
13. Steel, M.: Tools to Construct and Study Big Trees: A Mathematical Perspective. In: Hodkinson, T., Parnell, J., Waldren, S. (eds.) Reconstructing the Tree of Life: Taxonomy and Systematics of Species Rich Taxa, pp. 97–112. CRC Press (2007)
14. Tsirogiannis, C., Sandel, B.: Computing the skewness of the phylogenetic mean pairwise distance in linear time. In: Darling, A., Stoye, J. (eds.) WABI 2013. LNCS, vol. 8126, pp. 170–184. Springer, Heidelberg (2013)
15. Tsirogiannis, C., Sandel, B., Cheliotis, D.: Efficient computation of popular phylogenetic tree measures. In: Raphael, B., Tang, J. (eds.) WABI 2012. LNCS, vol. 7534, pp. 30–43. Springer, Heidelberg (2012)
16. Vellend, M., Cornwell, W.K., Magnuson-Ford, K., Mooers, A.Ø.: Measuring Phylogenetic Biodiversity. In: Magurran, A., McGill, B. (eds.) Biological Diversity: Frontiers in Measurement and Assessment, Oxford University Press (2010)
17. Webb, C.O., Ackerly, D.D., McPeek, M.A., Donoghue, M.J.: Phylogenies and Community Ecology. Annual review of ecology and systematics 33, 475–505 (2002)

The Divisible Load Balance Problem
and Its Application to Phylogenetic Inference

Kassian Kobert[1], Tomáš Flouri[1], Andre Aberer[1], and Alexandros Stamatakis[1,2]

[1] Heidelberg Institute for Theoretical Studies, Germany
[2] Karlsruhe Institute of Technology, Institute for Theoretical Informatics,
Postfach 6980, 76128 Karlsruhe

Abstract. Motivated by load balance issues in parallel calculations of the phylogenetic likelihood function we address the problem of distributing divisible items to a given number of bins. The task is to balance the overall sum of (fractional) item sizes per bin, while keeping the maximum number of unique elements in any bin to a minimum. We show that this problem is NP-hard and give a polynomial time approximation algorithm that yields a solution where the sums of (possibly fractional) item sizes are balanced across bins. Moreover, the maximum number of unique elements in the bins is guaranteed to exceed the optimal solution by at most one element. We implement the algorithm in two production-level parallel codes for large-scale likelihood-based phylogenetic inference: ExaML and ExaBayes. For ExaML, we observe best-case runtime improvements of up to a factor of 5.9 compared to the previously implemented data distribution algorithms.

1 Introduction

Maximizing the efficiency of parallel codes by distributing the data in such a way as to optimize load balance is one of the major objectives in high performance computing.

Here, we address a specific case of job scheduling (data distribution) which, to the best of our knowledge, has not been addressed before. We have a list of N divisible jobs, each of which consists of s_i atomic tasks, where $1 \leq i \leq N$, and B processors (or bins). All jobs have an equal, constant startup latency α, and each task, regardless of the job it appears in, requires a constant amount of time β to be processed. Although these times are constant, they depend on the available hardware architecture, and hence are not known a priori. Moreover, the jobs are independent of one another. We also assume that processors are equally fast. Therefore, any task takes time β to execute, independently of the processor it is scheduled to run on. Any job can be partitioned (or decomposed) into disjoint sets of its original tasks, which can then be distributed to different processors. However, each such set incurs its own startup latency α on the processor on which it is scheduled to run. Thus, a job of k tasks takes time $k \cdot \beta + \alpha$ to execute on any processor. The tasks (even of the same job) are independent of each other, that is, they can be executed in any order, and the sole purpose of the job

D. Brown and B. Morgenstern (Eds.): WABI 2014, LNBI 8701, pp. 204–216, 2014.

configuration is to group together the tasks that require the same initialization step and hence minimize the overall startup latency.

- Our work is motivated by parallel likelihood computations in phylogenetics (see [4,9] for an overview). There, we are given a multiple sequence alignment that is typically subdivided into distinct partitions (e.g., gene partitions; jobs in our context). Given the alignment and a partition scheme, the likelihood on a given candidate tree can be calculated. To this end, transition probabilities for the statistical nucleotide substitution model need to be calculated (start-up cost α in our context) for each partition separately because they are typically considered to evolve under different models. Note that, all alignment sites (job size) that belong to the same partition have identical model parameters.

The partitions are the divisible jobs to be distributed among processors. Each partition has a fixed number of *sites* (columns from the alignment), which denote the size of the partition. The sites represent the independent tasks a job (partition) consists of. Since alignment sites are assumed to evolve independently in the likelihood model, the calculations on a single site can be performed independently of all other sites. Thus, a single partition can easily be split among multiple processors. Finally, note that, parallel implementations of the phylogenetic likelihood function now form part of several widely-used tools and the results presented in this paper are generally applicable to all tools.

Related Work. A related problem is *bin-packing* with item fragmentation. Here, items may be fragmented, which can potentially reduce the total number of bins needed for packing the instance. However, since fragmentation incurs overhead, unnecessary fragmentations should be avoided. The goal is to pack all items in a minimum number of bins. For an overview of the fractional bin packing problem see [5, Chapter 33]. However, in contrast to our problem, the number of bins is not part of the input but is the objective function. The most closely related domain of research is *divisible load theory* (DLT). Here, the goal is to distribute optimal fractions of the total load among several processors such that the entire load is processed in a minimal amount of time. For a review on DLT, see [1]. However, in general DLT can accommodate more complex models, taking into account a number of factors, such as network parameters or processor speeds. Our problem falls into the category of scheduling divisible loads with start-up costs (see for instance [2,8]). To our knowledge the problem we present has not been solved before. Finally, there exists previous work by our group on improving the load-balance in parallel phylogenetic likelihood calculations. There, we considered, mostly for the sake of code simplicity, that single partitions/jobs are indivisible. Thus, the scheduling problem we addressed in this work was equivalent to the "classic" multi-processor scheduling problem. The paper also provides a detailed rationale as to why the calculation of transition probabilities (the overhead α) can become performance-critical [10].

Overview. In Section 2 we formally define two variations of the problem. We then prove that the problem is NP-hard (Section 3). The main contribution of this paper can be found in Section 4, where we give a polynomial-time approximation algorithm which yields solutions that assign at most one element more to any processor (or bin) than the optimal solution. We analyze the algorithm complexity and prove the $OPT+1$ approximation in (Section 5). Unless $\mathsf{P} = \mathsf{NP}$ [3,6], no polynomial time algorithm can guarantee a better worst case approximation. Finally, in Section 6, we present the performance gains we obtain, when employing our algorithm for distributing partitions in ExaML[1] [7]

2 Problem Definition

Assume we have N divisible items of sizes $s_1, s_2, \ldots, s_N$, and B available bins. Our task is to find an assignment of the N items to the B bins, by allowing an item to be partitioned into several sub-items whose total size is the size of the original item, in order to achieve the following two goals:

1. The sum of sizes of the (possibly partitioned) items assigned to each bin is well-balanced.
2. The maximum load over all bins is minimal with respect to the number of items added.

In the rest of the text we will use the term *solid* for the items that are not partitioned, and *fractional* for those that are partitioned.

We can now formally introduce two variations of the problem; one where we only allow items of integer sizes, and one where the sizes can be represented by real numbers. In the case of integers, the problem can be formulated as the following integer program.

Problem 1 (LBℕ). *Given a sequence of positive integers $s_1, s_2, \ldots, s_N$ and a positive integer B,*

$$minimize \qquad \max\{ \textstyle\sum_{j=1}^{N} x_{i,j} \mid i = 1, 2, \ldots, B \}$$

subject to

$$\sum_{i=1}^{B} q_{i,j} = s_j, \qquad 1 \leq j \leq N$$

$$\sum_{j=1}^{N} q_{i,j} \geq \lfloor \sigma/B \rfloor, \qquad 1 \leq i \leq B$$

$$\sum_{j=1}^{N} q_{i,j} \leq \lceil \sigma/B \rceil, \qquad 1 \leq i \leq B$$

$$\sigma = \sum_{i=1}^{N} s_i$$

$$0 \leq q_{i,j} \leq x_{i,j} \cdot s_j, \qquad 1 \leq i \leq B,\ 1 \leq j \leq N$$

$$q \in \mathbb{N}_{\geq 0}^{B \times N}$$

$$x \in \{0, 1\}^{B \times N}$$

---

[1] Available at http://www.exelixis-lab.org/web/software/examl/index.html.

Variable $x_{i,j}$ is a boolean value indicating whether bin i contains part of item j and if it does, $q_{i,j}$ denotes the amount. By removing the imposed restriction of integer sizes, and hence allowing for positive real values as the sizes of both solid and fractional items, we obtain the following mixed integer program.

Problem 2 (LBℝ). *Given a sequence of positive real values $s_1, s_2, \ldots, s_N$ and a positive integer value B,*

$$minimize \qquad \max\{ \textstyle\sum_{j=1}^{N} x_{i,j} \mid i = 1, 2, \ldots, B \}$$

subject to

$$\sum_{i=1}^{B} q_{i,j} = s_j, \qquad 1 \le j \le N$$

$$\sum_{j=1}^{N} q_{i,j} = \sigma/B, \qquad 1 \le i \le B$$

$$\sigma = \sum_{i=1}^{N} s_i$$

$$0 \le q_{i,j} \le x_{i,j} \cdot s_j, \qquad 1 \le i \le B,\ 1 \le j \le N$$

$$q \in \mathbb{R}^{B \times N}$$

$$x \in \{0,1\}^{B \times N}$$

If for some bin i and element j we get a solution with $q_{i,j} < s_j$, we say that element j is only assigned to bin i partially, or that only a fraction of element j is assigned to bin i. If $q_{i,j} = s_j$ we say that element j is fully assigned to bin i.

3   NP-hardness

We now show that problems LBN and LBℝ are NP-hard by reducing the well-known PARTITION [6] problem. We reduce it to another decision problem called *Equal Cardinality Partition* (ECP) that decides whether a set can be broken into disjoint sets of equal cardinality and equal sum of elements (see Def. 2), which can be solved by the two flavors of our problem.

Definition 1 (Partition). *Is it possible to partition a set S of positive integers into two disjoint subsets Q and R, such that $Q \cup R = S$ and $\sum_{q \in Q} q = \sum_{r \in R} r$?*

Definition 2 (ECP). *Let p, k be two positive integers and S a set of $p \cdot k$ positive integers. Can we partition S into p disjoint sets $S_1, S_2, \ldots, S_p$ of k elements each, such that $\bigcup_{i=1}^{p} S_i = S$ and $\sum_{s \in S_i} s = \sum_{s \in S_j} s$, for all $1 \le i, j \le p$?*

Clearly, if we can solve our original optimization problems LBN and LBℝ for any S exactly, we can also answer whether ECP returns *true* or *false* for the same set S. Thus, if we can show that ECP is NP-Complete we know that the original problems are NP-hard.

To show that ECP is NP-Complete, it is sufficient to show that ECP is in NP, that is the set of polynomial time verifiable problems, and some NP-Complete problem (here PARTITION) reduces to it.

Lemma 1. ECP *is* NP-*Complete.*

Proof. The first part, i.e., ECP $\in$ NP, is trivial. Given a solution (that is, the sets $S_1,\ldots,S_p$), we are able to verify, in polynomial time to p, that the conditions for problem ECP hold, by summing the elements of each set.

For the reduction of PARTITION to ECP consider the set S to be an instance of PARTITION. We derive an instance $\hat{S}$ of ECP from S, such that PARTITION(S) is true *iff* ECP($\hat{S}$) is true for 2 bins (that is $p = 2$). We define $\hat{S} = S \cup (a \cdot S)$ a set of integers, with $a = (1 + \sum_{s \in S} s)$ and $(a \cdot S) = \{a \cdot s \mid s \in S\}$. Clearly, if there is a solution for PARTITION given S, there must also be a solution for ECP given $\hat{S}$. If $Q, R \subset S$ is a solution for PARTITION, then $Q \cup (a \cdot R)$, $R \cup (a \cdot Q)$ is a solution for ECP.

Similarly, let $\hat{Q}, \hat{R}$ be a solution for ECP given $\hat{S}$. Let $Q = \hat{Q} \cap S$, $R = \hat{R} \cap S$, $(a \cdot Q) = \hat{Q} \cap (a \cdot S)$ and $(a \cdot R) = \hat{R} \cap (a \cdot S)$. Trivially, it holds that $Q = \{q \in \hat{Q} \mid q < a\}$, $R = \{r \in \hat{R} \mid r < a\}$ and $(a \cdot Q) = \hat{Q} \setminus Q$, $(a \cdot R) = \hat{R} \setminus R$. Thus, we obtain $Q \cup R = S$ and $(a \cdot Q) \cup (a \cdot R) = (a \cdot S)$. We also obtain that $\sum_{q \in Q} q = \sum_{r \in R} r$ (and $\sum_{q \in (a \cdot Q)} q = \sum_{r \in (a \cdot R)} r$). We prove that the equations hold by contradiction: suppose this was not the case for some solution of ECP, that is $\sum_{q \in Q} q \neq \sum_{r \in R} r$ and hence $\sum_{q \in (a \cdot Q)} q \neq \sum_{r \in (a \cdot R)} r$. By definition, $(a \cdot Q)$ and $(a \cdot R)$, q/a and r/a are integer values for any $q \in (a \cdot Q)$ and $r \in (a \cdot R)$, and therefore:

$$\left| \sum_{q \in (a \cdot Q)} q - \sum_{r \in (a \cdot R)} r \right| = \left| \sum_{q \in (a \cdot Q)} a \cdot q/a - \sum_{r \in (a \cdot R)} a \cdot r/a \right|$$

$$= a \cdot \overbrace{\left| \sum_{q \in (a \cdot Q)} q/a - \sum_{r \in (a \cdot R)} r/a \right|}^{\geq 1} \geq a$$

However, $\sum_{s \in S} s < a$. Thus, $\sum_{q \in \hat{Q}} q \neq \sum_{r \in \hat{R}} r$ which contradicts the assumption of $\hat{Q}, \hat{R}$ being a solution for ECP($\hat{S}$,2). Therefore, PARTITION reduces to ECP, which means that ECP is NP-Complete. $\square$

Corollary 1. *The optimization problems* LBN *and* LBR *are* NP-*hard.*

This follows directly from Lemma 1 and the fact that an answer for ECP can be obtained by solving the optimization problem.

4 Algorithm

As seen in Section 3, finding an optimal solution to this problem is hard. To overcome this hurdle, we propose an approximation algorithm running in polynomial time that guarantees a near-optimal solution. For an in-depth analysis of the complexity of the algorithm, see Section 5.

The input for the algorithm is a list S of N integer weights and the number of bins B these elements must be assigned to. The idea of the algorithm can be explained by the following three steps:

1. Sort S in ascending order.
2. Starting from the first (solid) element in the sorted list S, assign elements from S to the B bins in a cyclic manner (at any time no two bins can have a difference of more than one element) until any bin can not entirely hold the proposed next item.
3. Break the remaining elements from S to fill the remaining space in the bins.

Fig. 1 presents the pseudocode for the first two phases, while Fig. 2 illustrates phase 3. The output of this algorithm is an assignment, $list = (list[1], \ldots, list[p])$, of –possibly fractional– elements to bins. Each entry in $list$ is a set of triplets that specify which portion of an integer sized element is assigned to a bin. Let $(j, i, k) \in list[l]$ be one such triplet for bin number l. We interpret this triplet as follows: bin l is assigned the fraction of element j that starts at i and ends at k (including i and k).

For the application in phylogenetics, each triplet specifies which portion (how many sites) of a partition is assigned to which processor. Again, let $(j, i, k) \in list[l]$ be one such triplet for some processor l. We interpret this triplet as follows: processor l is assigned sites i through k of partition j.

If $i \neq 1$ or $k \neq s_j$ (recall s_j is the size of element j), we say that element j is partially assigned to bin i, that is, only a fraction of element j is assigned to

```
LOADBALANCE(N, B, S)
▷ Phase 1 — Initialization
 1. Sort S in ascending order and let S = (s₁, s₂, ..., sₙ)
 2. σ = Σⁿᵢ←₁ sᵢ
 3. c ← ⌈σ/B⌉
 4. r ← c · B − σ
 5. for i ← 1 to B do
 6.     size[b] ← 0; items[b] ← 0; list[b] ← ∅
 7. full_bins ← 0; b ← 0;
▷ Phase 2 — Initial filling
 8. for i ← 1 to N do
 9.     if size[b] + sᵢ ≤ c then
10.         size[b] ← size[b] + sᵢ
11.         items[b] = items[b] + 1
12.         ENQUEUE(list[b], (i, 1, sᵢ))
13.         if size[b] = c then
14.             full_bins ← full_bins + 1
15.             if full_bins = B − r then c ← c − 1
16.     else
17.         add ← sᵢ
18.         break
19.     b ← (b + 1)  mod B
```

Fig. 1. The algorithm accepts three arguments N, B and S, where N is the number of items in list S, and B is the number of bins

```
▷ Phase 3 — Partitioning items into bins
20. low ← B; ℓ ← B; high ← 1; h ← 1
21. while i ≤ N do
22.       while size[ℓ] ≥ c do
23.              low ← low − 1; ℓ ← low
24.       while size[h] ≥ c do
25.              high ← high + 1; h ← high
26.       if size[h] + add ≥ c then
27.              items[h] ← items[h] + 1
28.              ENQUEUE(list[h], (i, sᵢ − add + 1, sᵢ − add − size[d] + c))
29.              add ← size[h] + add − c
30.              size[h] ← c
31.              full_bins ← full_bins + 1
32.              if full_bins = B − r then c ← c − 1
33.       else
34.              items[ℓ] ← items[ℓ] + 1
35.              if size[ℓ] + add < c then
36.                     size[ℓ] ← size[ℓ] + add
37.                     ENQUEUE(list[ℓ], (i, sᵢ − add + 1, sᵢ))
38.                     add ← 0
39.                     high ← high − 1; h ← ℓ
40.                     low ← low − 1; ℓ ← low
41.              else
42.                     ENQUEUE(list[ℓ], (i, sᵢ − add + 1, sᵢ − add − size[d] + c))
43.                     add ← size[ℓ] + add − c
44.                     size[ℓ] ← c
45.                     full_bins ← full_bins + 1
46.                     if full_bins = B − r then c ← c − 1
47.       if add = 0 then
48.              i ← i + 1; add ← sᵢ
```

Fig. 2. Phase 3 of the algorithm

bin i. Otherwise, if $i = 1$ and $k = s_j$, then the triplet represents a solid element, i.e., element j is fully assigned to bin i.

For applications that allow any fraction of an integer to be assigned to a bin, not just whole integer values (that is, problem LB$\mathbb{R}$), we redefine the variable c, i.e. the maximum capacity of the bins, to be exactly σ/B, without rounding. Additionally, the output (*list*) must correctly state which ranges of the elements are assigned to which bin and not give integer lower and upper bounds.

We give two examples of how algorithm LOADBALANCE works on a specific set of integers.

Example 1. Consider the set $\{2, 2, 3, 5, 9\}$ and three bins. During initialization (phase 1) we have $c = 7$ and $r = 0$. Phase 2 makes the following assignments: $list[1] = \{(1, 1, 2), (4, 1, 5)\}$, $list[2] = \{(2, 1, 2)\}$, $list[3] = \{(3, 1, 3)\}$. Adding the next element of size 9 is not possible since $size[2] + 9 = 2 + 9 = 11 > c$. Thus, phase 2 ends. Phase 3 splits the last element of size 9 among bins 2 and

3, and the solution is $list[1] = \{(1,1,2),(4,1,5)\}$, $list[2] = \{(2,1,2),(5,1,5)\}$, $list[3] = \{(3,1,3),(5,6,9)\}$. With $\max\{|list[1]|,|list[2]|,|list[3]|\} = 2$. This is also an optimal solution.

Example 2. Consider the set $\{1,1,2,3,3,6\}$ and two bins. During the initialization (phase 1) we have $c = 8$ and $r = 0$. Phase 2 generates the following assignments: $list[1] = \{(1,1,1),(3,1,2),(5,1,3)\}$, $list[2] = \{(2,1,1),(4,1,3)\}$. The last element of size 6 can not be fully assigned to bin 2, thus phase 2 terminates. Finally, phase 3 splits the last element of size 6 among the two bins, and the solution is $list[1] = \{(1,1,1),(3,1,2),(5,1,3),(6,1,2)\}$, $list[2] = \{(2,1,1),(4,1,3),(6,3,6)\}$. We get $\max\{|list[1]|,|list[2]|\} = 4$. However, an optimal solution $list_1^\star = \{(1,1,1),(2,1,1),(6,1,6)\}$, $list_2^\star = \{(3,1,2),(4,1,3),(5,1,3)\}$ with $\max\{|list_1^\star|,|list_2^\star|\} = 3$ exists.

As we can see in Example 2, algorithm LOADBALANCE fails to find the optimal solution in certain cases. However in the next section we show that the difference of 1, as observed in Example 2, already represents the worst case scenario.

5 Algorithm Analysis

We now show that the score obtained by algorithm LOADBALANCE, for any given set of integers and any number of bins, is at most one above the optimal solution. We then give the asymptotic time and space complexities.

5.1 Near-Optimal Solution

Before we start with the proof, we make three observations associated with the algorithm that facilitate the proof. We use the same notation as in the description of the algorithm. That is, $items[i]$ indicates the number of items in bin i, $size[i]$ the sum of sizes of items in bin i, and $list[i]$ is a list of records per item in bin i, describing which fraction of the particular item is assigned to bin i.

Observation 1. *During phase 2 of algorithm* LOADBALANCE, *it holds that*

$$size[i] > size[j]$$

for any two bins j and i, such that $items[i] = items[j] + 1$.

The list of integers was sorted in Phase 1 of the algorithm to a non-decreasing sequence. Hence, any item added to a bin during the i-th cyclic iteration over bins, must be smaller or equal to an item that is added during iteration $i + 1$.

Observation 2. *For all bins i and j during phase 2 of algorithm* LOADBALANCE, *it holds that*

$$items[j] \leq items[i] + 1.$$

This follows directly from Observation 1.

Observation 3. *Phase 3 appends at most 2 more (fractional) items to a bin.*

Any remaining (unassigned) item of size s in this phase satisfies the condition $size[j] + s > c$, for any bin j and capacity c as computed in Fig. 1. Therefore, each bin will be assigned at most one fractional item that does not fill it completely, and one new element that is guaranteed to fill it up.

Lemma 2. *Let $OPT(S, B)$ be the score for the optimal solution for a set S distributed to B bins. Let list be the solution produced by algorithm LOADBALANCE for the same set S and B bins. Then:*

$$\max\{ |list[i]| \mid i = 1, 2, \ldots, B \} \leq OPT(S, B) + 1$$

Proof. Let $\hat{j}$ be the bin that terminates phase 2. That is, $\hat{j}$ is the last bin considered for any assignment in phase 2. After phase 2, if there exists a bin j with $items[j] = items[\hat{j}] + 1$ we get, by Observation 1 and the *pigeonhole principle*, that $OPT(S, B) \geq items[\hat{j}] + 1$. Otherwise, if no such bin exists, $OPT(S, B) \geq items[\hat{j}]$. Let K be the number of unassigned elements at the beginning of phase 3. Let J be the number of bins j with $items[j] = items[\hat{j}]$. We distinguish between three cases. First assume that $items[j] = items[\hat{j}]$ (after phase 2) for all bins j and $K > 0$. Clearly, $OPT(S, B) \geq items[\hat{j}] + 1$. By observation 3 we know that $items[j] \leq items[\hat{j}] + 2$ (after phase 3). Thus the lemma holds for this case. Now consider $K > J$ and $items[j] \neq items[\hat{j}]$ for some bin j, that is, there are more unassigned elements than there are bins with only $items[\hat{j}]$ elements assigned to them. By the pigeonhole principle, $OPT(S, B) \geq items[\hat{j}] + 2$. By observation 3 we get that $items[j] \leq items[\hat{j}] + 1 + 2 = items[\hat{j}] + 3$ for all j. Thus the lemma holds for this case as well. For the last case assume $K \leq J$ and $items[j] \neq items[\hat{j}]$ for some bin j. After a bin is assigned a fractional element that does not fill it completely, it is immediately filled up with the next element. Since preference is given to any bin j with $items[j] = items[\hat{j}]$ and there are at least as many such bins as remaining elements to be added ($K \leq J$), we get that $items[j] \leq items[\hat{j}] + 2$. Since we have seen above that $OPT(S, B) \geq items[\hat{j}] + 1$, the lemma holds. As this covers all cases, the lemma is proven.   □

5.2   Run-Time

The runtime analysis is straight forward. Phase 1 of the algorithm consists of initializing variables, sorting N items by size in ascending order and computing their sum. Using an algorithm such as MERGE-SORT, Phase 1 requires $\mathcal{O}(N \log(N))$ time. Phase 2 requires $\mathcal{O}(N)$ time to consider at most N items, and assign them to B bins in a cyclic manner. Phase 3 appends at most 2 items to a bin (see Observation 3), and hence has a time complexity of $\mathcal{O}(B)$. This yields an overall asymptotic run-time complexity of $\mathcal{O}(N \log(N) + B)$. Note that, if we are already given a sorted list of partitions, the algorithm runs in linear time $\mathcal{O}(N + B)$. Finally, LOADBALANCE requires $\mathcal{O}(B)$ space due to the arrays *items*, *size* and *list*, that are each of size B.

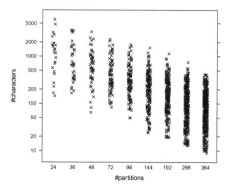

Fig. 3. Number of characters/sites in each partition for the partitioning schemes

6   Practical Application

As mentioned before, the scheduling problem arises for parallel phylogenetic likelihood calculations on large partitioned multi-gene or whole-genome datasets.
This type of partitioned analyses represent common practice at present. The
number of multiple sequence alignment partitions, the number of alignment sites
per partition, and the number of available processors are the input to our algorithm. The production-level maximum likelihood based phylogenetic inference
software ExaML for supercomputers implements two different data distribution
approaches: The *cyclic data distribution* scheme that does not balance the number of unique partitions per processor, but just assigns single sites to processors
in a cyclic fashion. The second approach is the *whole-partition data distribution*
scheme. Here, the individual partitions are not considered divisible and are assigned monolithically to processors using the longest processing time heuristic
for the 'classic' multi-processor scheduling problem [10]. This ensures that the
total and maximum number of initialization steps (substitution matrix calculations) is minimized, at the cost of not being balanced with respect to the sites
per processor. Nonetheless, using this scheme instead of the cyclic distribution
already yielded substantial performance improvements. In order to evaluate the
new distribution scheme, we compare it to these two previous schemes, in terms
of total ExaML runtime. Note that, our algorithm has also been implemented
in ExaBayes[2] which is a code for large-scale Bayesian phylogenetic inference.

6.1   Methods

We performed runtime experiments on a real-world alignment. The alignment
consists of 144 species and 38 400 amino acid characters[3] used the alignment to

---

[2] Available at http://www.exelixis-lab.org/web/software/exabayes/index.html
[3] Data from the 1KITE (www.1kite.org) project.

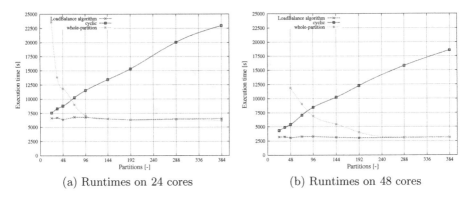

(a) Runtimes on 24 cores   (b) Runtimes on 48 cores

Fig. 4. Runtime comparison for ExaML employing algorithm LOADBALANCE, the cyclic data distribution scheme, or the whole-partition data distribution scheme

create 9 distinct partitioning schemes with an increasing number of partitions. For each scheme, partition lengths were drawn at random, while the number of partitions per scheme was fixed to 24, 36, 48, 72, 96, 144, 192, 288, 384, and 768, respectively. To generate n partition lengths, we drew n random numbers $x_1, \ldots, x_n$ from an exponential distribution $\exp(1) + 0.1$. For a partition p, the value of $x_p / \sum_{i=1..n} x_i$ then specifies the proportion of characters that belong to partition p. The offset of 0.1 was added to random numbers to prevent partition lengths from becoming unrealistically small, since the exponential distribution strongly favors small values. Fig. 3 displays the distributions of the partition lengths for each of the 9 partition schemes. As expected, partition lengths are distributed uniformly on the log-scale.

We executed ExaML using 24 and 48 processes, respectively, to assess performance with our new data distribution algorithm and compare it with the cyclic site and whole-partition data distribution performance. We used a cluster equipped with Intel SandyBridge nodes (2×6 cores per node) and an Infiniband interconnect. Thus, a total of 2 nodes was needed for runs with 24 processes and 4 nodes for runs with 48 processes (inducing higher inter-node communication costs). In Fig. 4.b, the run-times for the whole-partition distribution approach with less than 48 partitions are omitted, since they are identical to executing the runs on 24 processes. The reason is that this method does not divide partitions and thus, in case the number of partitions is smaller than the number of available processors, the extra processors will remain unused.

6.2   Results

As illustrated by Fig. 4, with algorithm LOADBALANCE ExaML always runs at least as fast as the two previous data distribution strategies with one minor exception. Compared to the cyclic data distribution, LOADBALANCE is 3.5× faster for 24 processes and up to 5.9× faster for 48 processes. Using LOADBALANCE,

ExaML requires up to 3.6× less runtime than with the whole partition distribution scheme for 24 processes and for 48 processes the runtime can be improved by a factor of up to 3.9×. For large numbers of partitions, the runtime of the whole partition distribution scheme converges against the runtime of LOADBALANCE. This is expected, since by increasing the number of partitions we break the alignment into smaller chunks and the chance of any heuristic to attain a near-optimal load/data distribution increases. However, if the same run is executed with more processes (i.e., 48 instead of 24), this break-even point shifts towards a higher number of partitions, as shown in Fig. 4.

The results show that, cyclic data distribution performance is acceptable for many processes and few partitions, whereas monolithic whole-partition data distribution is on par with our new heuristic for analyses with few processes and many partitions. Both figures show, that there exists a region where neither of the previous strategies exhibits acceptable performance compared to LOADBALANCE and that this performance gap widens, as parallelism increases.

Finally, employing LOADBALANCE, ExaML executes twice as fast with 48 processes than with 24 processes and thus exhibits an optimum scaling factor of about 2.07 in all cases. For comparison, under the cyclic data distribution, scaling factors ranged from 1.24 to 1.75 and under whole partition distribution, scaling factors ranged from 1.00 (i.e., no parallel runtime improvement) to 2.04. The slight superlinear speedups are due to increased cache efficiency.

7   Conclusion

We have introduced an approximation algorithm for solving a NP-hard scheduling problem with an acceptable worst-case performance guarantee. This theoretical work was motivated by our efforts to improve parallel efficiency of phylogenetic likelihood calculations. By implementing the approximation algorithm in ExaML, a dedicated code for large-scale maximum likelihood-based phylogenetic analyses on supercomputers, we show that (i) the data distribution is near-optimal, irrespective of the number of partitions, their lengths, and the number of processes used and (ii) substantial run time improvements can be achieved, thus saving scarce supercomputer resources. The data distribution algorithm is generally applicable to any code that parallelizes likelihood calculations.

References

1. Bharadwaj, V., Ghose, D., Robertazzi, T.: Divisible load theory: A new paradigm for load scheduling in distributed systems. Cluster Computing 6(1), 7–17 (2003)
2. Błażewicz, J., Drozdowski, M.: Distributed processing of divisible jobs with communication startup costs. Discrete Appl. Math. 76(1-3), 21–41 (1997)
3. Cook, S.A.: The complexity of theorem-proving procedures. In: STOC 1971 Proceedings of the Third Annual ACM Symposium on Theory of Computing, pp. 151–158 (1971)
4. Felsenstein, J.: Inferring phylogenies. Sinauer Associates (2003)

5. Gonzalez, T.F.: Handbook of Approximation Algorithms and Metaheuristics. Chapman & Hall/CRC (2007)
6. Karp, R.: Reducibility among combinatorial problems. Complexity of Computer Computations, 85–103 (1972)
7. Stamatakis, A., Aberer, A.J.: Novel parallelization schemes for large-scale likelihood-based phylogenetic inference. In: IPDPS, pp. 1195–1204 (2013)
8. Veeravalli, B., Li, X., Ko, C.C.: On the influence of start-up costs in scheduling divisible loads on bus networks. IEEE Transactions on Parallel and Distributed Systems 11(12), 1288–1305 (2000)
9. Yang, Z.: Computational Molecular Evolution. Oxford University Press (2006)
10. Zhang, J., Stamatakis, A.: The multi-processor scheduling problem in phylogenetics. In: IPDPS Workshops, pp. 691–698. IEEE Computer Society (2012)

Multiple Mass Spectrometry Fragmentation Trees Revisited: Boosting Performance and Quality

Kerstin Scheubert, Franziska Hufsky, and Sebastian Böcker

Lehrstuhl für Bioinformatik, Friedrich-Schiller-Universität Jena,
Ernst-Abbe-Platz 2, Jena, Germany
{kerstin.scheubert,franziska.hufsky,sebastian.boecker}@uni-jena.de

Abstract. Mass spectrometry (MS) in combination with a fragmentation technique is the method of choice for analyzing small molecules in high throughput experiments. The automated interpretation of such data is highly non-trivial. Recently, fragmentation trees have been introduced for *de novo* analysis of tandem fragmentation spectra (MS^2), describing the fragmentation process of the molecule. Multiple-stage MS (MS^n) reveals additional information about the dependencies between fragments. Unfortunately, the computational analysis of MS^n data using fragmentation trees turns out to be more challenging than for tandem mass spectra.

We present an Integer Linear Program for solving the COMBINED COLORFUL SUBTREE problem, which is orders of magnitude faster than the currently best algorithm which is based on dynamic programming. Using the new algorithm, we show that correlation between structural similarity and fragmentation tree similarity increases when using the additional information gained from MS^n. Thus, we show for the first time that using MS^n data can improve the quality of fragmentation trees.

Keywords: metabolomics, computational mass spectrometry, multiple-stage mass spectrometry, fragmentation trees, Integer Linear Programming.

1 Introduction

Studying metabolites and other small biomolecules with mass below 1000 Da, is relevant, for example, in drug design and the search for new signaling molecules and biomarkers [14]. Since such molecules cannot be predicted from the genome sequence, high-throughput *de novo* identification of metabolites is highly sought. Mass spectrometry (MS) in combination with a fragmentation technique is commonly used for this task. In liquid chromatography MS, a selected molecule can be fragmented in a second step typically using collision-induced dissociation (CID). The resulting fragment ions are recorded in tandem mass spectra (MS^2 spectra). For metabolites, the understanding of CID fragmentation is still in its infancy.

D. Brown and B. Morgenstern (Eds.): WABI 2014, LNBI 8701, pp. 217–231, 2014.

Multiple-stage MS (MS^n) allows to select the product ions of the initial fragmentation step (manually or automatically) and subject them to another fragmentation reaction. This reveals additional information about the dependencies between the fragments. The resulting fragment ions can, in turn, again be selected as precursor ions for further fragmentation. Typically, with each additional fragmentation reaction, the quality of mass spectra is reduced and measuring time increases. Thus, analysis is usually limited to a few fragmentation reactions beyond MS^2.

CID mass spectra (both MS^2 and MS^n) are limited in their reproducibility on different instruments, making spectral library search a non-trivial task [16]. Furthermore, spectral libraries are vastly incomplete. Recent approaches tend to replace searching in spectral libraries by searching in the more comprehensive molecular structure databases [1, 9–11, 26, 31]. However, many metabolites even remain uncharacterized with respect to their structure and function [17].

For the *de novo* interpretation of tandem mass spectra of small molecules, Böcker and Rasche [5] introduced fragmentation trees to identify the molecular formula of an unknown and its fragments. Moreover, fragmentation trees are reasonable descriptions of the fragmentation process and hence can also be used to derive further information about the unknown molecule [19]. Scheubert *et al.* [23, 24] adjusted the fragmentation tree concept to MS^n data to reflect the succession of fragmentation reactions.

Adjusting the fragmentation tree concept to MS^n data, results in the NP-hard COLORFUL SUBTREE CLOSURE problem [24] which has to be solved in conjunction with the original NP-hard MAXIMUM COLORFUL SUBTREE problem [5], resulting in the COMBINED COLORFUL SUBTREE problem [24]. To solve this problem, Scheubert *et al.* [24] presented a fixed-parameter algorithm based on dynamic programming (DP) with worst-case running time depending exponentially on the number of peaks in the spectrum.

To compare two molecules based on their fragmentation spectra, Rasche *et al.* [18] introduced fragmentation tree alignments. By this, similar fragmentation cascades in the two trees are identified and scored. This allows us to use fragmentation trees in applications such as database searching, assuming that structural similarity is inherently coded in the CID spectra fragments. Improving the quality of the fragmentation trees using the additional information provided by MS^n, may improves this downstream analysis.

Here, we present a novel exact algorithm for solving the COMBINED COLORFUL SUBTREE problem. This Integer Linear Program (ILP) is faster than the DP algorithm. Further, we demonstrate the impact of the additional information from MS^n data for the downstream analysis: We compute fragmentation tree alignments [18] and find that correlation between the similarity score of two fragmentation trees and the structural similarity score of the corresponding molecules increases when using the additional information gained from the succession of fragments in multiple MS.

2   Constructing Fragmentation Trees

Given the molecular structure of a molecule and the measured fragmentation spectrum, an MS expert can assign peaks to fragments of the molecule and derive a "fragmentation diagram". *Fragmentation trees* are similar to experts' "fragmentation diagrams" but are extracted directly from the data, without knowledge about a molecule's structure. A fragmentation tree consists of vertices annotated with the molecular formulas of the precursor ion and fragment ions, and directed edges representing the fragmentation steps. Fragmentation trees must not be confused with *spectral trees* for multiple stage mass spectrometry [22,25]. Spectral trees are a formal representation of the MS setup and describe the relationship between the MS^n spectra, but do not contain any additional information.

For the computation of fragmentation trees [5], a fragmentation graph is constructed (see Fig. 1): vertices represent all fragment molecular formulas with mass sufficiently close to the peak mass [3,4]; and weighted edges represent the fragmentation steps leading to those formulas. Two vertices u, v are connected by a directed edge if the molecular formula of v is a sub-molecule of the molecular formula of u. We assume the molecular formula of the full molecule to be given (see [19] for details). The resulting graph is a directed acyclic graph (DAG) $G = (V, E)$, since fragments can only lose, never gain, weight. Vertices in the graph are colored $c : V \rightarrow C$, such that vertices that explain the same peak receive the same color. Edges are weighted, reflecting that some fragmentation steps are more likely. Common fragmentation steps get a higher weight than implausible fragmentation steps. Also peak intensities and mass deviations are taken into account in these weights. The resulting fragmentation graph contains all possible fragmentation trees as subgraphs. The *weight* of an induced tree $T = (V_T, E_T)$ is defined as the sum of its edge weights: $w(T) := \sum_{(u,v) \in E_T} w(u, v)$.

The MS^n data does not only hint to direct but also to indirect successions, that is a fragment is not only scored based on its direct ancestor (its parent node), but also on indirect ancestors (grandparent node etc). Thus, we also have to score the transitive closure of the induced subtrees [24]. The *transitive closure* $G^+ = (V, E^+)$ of a DAG $G = (V, E)$ contains the edge $(u, v) \in E^+$ if and only if there is a directed path in G from u to v. As MS^n data does not differentiate between different explanations of the peaks, we score pairs of colors: $w^+ : C^2 \rightarrow \mathbb{R}$. The *transitive weight* of an induced tree $T = (V_T, E_T)$ with transitive closure $T^+ = (V_T, E_T^+)$ is defined as

$$w^+(T) := \sum_{(u,v) \in E_T^+} w^+\big(c(u), c(v)\big) \tag{1}$$

Scheubert *et al.* [24] introduced three parameters σ_1, σ_2 and σ_3 to score the transitive closure. Parameter σ_1 rewards fragments of an MS^n spectrum that are successors of its parent fragment ($\sigma_1 \geq 0$). Parameter σ_2 penalizes fragments that are successors of a parent fragment of an MS^n spectrum although the corresponding peak is not contained in this spectrum ($\sigma_2 \leq 0$). Parameter σ_3 penalizes direct and indirect fragmentation steps that occur at high collision

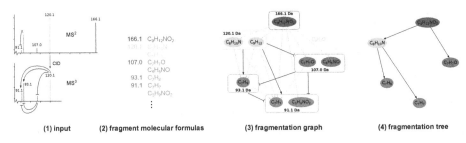

Fig. 1. (1) As input we use MS^n spectra that contain additional information on the succession of fragments. (2) For each peak, we compute all fragment molecular formulas with mass sufficiently close to the peak mass. (3) A fragmentation graph is constructed with vertices for all fragment molecular formulas and edges (grey) for all possible fragmentation steps. Explanations of the same peak receive the same color. The transitive closure of the graph is scored based on pairs of colors. To simplify the drawing, we only show non zero edges of the transitive closure (black). (4) The colorful subtree with maximum combined weight of the edges and the transitive closure is the best explanation of the observed fragments.

energy but not at low collision energy ($\sigma_3 \leq 0$). For a more detailed description of the parameters see [24].

Now, each subtree of the fragmentation graph corresponds to a possible fragmentation tree. Considering trees, every fragment is explained by a unique fragmentation pathway. To avoid the case that one peak is explained by more than one molecular formula, we limit our search to *colorful* trees, where each color is used at most once. In practice, it is very rare that a peak is indeed created by two different fragments. Searching for a colorful subtree of maximum sum of edge weights is known as the MAXIMUM COLORFUL SUBTREE problem, which is NP-hard [5,8]. Searching for a colorful subtree of maximum weight of the transitive closure is known as the COLORFUL SUBTREE CLOSURE problem, which is again NP-hard (even for unit weights) [24]. In addition, both problems are even hard to approximate [6,24,27]. The problem we are interested in combines the two above problems, that is searching for a colorful subtree of maximum combined weight of the edges and the transitive closure, which is the best explanation of the observed fragments [24]:

Combined Colorful Subtree Problem. Given a vertex-colored DAG $G = (V, E)$ with colors $\mathcal{C}$, edge weights $w : E \to \mathbb{R}$, and transitive weights $w^+ : \mathcal{C}^2 \to \mathbb{R}$. Find the induced colorful subtree T of G of maximum weight $w^*(T) = w(T) + w^+(T)$.

3   Integer Linear Programming for Fragmentation Trees

For the computation of fragmentation trees from tandem MS data, several exact and heuristic algorithms to solve the MAXIMUM COLORFUL SUBTREE problem have been proposed and evaluated [5,19,20], inter alia a fixed-parameter algorithm using dynamic programming (DP) over vertices and color subsets [5,7], and an Integer Linear Program (ILP) [20] (see below) – both computing an exact solution. For multiple MS data, Scheubert *et al.* [24] presented an exact DP algorithm for the COMBINED COLORFUL SUBTREE problem, which is parameterized by the number of colors k in the graph. Here, we present an ILP for solving the COMBINED COLORFUL SUBTREE problem. ILPs are a classical approach for finding exact solutions of computationally hard problems.

3.1   ILP for Tandem MS

We first repeat the ILP introduced by Rauf *et al.* [20] for tandem MS data. By mapping all peaks into a single "pseudo tandem MS" spectrum we can also use this ILP to find a fragmentation tree for multiple MS data. However, by doing so, we ignore the additional information gained from the succession of fragments in multiple MS.

Let $G = (V, E)$ be the input graph, and let $\mathcal{C} : V \to C$ denote the vertex coloring of G. We assume that G has a unique source r that will be the root of the subtree. For each color $c \in C$ let $V(c)$ be the set of all vertices in G which are colored with c. We introduce binary variables x_{uv} for each edge $uv \in E$, where $x_{uv} = 1$ if and only if uv is part of the subtree.

$$\max \sum_{uv \in E} w(u, v) \cdot x_{uv} \tag{2}$$

$$\text{s.t.} \quad \sum_{u \text{ with } uv \in E} x_{uv} \leq 1 \qquad \qquad \text{for all } v \in V \setminus \{r\}, \tag{3}$$

$$x_{vw} \leq \sum_{u \text{ with } uv \in E} x_{uv} \qquad \text{for all } vw \in E \text{ with } v \neq r, \tag{4}$$

$$\sum_{uv \in E \text{ with } v \in V(c)} x_{uv} \leq 1 \qquad \qquad \text{for all } c \in C, \tag{5}$$

$$x_{uv} \in \{0, 1\} \qquad \qquad \text{for all } uv \in E. \tag{6}$$

Constraints (3) ensure that the feasible solution is a tree, whereas constraints (5) make sure that there is at most one vertex of each color present in the solution. Finally, constraints (4) require the solution to be connected. Note that in general graphs, we would have to ensure for every cut of the graph to be connected to some parent vertex. That would require an exponential number of constraints [15]. But since our graph is directed and acyclic, a linear number of constraints suffice. White *et al.* [30] pointed out that constraints (3) are redundant due to constraints (5). However, in the following we will refer to the original ILP from [20].

3.2   ILP for Multiple MS Allowing Transitivity Penalties Only

A rather simple ILP for solving the COMBINED COLORFUL SUBTREE problem extends the ILP from Rauf *et al.* [20] by adding constraints similar to [2] to capture the transitivity of the closure. To this end, we will introduce additional variables that capture the edges of the transitive closure of the tree. Unfortunately, this simple approach is only working for negative weights for all edges of the transitive closure and cannot be generalized to arbitrary transitivity scores.

Let $G^+ = (V, E^+)$ be the transitive closure of the input graph G. We assume that $w^+(c(u), c(v)) \leq 0$ holds for all edges uv of the transitive closure. Let us define binary variables x_{uv} for each edge $uv \in E$, and z_{uv} for each edge $uv \in E^+$. We assume $x_{uv} = 1$ if and only if uv is part of the subtree; and $z_{uv} = 1$ if uv is part of the closure of the subtree. We can formulate the following ILP:

$$\max \sum_{uv \in E} w(u, v) \cdot x_{uv} + \sum_{uv \in E^+} w^+(c(u), c(v)) \cdot z_{uv} \tag{7}$$

satisfying constraints (3), (4), (5) and, in addition:

$$x_{uv} \leq z_{uv} \qquad\qquad \text{for all } uv \in E, \tag{8}$$

$$z_{uv} + z_{vw} - z_{uw} \leq 1 \qquad\qquad \text{for all } uv, vw \in E^+, \tag{9}$$

$$x_{uv} \in \{0, 1\} \qquad\qquad \text{for all } uv \in E, \tag{10}$$

$$z_{uv} \in \{0, 1\} \qquad\qquad \text{for all } uv \in E^+. \tag{11}$$

As $w^+(c(u), c(v)) \leq 0$ for all $uv \in E^+$ we may assume that $z_{uv} = 0$ holds unless required otherwise by (8) or (9). Constraint (8) requires that all edges of the subtree are also edges of the closure; constraint (9) results in the transitivity of the closure.

Unfortunately, the above ILP cannot be generalized to arbitrary transitivity scores, demonstrated by the example that $z_{uv} = 1$ for all $uv \in E^+$ satisfies both constraints (8) and (9), independently of the actual assignment of variables x_{uv}.

3.3   ILP for Multiple MS Using General Transitivity Scores

Here, we present an ILP for solving the COMBINED COLORFUL SUBTREE problem using general transitivity scores. Let $G = (V, E)$ be the input graph, and let $\mathcal{C} : V \to C$ denote the vertex coloring of G. For each color $c \in C$ let $V(c)$ be the set of all vertices in G which are colored with c. Let $H = (U, F)$ be the color version of G with

$$U := \mathcal{C}(V) \quad \text{and} \quad F := \{\mathcal{C}(u)\mathcal{C}(v) : uv \in E\}.$$

We may assume $U = C$, but for the sake of clarity we will use U whenever we refer to the vertices of the color graph H.

Let us define binary variables x_{uv} for each edge $uv \in E$, and z_{ab} and y_{ab} for each edge $ab \in F$. We assume $x_{uv} = 1$ if and only if uv is part of the subtree, and $y_{ab} = 1$ if there exist $u \in V(a)$ and $v \in V(b)$ such that uv is part of the subtree, that is, $x_{uv} = 1$. Variables y_{ab} are merely helper variables that map the subtree to the color space. Finally, we assume $z_{ab} = 1$ if ab is part of the closure of the subtree in color space. The following ILP captures the maximum colorful subtree problem as well as the COLORFUL SUBTREE CLOSURE problem using arbitrary transitivity scores:

$$\max \sum_{uv \in E} w(u,v) \cdot x_{uv} + \sum_{ab \in F} w^+(a,b) \cdot z_{ab} \tag{12}$$

satisfying constraints (3), (4), (5) and, in addition:

$$x_{uv} \leq y_{C(u)C(v)} \qquad\qquad \text{for all } uv \in E, \tag{13}$$

$$y_{ab} \leq \sum_{u \in V(a), v \in V(b)} x_{uv} \qquad\qquad \text{for all } ab \in F, \tag{14}$$

$$y_{ab} \leq z_{ab} \qquad\qquad \text{for all } ab \in F, \tag{15}$$

$$z_{ab} + y_{bc} - 1 \leq z_{ac} \qquad\qquad \text{for all } bc \in F, a \in U, \tag{16}$$

$$z_{ab} - y_{bc} + 1 \geq z_{ac} \qquad\qquad \text{for all } bc \in F, a \in U, \tag{17}$$

$$z_{ac} \leq \sum_{\substack{b \in U \text{ with} \\ bc \in F}} y_{bc} \qquad\qquad \text{for all } ac \in F, \tag{18}$$

$$x_{uv} \in \{0,1\} \qquad\qquad \text{for all } uv \in E, \tag{19}$$

$$y_{ab}, z_{ab} \in \{0,1\} \qquad\qquad \text{for all } ab \in F. \tag{20}$$

Constraints (13) and (14) ensure that there is an edge in the color version of the tree if and only if there is an edge between vertices of the corresponding colors. Constraints (15) guarantee that for each edge that is part of the solution, also its transitive edge is part of the solution. Constraints (16) and (17) ensure the transitivity of the transitive closure of the solution: For a given edge y_{bc} in the color version of the tree and an arbitrary color a, a is either an ancestor of b (and thus also of c), or not. The first case implies that there must be transitive edges from a to b as well as from a to c. In the second case, transitive edges from a to b as well as from a to c are prohibited. Constraints (18) guarantee that only the transitive closure of the solution tree is part of the solution, and not the transitive closure of other subgraphs.

4   Correlation with Structural Similarity

Rasche et al. [18] presented the comparison of fragmentation trees using fragmentation tree alignments. One important application of this approach is searching in a database for molecules that are similar to the measured unknown molecule. Two structurally similar molecules have similar fragmentation trees and vice versa [18]. Hence, the similarity of high quality fragmentation trees correlates with the structural similarity of the corresponding molecules. We will use the correlation coefficient to optimize the parameters of the transitivity score and to evaluate the benefit of MS^n data compared to MS^2 data.

Fragmentation tree similarity is defined via edges, representing fragmentation steps, and vertices, representing fragments. A local fragmentation tree alignment contains those parts of the two trees where similar fragmentation cascades occurred [18]. To compute fragmentation tree alignments we use the sparse DP introduced by Hufsky *et al.* [12] which is very fast in practice.

For the comparison of molecular structures, many different similarity scores have been developed [13]. Molecular structures can be represented as binary fingerprints. Here, we use two of those fingerprint representations, that is the fingerprints from PubChem database [29] accessed via the Chemistry Development Toolkit version 1.3.37 [28][1], and Molecular ACCess System (MACCS) fingerprints implemented in OpenBabel[2]. We use Tanimoto similarity scores (Jaccard indices) [21] to compare those binary vectors.

To assess the correlation between fragmentation tree similarity and structural similarity, we use the well-known Pearson correlation coefficient r which measures the linear dependence of two variables, as well as the Spearman's rank correlation coefficient ρ that is the Pearson correlation coefficient between the ranked variables. The coefficient of determination, r^2, measures how well a model explains and predicts future outcomes. Fragmentation tree alignment scores and structural similarity scores are two measures where one would not expect a linear dependence. This being said, we argue that any Pearson correlation coefficients $r > 0.5$ ($r^2 > 0.25$) can be regarded as strong correlation.

5 Results

To evaluate our work, we analyze spectra from a dataset introduced in [24]. It contains 185 mass spectra of 45 molecules, mainly representing plant secondary metabolites. All spectra were measured on a Thermo Scientific Orbitrap XL instrument using direct infusion. For more details of the dataset see [24].

For the construction of the fragmentation graph, we use a relative mass error of 20 ppm and the standard alphabet – that is carbon, hydrogen, nitrogen, oxygen, phosphorus, and sulfur – to compute the fragment molecular formulas. For weighting the fragmentation graph, we use the scoring parameters from [19]. For scoring the transitive closure, we evaluate the influence of parameters σ_1, σ_2 and σ_3 on the quality of fragmentation trees. We assume the molecular formula of the unfragmented molecule to be given (for details, see [18, 19, 24]).

For the computation of fragmentation trees from tandem MS data, we use the DP algorithm from [5] (called DP-MS^2 in the following) and the ILP from [20] (ILP-MS^2). Recall, that we can convert MS^n data to "pseudo MS^2" data by mapping all peaks into a single spectrum and ignoring the additional information gained from the succession of fragments in MS^n. For the computation of fragmentation trees from multiple MS data, we use the DP algorithm from [24] (DP-MS^n) as well as our novel ILP (ILP-MS^n). Both DP algorithms are restricted by memory and time consumptions. Thus, exact calculations are limited to the k' most intense peaks. The remaining peaks are added in descending intensity order by a greedy heuristic (see the tree completion heuristic from [20, 24]).

---

[1] https://sourceforge.net/projects/cdk/
[2] http://openbabel.sourceforge.net/

For solving the ILPs we use Gurobi 5.6[3]. The experiments were run on a cluster with four nodes each containing 2x Intel XEON 6 Cores E5645 at 2.40 GHz with 48 GB RAM. Each instance is started on a single core.

For computing fragmentation tree alignments, we use the sparse DP from [12] and the scoring from [18]. Estimation of Pearson and Spearman correlation coefficients was done using the programming language R.

Running Time Comparison. For the evaluation of running times depending on the number of peaks in the spectrum, we calculate the exact solution (using all four algorithms) for the k' most intense peaks for each molecule. Afterwards, remaining peaks are added heuristically. For each k', we exclude instances with less than k' peaks in the spectrum. For very small instances, the DP algorithms are slightly faster than the ILPs (see Fig. 2 (left)). On moderate large instances (e.g. $k' = 17$), the ILPs clearly outperform the DP algorithms. For $k' > 20$ it is not possible to calculate fragmentation trees with the DP due to memory and time constraints. On huge instances ($k' > 30$) the ILP-MS^n is slower than the ILP-MS^2.

To get an overview of differences in the running times between hard and easy fragmentation tree computations for tandem MS and multiple MS data, we sort the instances by their running times in increasing order. This is done separately for the ILP-MS^2 and the ILP-MS^n algorithm (see Fig. 2 (right)). We find that solving the COMBINED COLORFUL SUBTREE problem using the ILP-MS^n is still very fast on most instances. Further, we find that for the ILP-MS^n, there is one molecule for which the calculation of the fragmentation tree takes nearly as much time as for the remaining 39 molecules together.

Parameter Estimation. In [24] the estimation of parameters was based on the assumption that fragmentation trees change when using the additional scoring of the transitive closure. Here, we want to optimize the scoring of the transitive closure by maximizing the correlation of fragmentation tree alignment scores and the structural similarity scores of the corresponding molecules. For three of the 45 molecules, it was not possible to calculate fragmentation tree alignments due to memory and time constraints. Those compounds were excluded from the analysis.

For estimating the optimal scoring parameters σ_1, σ_2 and σ_3 of the transitive closure, we compute exact fragmentation trees using the $k' = 20$ most intense peaks and attach the remaining peaks by the tree completion heuristic. For scoring the transitive closure of the fragmentation graph, we separately vary $0 \leq \sigma_1 \leq 6$, $-3 \leq \sigma_2 \leq 0$ and $-3 \leq \sigma_3 \leq 0$. We compute fragmentation tree alignments and analyze the resulting PubChem/Tanimmoto as well as MACCS/Tanimoto Pearson correlation coefficients (see Fig. 3). Increasing σ_1 the correlation coefficient increases and converges at approximately $\sigma_1 = 3$. For σ_2 and σ_3 the highest correlation is reached around -0.5. For the further evaluation, we set $\sigma_1 = 3$, $\sigma_2 = -0.5$ and $\sigma_3 = -0.5$. We find, that this result agrees

---

[3] Gurobi Optimizer 5.6. Houston, Texas: Gurobi Optimization, Inc.

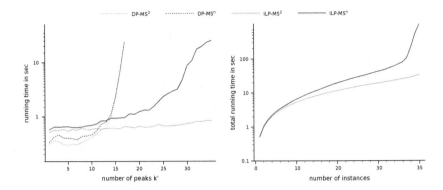

Fig. 2. Running times for calculating fragmentation trees. Times are averaged on 10 repetitive evaluations and given in seconds. Note the logarithmic y-axis. Left: Average running times for calculating one fragmentation tree with exact solution for the k' most intense peaks. The remaining peaks are attached by tree completion heuristic. Right: Total running times for instances of size $k' = 35$. Again, the remaining peaks are attached heuristically. We calculate the total running time of the x instances for which the tree was computed faster than for any of the remaining instances. For each algorithm, instances were sorted separately.

with the original scoring parameters from [24]. Although they were chosen ad hoc, they seem to work very well in practice. We further find, that σ_1 has a larger effect on the correlation than σ_2 and σ_3 (see Fig. 3). This was expected, as the requirement that a fragments is placed below its parent fragment is very strong.

Further, we evaluate the effect of using more peaks for the exact fragmentation tree computation on the correlation. We set $\sigma_1 = 3$, $\sigma_2 = -0.5$ and $\sigma_3 = -0.5$, and vary the number of peaks from $10 \leq k' \leq 35$. We find that the highest PubChem/Tanimoto correlation coefficient $r = 0.5643137$ ($r^2 = 0.31844500$) is achieved for $k' = 21$ (see Fig. 4).

Note that the DP-MS^n is not able to solve problems of size $k' = 21$ with acceptable running time and memory consumption. Thus, only by help of the ILP-MS^n it is possible to compute trees with best quality.

The optimum of k' remains relatively stable in a *leave-one-out* validation experiment: For each compound, we delete the corresponding fragmentation tree from the dataset and repeat the former analysis to determine the best k'. For 30 of the 42 sub-datasets $k' = 21$ achieves the best correlation. For the remaining 11 sub-datasets $k' = 14$, $k' = 20$ or $k' = 25$ are optimal.

Due to the small size of the dataset, it is hard to determine best parameters without overfitting. Hence, these analyzes should not be seen as perfect parameter estimation, but more as a rough estimation until a bigger dataset becomes available.

Comparison between Trees from MS^2, Pseudo-MS^2 and MS^n Data. To evaluate the benefit of scoring the additional information from MS^n data, we compare the

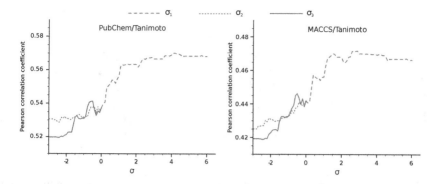

Fig. 3. Pearson correlation coefficients of PubChem/Tanimoto (left) and MACCS/Tanimoto (right) scores with fragmentation tree alignment scores, separately varying the scoring parameters σ_1, σ_2 and σ_3 of the transitive closure for fragmentation tree computation. When varying σ_1, we set $\sigma_2 = 0$ and $\sigma_3 = 0$ and vice versa

correlation coefficients of using only the MS^2 spectra, using Pseudo-MS^2 data, and using MS^n data. As mentioned above, Pseudo-MS^2 data means mapping all peaks into a single spectrum and ignoring the additional information gained from the succession of fragments in MS^n, that is not scoring the transitive closure. For fragmentation tree computation from MS^2 and Pseudo-MS^2 data we use the ILP-MS^2, for MS^n data we use the ILP-MS^n. For a fair evaluation, we again vary the number of peaks from $10 \leq k' \leq 35$ to choose the k' with the highest correlation coefficient. The highest Pearson correlation coefficient with PubChem/Tanimoto fingerprints for MS^2 data is $r = 0.3860874$ ($r^2 = 0.1490635$) with $k' = 21$ and for Pseudo-MS^2 data $r = 0.5477199$ ($r^2 = 0.2999970$) with $k' = 25$ (see Fig. 4).

Further, we compare the Pearson correlation coefficients between the three datasets MS^2, Pseudo-MS^2 and MS^n (see Table 1). We find that the benefit of MS^n data is huge in comparison to using only MS^2 data, which is expected since the MS^2 spectra contain too few peaks. The question that is more intriguing is whether scoring the transitive closure improves correlation results. Comparing Pseudo-MS^2 with MS^n data, we get an increase in the coefficient of determination r^2 by up to 6.7 % for PubChem fingerprints and 6.3 % for MACCS fingerprints. The results for Spearman correlation coefficients look similar. When restricting the evaluation to large trees (at least three edges, five edges, seven edges), we cannot observe an increase in correlation.

When fragmentation trees are used in database search the relevant accuracy measure is not Pearson correlation, but identification accuracy. The dataset used in this paper is small and there is only one measurement per compound. Thus we cannot evaluate the identification accuracy. Instead we analyze the Tanimoto scores $T(h)$ of the first h hits with h ranging from one to the number of compounds (see Fig. 5). We exclude the identical compound from the hitlist and then

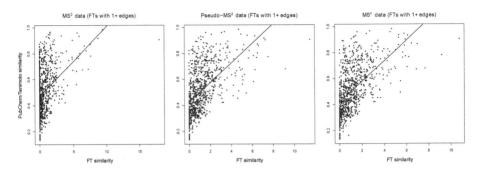

Fig. 4. Correlation and regression line for the complete datasets. Fragmentation tree similarity (x-axis) plotted against structural similarity measured by Pub-Chem/Tanimoto score (y-axis). (a) Fragmentation trees for MS^2 data ($k' = 21$). Pearson correlation is $r = 0.386$. Spearman correlation is $\rho = 0.364$ (b) Fragmentation trees for Pseudo-MS^2 data ($k' = 25$). Pearson correlation is $r = 0.548$. Spearman correlation is $\rho = 0.615$ (c) Fragmentation trees for MS^n data ($k' = 21$). Pearson correlation is $r = 0.564$. Spearman correlation is $\rho = 0.624$.

Table 1. Pearson correlation r and coefficient of determination r^2 (in brackets) of structural similarity (PubChem/Tanimoto and MACCS/Tanimoto) with fragmentation tree similarity, for all three datasets and different minimum tree sizes (at least one edge, three edges, five edges, seven edges). We report the number of alignments (molecule pairs) N for each set. The subsets with different minimum tree sizes are determined by the tree sizes of the MS^n trees (that is, the MS^2 and Pseudo-MS^2 subsets contain the same molecules).

| fingerprint | dataset | only molecules with at least | | | |
		1 edge	3 edges	5 edges	7 edges
PubChem	MS^2	0.386 (0.149)	0.386 (0.149)	0.374 (0.140)	0.384 (0.147)
	Pseudo-MS^2	0.548 (0.300)	0.549 (0.301)	0.530 (0.281)	0.549 (0.301)
	MS^n	0.564 (0.318)	0.567 (0.321)	0.547 (0.299)	0.565 (0.319)
MACCS	MS^2	0.379 (0.143)	0.371 (0.138)	0.371 (0.138)	0.373 (0.139)
	Pseudo-MS^2	0.453 (0.206)	0.445 (0.198)	0.438 (0.192)	0.439 (0.193)
	MS^n	0.466 (0.217)	0.456 (0.210)	0.449 (0.202)	0.449 (0.201)
	no. molecule pairs N	861	820	630	561

average over the hitlists of all compounds in the dataset. We compare the results from MS^2, Pseudo-MS^2 and MS^n data with pseudo hitlists containing randomly ordered compounds (minimum value, RANDOM) and compounds arranged in descending order in accordance with the Tanimoto scores (upper limit, BEST). There is a significant increase of average Tanimoto scores from MS^2 data to MS^n data, and a slight increase from Pseudo-MS^2 data to MS^n data especially for the first $h = 5$ compounds.

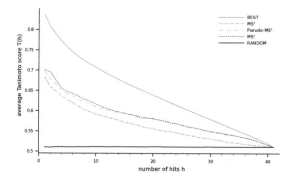

Fig. 5. Average Tanimoto scores $T(h)$ between query structures and the first h structures from hitlists obtained by FT alignments (MS^2, Pseudo-MS^2, MS^n data), pseudo hitlists containing the structures with maximum Tanimoto score to query structure (BEST) and randomly selected pseudo hitlists (RANDOM)

6   Conclusion

In this work, we have presented an Integer Linear Program for the COMBINED COLORFUL SUBTREE problem, that outperforms the Dynamic Programming algorithm that has been presented before [24]. Solving this problem is relevant for calculating fragmentation trees from multistage mass spectrometry data.

Quality of fragmentation trees is measured by correlation of tree alignment scores with structural similarity scores of the corresponding compounds. Experiments on a dataset with 45 compounds revealed that trees computed with transitivity scores $\sigma_1 = 3$, $\sigma_2 = -0.5$ and $\sigma_3 = -0.5$ achieve the best quality. The highest correlation of $r = 0.564$ was achieved when computing exact fragmentation trees for the $k' = 21$ most intense peaks and attaching the remaining peaks heuristically. Using the additional information provided by multiple MS data, the coefficient of determination r^2 increases by up to 6.7 % compared to trees computed without transitivity scores. Thus, we could show for the first time that additional information from MS^n data can improve the quality of fragmentation trees.

For the computation of those trees with highest quality ($k' = 21$), our ILP needs 1.3 s on average. In contrast, the original DP is not able to solve those instances with acceptable running time and memory consumption. The ILP for MS^n is, however, slower than the ILP for MS^2 that has been presented before [20]. This is due to the number of constraints which increases by an order of magnitude from MS^2 to MS^n. White *et al.* [30] suggested rules to speed up computations for the ILP on MS^2 data. These rules may also improve the running time of our algorithm.

Acknowledgements. We thank Aleš Svatoš and Ravi Kumar Maddula from the Max Planck Institute for Chemical Ecology in Jena, Germany for supplying us with the test data. We thank Kai Dührkop for helpful discussions on the ILP. F. Hufsky was funded by Deutsche Forschungsgemeinschaft, project "IDUN".

References

1. Allen, F., Wilson, M., Pon, A., Greiner, R., Wishart, D.: CFM-ID: a web server for annotation, spectrum prediction and metabolite identification from tandem mass spectra. Nucleic. Acids Res. (2014)
2. Böcker, S., Briesemeister, S., Klau, G.W.: On optimal comparability editing with applications to molecular diagnostics. BMC Bioinformatics 10(suppl. 1), S61 (2009); Proc. of Asia-Pacific Bioinformatics Conference (APBC 2009)
3. Böcker, S., Lipták, Z.: Efficient mass decomposition. In: Proc. of ACM Symposium on Applied Computing (ACM SAC 2005), pp. 151–157. ACM Press, New York (2005)
4. Böcker, S., Lipták, Z.: A fast and simple algorithm for the Money Changing Problem. Algorithmica 48(4), 413–432 (2007)
5. Böcker, S., Rasche, F.: Towards de novo identification of metabolites by analyzing tandem mass spectra. Bioinformatics 24, I49–I55 (2008); Proc. of European Conference on Computational Biology (ECCB 2008)
6. Dondi, R., Fertin, G., Vialette, S.: Complexity issues in vertex-colored graph pattern matching. J. Discrete Algorithms 9(1), 82–99 (2011)
7. Dreyfus, S.E., Wagner, R.A.: The Steiner problem in graphs. Networks 1(3), 195–207 (1972)
8. Fellows, M.R., Gramm, J., Niedermeier, R.: On the parameterized intractability of motif search problems. Combinatorica 26(2), 141–167 (2006)
9. Gerlich, M., Neumann, S.: MetFusion: integration of compound identification strategies. J. Mass Spectrom 48(3), 291–298 (2013)
10. Heinonen, M., Shen, H., Zamboni, N., Rousu, J.: Metabolite identification and molecular fingerprint prediction via machine learning. Bioinformatics 28(18), 2333–2341 (2012); Proc. of European Conference on Computational Biology (ECCB 2012)
11. Hill, D.W., Kertesz, T.M., Fontaine, D., Friedman, R., Grant, D.F.: Mass spectral metabonomics beyond elemental formula: Chemical database querying by matching experimental with computational fragmentation spectra. Anal. Chem. 80(14), 5574–5582 (2008)
12. Hufsky, F., Dührkop, K., Rasche, F., Chimani, M., Böcker, S.: Fast alignment of fragmentation trees. Bioinformatics 28, i265–i273 (2012); Proc. of Intelligent Systems for Molecular Biology (ISMB 2012)
13. Leach, A.R., Gillet, V.J.: An Introduction to Chemoinformatics. Springer, Berlin (2005)
14. Li, J.W.-H., Vederas, J.C.: Drug discovery and natural products: End of an era or an endless frontier? Science 325(5937), 161–165 (2009)
15. Ljubić, I., Weiskircher, R., Pferschy, U., Klau, G.W., Mutzel, P., Fischetti, M.: Solving the prize-collecting Steiner tree problem to optimality. In: Proc. of Algorithm Engineering and Experiments (ALENEX 2005), pp. 68–76. SIAM (2005)
16. Oberacher, H., Pavlic, M., Libiseller, K., Schubert, B., Sulyok, M., Schuhmacher, R., Csaszar, E., Köfeler, H.C.: On the inter-instrument and inter-laboratory transferability of a tandem mass spectral reference library: 1. Results of an Austrian multicenter study. J. Mass Spectrom. 44(4), 485–493 (2009)

17. Patti, G.J., Yanes, O., Siuzdak, G.: Metabolomics: The apogee of the omics trilogy. Nat. Rev. Mol. Cell Biol. 13(4), 263–269 (2012)
18. Rasche, F., Scheubert, K., Hufsky, F., Zichner, T., Kai, M., Svatoš, A., Böcker, S.: Identifying the unknowns by aligning fragmentation trees. Anal. Chem. 84(7), 3417–3426 (2012)
19. Rasche, F., Svatoš, A., Maddula, R.K., Böttcher, C., Böcker, S.: Computing fragmentation trees from tandem mass spectrometry data. Anal. Chem. 83(4), 1243–1251 (2011)
20. Rauf, I., Rasche, F., Nicolas, F., Böcker, S.: Finding maximum colorful subtrees in practice. In: Chor, B. (ed.) RECOMB 2012. LNCS, vol. 7262, pp. 213–223. Springer, Heidelberg (2012)
21. Rogers, D.J., Tanimoto, T.T.: A computer program for classifying plants. Science 132(3434), 1115–1118 (1960)
22. Rojas-Chertó, M., Kasper, P.T., Willighagen, E.L., Vreeken, R.J., Hankemeier, T., Reijmers, T.H.: Elemental composition determination based on MS^n. Bioinformatics 27, 2376–2383 (2011)
23. Scheubert, K., Hufsky, F., Rasche, F., Böcker, S.: Computing fragmentation trees from metabolite multiple mass spectrometry data. In: Bafna, V., Sahinalp, S.C. (eds.) RECOMB 2011. LNCS, vol. 6577, pp. 377–391. Springer, Heidelberg (2011)
24. Scheubert, K., Hufsky, F., Rasche, F., Böcker, S.: Computing fragmentation trees from metabolite multiple mass spectrometry data. J. Comput. Biol. 18(11), 1383–1397 (2011)
25. Sheldon, M.T., Mistrik, R., Croley, T.R.: Determination of ion structures in structurally related compounds using precursor ion fingerprinting. J. Am. Soc. Mass. Spectrom 20(3), 370–376 (2009)
26. Shen, H., Dührkop, K., Böcker, S., Rousu, J.: Metabolite identification through multiple kernel learning on fragmentation trees. Bioinformatics (2014) Accepted Proc. of Intelligent Systems for Molecular Biology (ISMB 2014)
27. Sikora, F.: Aspects algorithmiques de la comparaison d'éléments biologiques. PhD thesis, Université Paris-Est (2011)
28. Steinbeck, C., Hoppe, C., Kuhn, S., Floris, M., Guha, R., Willighagen, E.L.: Recent developments of the Chemistry Development Kit (CDK) - an open-source Java library for chemo- and bioinformatics. Curr. Pharm. Des. 12(17), 2111–2120 (2006)
29. Wang, Y., Xiao, J., Suzek, T.O., Zhang, J., Wang, J., Bryant, S.H.: PubChem: A public information system for analyzing bioactivities of small molecules. Nucleic Acids Res. 37(Web Server issue), W623–W633 (2009)
30. White, W.T.J., Beyer, S., Dührkop, K., Chimani, M., Böcker, S.: Speedy colorful subtrees. Submitted to European Conference on Computational Biology, ECCB 2014 (2014)
31. Wolf, S., Schmidt, S., Müller-Hannemann, M., Neumann, S.: In silico fragmentation for computer assisted identification of metabolite mass spectra. BMC Bioinformatics 11, 148 (2010)

An Online Peak Extraction Algorithm for Ion Mobility Spectrometry Data

Dominik Kopczynski[1,2] and Sven Rahmann[1,2,3]

[1] Collaborative Research Center SFB 876, TU Dortmund, Germany
[2] Bioinformatics, Computer Science XI, TU Dortmund, Germany
[3] Genome Informatics, Institute of Human Genetics, Faculty of Medicine,
University Hospital Essen, University of Duisburg-Essen, Germany

Abstract. Ion mobility (IM) spectrometry (IMS), coupled with multi-capillary columns (MCCs), has been gaining importance for biotechnological and medical applications because of its ability to measure volatile organic compounds (VOC) at extremely low concentrations in the air or exhaled breath at ambient pressure and temperature. Ongoing miniaturization of the devices creates the need for reliable data analysis on-the-fly in small embedded low-power devices. We present the first fully automated online peak extraction method for MCC/IMS spectra. Each individual spectrum is processed as it arrives, removing the need to store a whole measurement of several thousand spectra before starting the analysis, as is currently the state of the art. Thus the analysis device can be an inexpensive low-power system such as the Raspberry Pi.

The key idea is to extract one-dimensional peak models (with four parameters) from each spectrum and then merge these into peak chains and finally two-dimensional peak models. We describe the different algorithmic steps in detail and evaluate the online method against state-of-the-art peak extraction methods using a whole measurement.

1 Introduction

Ion mobility (IM) spectrometry (IMS), coupled with multi-capillary columns (MCCs), MCC/IMS for short, has been gaining importance for biotechnological and medical applications. With MCC/IMS, one can measure the presence and concentration of volatile organic compounds (VOCs) in the air or exhaled breath with high sensitivity; and in contrast to other technologies, such as mass spectrometry coupled with gas chromatography (GC/MS), MCC/IMS works at ambient pressure and temperature. Several diseases like chronic obstructive pulmonary disease (COPD) [1], sarcoidosis [4] or lung cancer [20] can potentially be diagnosed early. IMS is also used for the detection of drugs [11] and explosives [9]. Constant monitoring of VOC levels is of interest in biotechnology, e.g., for watching fermenters with yeast producing desired compounds [12] and in medicine, e.g., monitoring propofol levels in the exhaled breath of patients during surgery [15].

IMS technology is moving towards miniaturization and small mobile devices. This creates new challenges for data analysis: The analysis should be possible

D. Brown and B. Morgenstern (Eds.): WABI 2014, LNBI 8701, pp. 232–246, 2014.
© Springer-Verlag Berlin Heidelberg 2014

within the measuring device without requiring additional hardware like an external laptop or a compute server. Ideally, the spectra can be processed on a small embedded chip or small device like a Raspberry Pi or similar hardware with restricted resources. Algorithms in small mobile hardware face constraints, such as the need to use little energy (hence little random access memory), while maintaining prescribed time constraints.

The basis of each MCC/IMS analysis is *peak extraction*, by which we mean a representation of all high-intensity regions (peaks) in the measurement by using a few descriptive parameters per peak instead of the full measurement data. State-of-the-art software (like IPHEx [3], Visual Now [2], PEAX [5]) only extracts peaks when the whole measurement is available, which may take up to 10 minutes because of the pre-separation of the analytes in the MCC. Our own PEAX software in fact defines modular pipelines for fully automatic peak extraction and compares favorably with a human domain expert doing the same work manually when presented with a whole MCC/IMS measurement. However, storing the whole measurement is not desirable or possible when the memory and CPU power is restricted. Here we introduce a method to extract peaks and estimate a parametric representation while the measurement is being captured. This is called *online peak extraction*, and this article presents the first algorithm for this purpose on MCC/IMS data.

Section 2 contains background on the data produced in an MCC/IMS experiment, on peak modeling and on optimization methods used in this work. A detailed description of the novel online peak extraction method is provided in Section 3. An evaluation of our approach is presented in Section 4, while Section 5 contains a concluding discussion.

2 Background

We primarily focus on the data generated by an MCC/IMS experiment (Section 2.1) and related peak models. Ion mobility spectrometers and their functions are well documented [7], and we do not go into technical details. In Section 2.2 we describe a previously used parametric peak model, and in Section 2.3 we review two optimization methods that are being used as subroutines in this work.

2.1 Data from MCC/IMS Measurements

In an MCC/IMS experiment, a mixture of several unknown volatile organic compounds (VOCs) is separated in two dimensions: first by retention time r in the MCC (the time required for a particular compound to pass through the MCC) and second by drift time d through the IM spectrometer. Instead of the drift time itself, a quantity normalized for pressure and temperature called the *inverse reduced mobility* (IRM) t is used to compare spectra taken under different or changing conditions. Thus we obtain a time series of IM spectra (one each 100 ms at each retention time point), and each spectrum is a vector of ion concentrations (measured by voltage change on a Faraday plate) at each IRM.

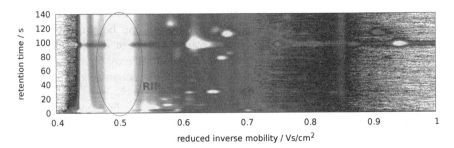

Fig. 1. Visualization of a raw measurement (IMSC) as a heat map; X-axis: inverse reduced mobility $1/K_0$ in Vs/cm^2; y-axis: retention time r in seconds; signal: white (lowest) < blue < purple < red < yellow (highest). The constantly present reactant ion peak (RIP) with mode at 0.48 Vs/cm^2 and exemplarily one VOC peak are annotated.

Let R be the set of (equidistant) retention time points and let T be the set of (equidistant) IRMs where a measurement is made. If D is the corresponding set of drift times (each $1/250000$ second for 50 ms, that is $12\,500$ time points), there exists a constant f_{ims} depending on external conditions such that $T = f_{ims} \cdot D$ [7]. Then the data is an $|R| \times |T|$ matrix $S = (S_{r,t})$ of measured ion intensities, which we call an *IM spectrum-chromatogram* (IMSC). The matrix can be visualized as a heat map (Figure 1). A row of S is a *spectrum*, while a column of S is a *chromatogram*.

Areas of high intensity in S are called peaks, and our goal is to discover these peaks. Comparing peak coordinates with reference databases may reveal the identity of the corresponding compound. A peak caused by a VOC occurs over several IM spectra. We have to mention some properties of MCC/IMS data that complicate the analysis.

– An IM spectrometer uses a carrier gas, which is also ionized. The ions are present in every spectrum, which is referred to as the reactant ion peak (RIP). In the whole IMSC it is present as high-intensity chromatogram at an IRM between 0.47 and 0.53 Vs/cm^2. When the MCC/IMS is in idle mode, no analytes are injected into the IMS, and the spectra contain only the RIP. These spectra are referred to as *RIP-only spectra*.
– Every spectrum contains a tailing of the RIP, meaning that it decreases slower than it increases; see Figure 2. To extract peaks, the effect of both RIP and its tailing must be estimated and removed.
– At higher concentrations, compounds can form dimer ions, and one may observe both the monomer and dimer peak from one compound. This means that there is not necessarily a one-to-one correspondence between peaks and compounds, and our work focuses on peak detection, not compound identification.
– An IM spectrometer may operate in positive or negative mode, depending on which type of ions (positive or negative) one wants to detect. In either case, signals are reported in positive units. All experiments described here were done in positive mode.

2.2   Peak Modeling

Peaks can be described by parametrized distribution functions, such as shifted Inverse Gaussian distributions g in both retention time and IRM dimension [13] with

$$g(x; \mu, \lambda, o) := [x > o] \cdot \sqrt{\frac{\lambda}{2\pi(x-o)^3}} \cdot \exp\left(-\frac{\lambda((x-o)-\mu)^2}{2\mu^2(x-o)}\right), \quad (1)$$

where o is the shift, $\mu + o$ is the mean and λ is a shape parameter. A peak is then given as the product of two shifted Inverse Gaussians, scaled by a volume factor v, i.e., by seven parameters, namely $P(r,t) := v \cdot g(r, \mu_r, \lambda_r, o_r) \cdot g(t, \mu_t, \lambda_t, o_t)$ for all $r \in R, t \in T$.

Since the parameters μ, λ, o of a shifted Inverse Gaussian may be very different, although the resulting distributions have a similar shape, it is more intuitive to describe the shifted Inverse Gaussian in terms of three different descriptors, the mean μ', the standard deviation σ and the mode m. There is a bijection between (μ, λ, o) and (μ', σ, m), described in [13] and shown in Appendix A.

2.3   Optimization Methods

The online peak extraction algorithm makes use of non-linear least squares (NLLS) parameter estimation and of the EM algorithm, summarized here.

Non-linear Least Squares. The non-linear least squares (NLLS) method is an iterative method to estimate parameters $\theta = (\theta_1, \dots, \theta_q)$ of a supposed parametric function f, given n observed data points $(x_1, y_1), \dots, (x_n, y_n)$ with $y_i = f(x_i; \theta)$. The idea is to minimize the quadratic error $\sum_{i=1}^n e_i(\theta)$ between the function and the observed data with $e_i(\theta) := r_i(\theta)^2$ and $r_i(\theta) := y_i - f(x_i; \theta)$ is the *residual*. The necessary optimality condition is $\sum_i r_i(\theta) \cdot \partial r_i(\theta)/\partial \theta_j = 0$ for all j. If f is linear in θ (e.g., a polynomial in x with θ being the polynomial coefficients, a setting called polynomial regression), then the optimality condition results in a linear system, which can be solved in closed form. However, often f is not linear in θ and we obtain a non-linear system, which is solved iteratively, given initial parameter values, by linearizing it in each iteration. Details and different algorithms for NLLS can be found in [18, Chapter 10].

The EM Algorithm for Mixtures with Heterogeneous Components. The idea of the expectation maximization (EM) algorithm [6] is that the observed data $x = (x_1, \dots, x_n)$ is viewed as a *sample* from a *mixture* of probability distributions, where the mixture density is specified by $f(x_i \mid \omega, \theta) = \sum_{c=1}^C \omega_c \, f_c(x_i \mid \theta_c)$. Here c indexes the C different component distributions f_c, where θ_c denotes the parameters of f_c, and $\theta = (\theta_1, \dots, \theta_C)$ is the collection of all parameters. The mixture coefficients satisfy $\omega_c \geq 0$ for all c, and $\sum_c \omega_c = 1$. Unlike in most applications, where all component distributions f_c are multivariate

Gaussians, here the f_c are of different types (e.g., uniform and Inverse Gaussian). The goal is to determine the parameters ω and θ such that the probability of the observed sample is maximal (maximum likelihood paradigm). Since the resulting optimization problem is non-convex in (ω, θ), the EM algorithm is an iterative method that converges towards a local optimum that depends on the initial parameter values. The EM algorithm consists of two repeated steps: The E-step (expectation) estimates the expected membership of each data point in each component and then the component weights ω, given the current model parameters θ. The M-step (maximization) estimates maximum likelihood parameters θ_c for each parametric component f_c individually, using the expected memberships as hidden variables that decouple the model.

E-Step. To estimate the expected membership $W_{i,c}$ of data point x_i in each component c, the component's relative probability at that data point is computed, such that $\sum_c W_{i,c} = 1$ for all i. Then the new component weight estimates ω_c^+ are the averages of $W_{i,c}$ across all n data points.

$$W_{i,c} = \frac{\omega_c f_c(x_i \mid \theta_c)}{\sum_k \omega_k f_k(x_i \mid \theta_k)}, \qquad \omega_c^+ = \frac{1}{n} \sum_{i=1}^{n} W_{i,c}, \tag{2}$$

Convergence. After each M-step of an EM cycle, we compare $\theta_{c,q}$ (old parameter value) and $\theta_{c,q}^+$ (updated parameter value), where q indexes the elements of θ_c, the parameters of component c. We say that the algorithm has converged when the relative change $\kappa_{c,q} := |\theta_{c,q}^+ - \theta_{c,q}| / \max\left(|\theta_{c,q}^+|, |\theta_{c,q}|\right)$ drops below a given threshold ε for all c, q. (If $\theta_{c,q}^+ = \theta_{c,q} = 0$, we set $\kappa_{c,q} := 0$.)

3   An Algorithm for Online Peak Extraction

The basic idea of our algorithm is to process each IM spectrum as soon as it arrives (and before the next one arrives) and convert it into a mixture of parametric one-dimensional peak models, described in Section 3.1. The resulting models of consecutive spectra then have to be aligned to each other to track peaks through time; this step is explained in Section 3.2. Finally, we discuss how the aligned models are merged into two-dimensional peak models (Section 3.3).

3.1   Single Spectrum Processing

The idea of processing a single IM spectrum S is to deconvolute it into its single components. Several components appear in each spectrum besides the peaks, namely the previously described RIP, the tailing described in Section 2.1 and background noise. We first erase background noise, then determine and remove the tailing function and finally determine the peak parameters.

Erasing Background Noise. Background noise intensities are assumed to follow a Gaussian distribution at small intensity values. We can determine its approximate mean μ_R and standard deviation σ_R by considering the ranges between 0 and 0.2175 Vs/cm^2 and between 1.2345 and 1.45 (end of a spectrum) Vs/cm^2 from a RIP-only spectrum (cf. Section 2.1) since these ranges typically contain only background noise. We now apply an idea that we previously used on the whole IMSC [14] to each single spectrum $S = (S_t)$: We use the EM algorithm to determine (a) the mean μ_n and standard deviation σ_n of the background noise, and (b) for each t the fraction of the signal intensity S_t that belongs to noise. We then correct for both quantities.

The EM algorithm receives a smoothed spectrum $A = (A_t)$ averaged with window radius ρ as input, $A_t := \frac{1}{2\rho+1} \cdot \sum_{t'=t-\rho}^{t+\rho} S_{t'}$ and deconvolves it into three components: noise (n), data (d), and background (b) that can be described by neither noise nor data. The noise intensities follow a Gaussian distribution with parameters (μ_n, σ_n), whereas the data intensities follow an Inverse Gaussian model with parameters (μ_d, λ_d). The background is uniformly distributed. The initial parameters for the noise are estimated from all data points whose intensity lies within $[\mu_R - 3\sigma_R, \mu_R + 3\sigma_R]$. The initial parameters for the data component are analogously estimated from all points whose intensity exceeds $\mu_R + 3\sigma_R$. The weights ω_n and ω_d are set according to the number of data points inside and outside the interval, respectively, times 0.999, and the background model initially has $\omega_b = 0.001$, hence the sum over all three weights equals 1. The choice of the precise background weight is not critical, since the EM algorithm re-estimates its optimal value, but the initial conditions should suggest a low value, so the background explains only those of the 12 500 points which cannot be explained by the two other components.

The EM algorithm alternates between E-step (Eq. (2)) and M-step, where the new parameters for the non-uniform component are estimated by the maximum likelihood estimators

$$\mu_c^+ = \frac{\sum_t W_{t,c} \cdot A_t}{\sum_t W_{t,c}}, \quad \sigma_n^+ = \sqrt{\frac{\sum_t W_{t,n} \cdot (A_t - \mu_n^+)^2}{\sum_t W_{t,n}}}, \quad \lambda_d^+ = \frac{\sum_t W_{t,d}}{\sum_t W_{t,d}\alpha_t}$$

for $c \in \{n, d\}$, where $\alpha_t = 1/A_t - 1/\mu_d^+$. These computations are repeated until convergence as defined in Section 2.3.

The denoised signal vector S^+ is computed as $S_t^+ := \max(0, (S_t - \mu_n) \cdot (1 - W_{t,n}))$ for all $t \leq |T|$, thus setting mean noise level to zero and erasing data points that belong exclusively to noise.

Determining the Tailing Function. The tailing function appears as a baseline in every spectrum (see Figure 2 for an example). Its shape and scale changes from spectrum to spectrum; so it has to be determined in each spectrum and subtracted in order to extract peaks from the remaining signal in the next step. Empirically, we observe that the tailing function $f(t)$ can be described by a scaled shifted Inverse Gaussian, $f(t) = v \cdot g(t; \mu, \lambda, o)$ with g given by (1). The

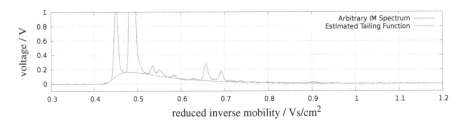

Fig. 2. A spectrum and its estimated tailing function

goal is to determine the parameters $\theta = (v, \mu, \lambda, o)$ such that $f_\theta(t)$ under-fits the given data $S = (S_t)$, as shown in Figure 2.

Let $r_\theta(t) := S(t) - f_\theta(t)$ be the residual function for a given choice θ of parameters. As we want to penalize $r(t) < 0$ but not (severely) $r(t) > 0$, maximum likelihood estimation of the parameters is not appropriate. Instead, we use a modified version of non-linear least squares (NLLS) estimation. In the standard NLLS method, the error function to be minimized is $e(\theta) = \sum_t e_t(\theta)$, where $e_t(\theta) := r_\theta(t)^2$. Instead, we use

$$e_t(\theta) := \begin{cases} r_\theta(t)^2/2 & \text{if } r_\theta(t) < \rho, \\ \rho \cdot r_\theta(t) - \rho^2/2 & \text{if } r_\theta(t) \geq \rho. \end{cases}$$

That is, the error is the residual squared when it has a negative or small positive value less than a given $\rho > 0$, but becomes a linear function for larger residuals. We refer to this modification (and corresponding algorithms) as the modified NLLS (MNLLS) method. To estimate the tailing function,

1. we determine reasonable initial values for the parameters (v, μ, λ, o); see below,
2. we use MNLLS to estimate the scaling factor v with $\rho = \sigma_R^2$, leaving the other parameters fixed,
3. we use MNLLS to estimate all four parameters with $\rho = \sigma_R^2$,
4. we use MNLLS to re-estimate the scaling factor v with $\rho = \sigma_R^2/100$.

The initial parameter values are determined as follows. The initial σ is set to the standard deviation of the whole RIP-only spectrum. An additional offset o' is set to the largest IRM left of the RIP mode where the signal is below σ_R. Having determined the IRM of the RIP mode T_R, the initial μ' can only range within the interval $]T_R, T_R + 0.7\sigma]$. To obtain appropriate model parameters, μ' is being increased in small steps within the interval. The descriptor set (μ', σ, T_R) is being recomputed into the parameter set (μ, λ, o) until $o \geq o'$. This parameter set contains the initial parameter values. For the scaling factor, we initially set $v = (1/2) \sum_{t \leq |T|} S_t$.

Extracting Peak Parameters from a Single Spectrum. To extract all peaks from a spectrum (from left to right), we repeat three sub-steps:

1. scanning for a potential peak, starting where the previous iteration stopped
2. determining peak parameters (Inverse Gaussian distribution)
3. subtracting the peak from the spectrum and continuing with the remainder

Scanning. The algorithm scans for peaks, starting at the left end of S, by sliding a window of width ρ across S and fitting a quadratic polynomial to the data points within the window. Assume that the window starts at index β and ends at $\beta + \rho$, the latter being not included. The value of ρ is determined by the grid opening time d_{grid}, the maximum drift time of the spectrum D_{last} and the number of data points in the spectrum, $\rho := d_{\text{grid}}/D_{\text{last}} \cdot |D|$ data point units. Let $f(x; \theta) = \theta_2 x^2 + \theta_1 x + \theta_0$ be the fitted polynomial. We call a window a *peak window* if the following conditions are fulfilled:

- the extreme drift time $D_x = \theta_1/(2\theta_2)$ ranges within the interval $[D_\beta, D_{\beta+\rho}]$.
- $f(D_x; \theta) \geq 2\sigma_R$
- There is a maximum at D_x, i.e., $f(D_x; \theta) > f(D_\beta; \theta)$

The first condition can be more restricted to achieve more reliable results, by shrinking the interval towards the center of the window. When no peak is found, the moving window is shifted one index forward. If a peak is detected, the window is shifted half the window length forward before the next scan begins, but first the peak parameters and the reduced spectrum are computed.

Determining Peak Parameters. Given the drift time D_x of maximal intensity in the window, the mode descriptor of the peak is simply $m = D_x \cdot f_{\text{ims}}$, where f_{ims} is the scaling constant that converts drift times (in ms) into IRMs (in Vs/cm^2; see Section 2.1). Given the mode, the other peak parameters can be inferred. Spangler *et al.* [19] empirically derived that the width $w_{1/2}$ of the drift time interval between the point of maximal intensity and the point of half the maximum intensity (assuming peak symmetry) is $w_{1/2} = \sqrt{(11.09\,\mathcal{D}\,d)/V_d^2 + d_{\text{grid}}^2}$, where $\mathcal{D}$ is the diffusion coefficient, d the mean drift time of the compound, V_d the drift velocity. Using the Einstein relation [8], $\mathcal{D}$ can be computed as $\mathcal{D} = k\mathcal{K}_B\mathcal{T}/q$, where k is the ion mobility, $\mathcal{K}_B$ the Boltzmann constant, $\mathcal{T}$ the absolute temperature and q the electric charge. From this, we can derive the standard deviation $\sigma = w_{1/2}/2.3548 \cdot f_{\text{ims}}$, remark that in a Gaussian curve $w_{1/2} \approx 2.3548\sigma$. Empirically, the mean is found to be $\mu' = \left(d + \sqrt{(4.246 \cdot 10^{-5})^2 + d^2/585048.1633}\right) \cdot f_{\text{ims}}$.

Having computed the peak descriptors, we convert them into the parameters (μ, λ, o) of the Inverse Gaussian parameterization (see Appendix A). The scaling factor v for the peak is $v = f(d; \theta)/g(d \cdot f_{\text{ims}}; \mu, \lambda, o)$. The model function is subtracted from the spectrum, and the next iteration is started with a window shifted by $\rho/2$ index units. For each spectrum, the output of this step is a *reduced spectrum*, which is a set of parameters for a mixture of weighted Inverse Gaussian models describing the peaks.

3.2 Aligning Two Consecutive Reduced Spectra

We now have, for each spectrum, a set of peak parameters, and the question arises how to merge the sets $P = (P_i)$ and $P^+ = (P_j^+)$ of two consecutive spectra. For each peak P_i, we have stored the Inverse Gaussian parameters μ_i, λ_i, o_i, the peak descriptors μ_i', σ_i, m_i (mean, standard deviation, mode) and the scaling factor v_i, and similarly so for the peaks P_j^+. The idea is to compute a global alignment similar to the Needleman-Wunsch method [17] between P and P^+. We need to specify how to score an alignment between P_i and P_j^+ and how to score leaving a peak unaligned (i.e., a gap). The score S_{ij} for aligning P_i to P_j^+ is chosen proportionally to the P_i's Inverse Gaussian density at the mode m_j^+, since we want to align new peaks to the current and not vice versa. The score γ_i in comparison to align P_i to a gap is proportional to P_i's density at $m_i + \delta$, where $\delta := d_{grid}/2.3548 \cdot f_{ims}$ corresponds approximate to a minimal standard deviation width in IRM units. In other words, with $g(x; \mu, \lambda, o)$ as in Eq. (1), let

$$S_{ij}' := g(m_j^+; \mu_i, \lambda_i, o_i), \quad \gamma_i' := g(m_i + \delta; \mu_i, \lambda_i, o_i), \quad Z := \gamma_i' + \sum_j S_{ij}',$$
$$S_{ij} := S_{ij}'/Z, \qquad\qquad \gamma_i := \gamma_i'/Z.$$

Accordingly, γ_j^+ refers to the corresponding value for P_j^+.

Computing the alignment now in principle uses the standard dynamic programming approach with an alignment matrix E, such that E_{ij} is the optimal score between the first i peaks of P and the first j peaks of P^+. However, while we use γ_i for *deciding* whether to align P_i to a gap, we *score* it as zero. Thus we initialize the borders of (E_{ij}) to zero and then compute, for $i \geq 1$ and $j \geq 1$,

$$\zeta_{i,j} = \max(E_{i-1,j-1} + S_{i,j}, \ E_{i-1,j} + \gamma_i, \ E_{i,j-1} + \gamma_j^+),$$

$$E_{i,j} = \begin{cases} \zeta_{i,j} & \text{if } \zeta_{i,j} = E_{i-1,j-1} + S_{i,j}, \\ E_{i-1,j} & \text{if } \zeta_{i,j} = E_{i-1,j} + \gamma_i, \\ E_{i,j-1} & \text{if } \zeta_{i,j} = E_{i,j-1} + \gamma_j^+. \end{cases}$$

The alignment is obtained with a traceback, recording the optimal case in each cell, as usual. There are three cases to consider.

- If P_j^+ is not aligned with a peak in P, potentially a new peak starts at this retention time. Thus model P_j^+ is put into a new peak chain.
- If P_j^+ is aligned with a peak P_i, the chain containing P_i is extended with P_j^+.
- All peaks P_i that are not aligned to any peak in P^+ indicate the end of a peak chain at the current retention time.

All completed peak chains are forwarded to the next step, two-dimensional peak model estimation.

3.3 Estimating 2-D Peak Models

Let $C = (P_1, \ldots, P_n)$ be a chain of one-dimensional Inverse Gaussian models. The goal of this step is to estimate a two-dimensional peak model (product of two one-dimensional Inverse Gaussians) from the chain, as described in Section 2.2, or to reject the chain if the chain does not fit such a model well. Potential problems are that a peak chain may contain noise 1-D peaks, or in fact consist of several consecutive 2-D peaks at the same drift time and successive retention times.

Empirically, we find that the retention time interval where a peak has more than half the height of its maximum height is at least 2.5 s. (This depends on the settings at the MCC and has been found for temperature 40°C and throughput 150 mL/min.) The time between two spectra is 0.1 s; so for a peak the number of chain entries should be $\geq 2.5\,\mathrm{s}\,/\,0.1\,\mathrm{s} = 25$. If it is lower, then the chain is discarded immediately.

We first collect the peak height vector $h = (h_i)_{i=1,\ldots,n}$ at the individual modes; $h_i := v_i \cdot g(m_i; \mu_i, \lambda_i, o_i)$ and in parallel the vector $r = (r_i)$ of corresponding retention times of the models.

To identify noise chains, we fit an affine function of r to h by finding (θ'_0, θ'_1) to minimize $\sum_i (h_i - \theta'_1 r_i - \theta'_0)^2$ (linear least squares). We then compute the cosine similarity between $\ell = (\ell_i)$ with $\ell_i := \theta'_1 r_i + \theta'_0$ and h, which is defined as normalized inner product $\langle h, \ell \rangle / (\|h\| \|\ell\|) \in [-1, 1]$. If it exceeds 0.99, there is no detectable concave peak shape and the chain is discarded as noise.

Otherwise we proceed similarly to the paragraph "Extracting Peak Parameters from A Single Spectrum" in Section 3.1 by fitting quadratic polynomials $h_i \approx \theta_2 r_i^2 + \theta_1 r_i + \theta_0$ to appropriately sized windows of retention times. We record the index $i^* := \arg\max_i(h_i)$ of the highest signal and check the peak conditions: (1) As a minimum peak height, we require $h_{i^*} \geq 5\sigma_R/2$. (2) As a minimum peak width ρ (size of the moving window), we use $\rho = (25 + 0.01 \cdot R_{i^*})$. The lower bound of 25 was explained above, but with increasing retention time R_i, the peaks become wider. This was found empirically by manually examining about 100 peaks in several measurements and noting a linear correlation of peak retention time and width.

When the peak conditions are satisfied, we compute the descriptors for an Inverse Gaussian as follows: $v = -\theta_2(\theta_1/(2\theta_2))^2 + \theta_0$, $\sigma = \sqrt{v/(2|\theta_2|)}$, $m = -\theta_1/(2a)$, $\mu' = m + 0.1\sqrt{m}$. Having computed the descriptors, we compute the model parameters (μ, λ, o). Ideally, we have now obtained a single shifted Inverse Gaussian model from the chain and are done. However, to optimize the model fit and to deal with the possibility of finding several peak windows in the chain, we use the EM algorithm to fit a mixture of Inverse Gaussians to (h_i) and finally only take the dominant model from the mixture as the resulting peak. To achieve descriptors in IRM dimension and the volume v, we take the weighted average over all models within the chain for every descriptor.

4   Evaluation

We tested different properties of our online algorithm: (1) the execution time, (2) the quality of reducing a single spectrum to peak models, (3) the correlation between manual annotations on full IMSCs by a computer-assisted expert and our automated online extraction method.

Table 1. Average processing time of both spectrum reduction and consecutive alignment on two platforms with different degrees of averaging

Platform	Average 1	Average 2	Average 5
Desktop PC	7.79 ms	3.10 ms	1.52 ms
Raspberry Pi	211.90 ms	85.49 ms	37.82 ms

Execution time. We tested our method on two different platforms, (1) a desktop PC with Intel(R) Core(TM) i5 2.80GHz CPU, 8GB memory, Ubuntu 12.04 64bit OS and (2) a Raspberry Pi[1] type B with ARM1176JZF-S 700MHz CPU, 512 MB memory, Raspbian Wheezy 32bit OS. The Raspberry Pi was chosen because it is a complete credit card sized low-cost single-board computer with low CPU and power consumption (3.5 W). This kind of device is appropriate for data analysis in future mobile measurement devices.

Recall that each spectrum contains 12 500 data points. It is current practice to analyze not the full spectra, but aggregated ones, where five consecutive values are averaged. Here we consider the full spectra, slightly aggregated ones (av. over two values, 6 250 data points) and standard aggregated ones (av. over five values, 2 500 data points). We measured the average execution time of the spectrum reduction and consecutive alignment. Table 1 shows the results. At the highest resolution (Average 1) only the desktop PC satisfies the time bound of 100 ms between consecutive spectra. At lower resolutions, the Raspberry Pi satisfies the time restrictions.

We found that in the steps that use the EM algorithm, on average 25–30 EM iterations were necessary for a precision of $\varepsilon := 0.001$ (i.e., 0.1%) (see Convergence in Section 2.3). Relaxing the threshold from 0.001 to 0.01 halved the number of iterations without noticeable difference in the results.

Quality of single spectrum reduction. In a second experiment we tested quality of the spectrum reduction method using an idea by Munteanu and Wornowizki [16] that determines the agreement between an observed set of data points, interpreted as an empirical distribution function F and a model distribution G (the reduced spectrum in our case). The approach works with the fairly general two-component mixture model $F = \tilde{s} \cdot G + (1 - \tilde{s}) \cdot H$ with $\tilde{s} \in [0, 1]$, where H is a non-parametric distribution whose inclusion ensures the fit of the model G to the data F. If the weight $\tilde{s}$ is close to 1.0, then F is a plausible sample from G. We compare the original spectra and reduced spectra from a previously used dataset [10]. This set contains 69 measurements preprocessed with a 5×5 average. Every measurement contains 1200 spectra. For each spectrum in all measurements, we computed the reduced spectrum model and determined $\tilde{s}$. Over 92% of all 82 000 models achieved $\tilde{s} = 1$ and over 99% reached $\tilde{s} \geq 0.9$. No $\tilde{s}$ dropped below 85%.

---

[1] http://www.raspberrypi.org/

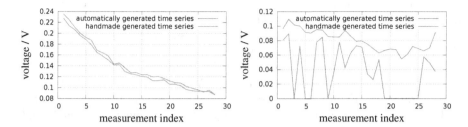

Fig. 3. Time series of discovered intensities of two peaks. **Left:** A peak with agreement between manual and automated online annotation. **Right:** A peak where the online method fails to extract the peak in several measurements. If one treated zeros as missing data, the overall trend would still be visible.

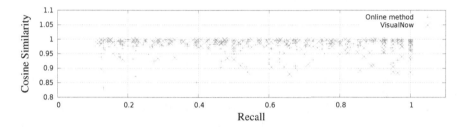

Fig. 4. Comparison of peak intensity time series from manual annotation of full measurements vs. our online algorithm. Each cross corresponds to a time series of one peak. Ideally, the online algorithm finds the peak in all measurements where it was manually annotated (high recall, x-axis), and the time series has a high cosine similarity (y-axis).

Similarity of online extracted peaks with manual offline annotation. The third experiment compares extracted peaks on a time series of measurements between expert manual annotation and our algorithm. Here 15 rats were monitored in 20 minute intervals for up to a day. Each rat resulted in 30–40 measurements (a time series) for a total of 515 measurements, all captured in positive mode. To track peaks within a single time series, we used the previously described EM clustering method [14] as well as the state-of-the-art software, namely VisualNow. As an example, Figure 3 shows time series of intensities of two peaks detected by computer-assisted manual annotation and using our online algorithm. Clearly, the sensitivity of the online algorithm is not perfect.

To obtain an overview over all time series, we computed the cosine similarity $\gamma_{X,Y} \in [-1, +1]$ between the intensities over time as discovered by manual annotation (X) and our online algorithm (Y). We also computed the recall of the online algorithm for each time series, that is, the relative fraction of measurements where the peak was found by the algorithms among those where it was found by manual annotation.

In summary, we outperform VisualNow in terms of sensitivity and computation time. Almost 27% of the points extracted by the online method exceed 90%

recall and 95% cosine similarity whereas only 7% of the time series extracted by VisualNow achieve that values. The peak detection of one measurement took about 2 seconds on average (when the whole measurement is available at once) with the online method and about 20 seconds with VisualNow on the previous described desktop computer. VisualNow only provides the position and signal intensity of the peak's maximum, whereas our method additionally provides shape parameters. Figure 4 shows generally good agreement between the online method and the manual one, and similarly good agreement between Visual-Now and the manual annotation. Problems of our online method stem from low-intensity peaks only slightly above the detection threshold, and resulting fragmentary or rejected peak chains.

5   Discussion and Conclusion

We presented the first approach to extract peaks from MCC/IMS measurements while they are being captured, with the long-term goal to remove the need for storing full measurements before analyzing them in small embedded devices. Our method is fast and satisfies the time restrictions even on a low-power CPU platform like a Raspberry Pi.

While performing well on single spectra, there is room for improvement in merging one-dimensional peak models into two-dimensional peak models. We currently ignore the fact that a peak chain may contain more than one peak. Our method has to be further evaluated in clinical studies or biotechnological monitoring settings. It also has not been tested with the negative mode of an IMS for lack of data. In general, the robustness of the method under adversarial conditions (high concentrations with formation of dimer ions, changes in temperature or carrier gas flow in the MCC) has to be evaluated and probably improved.

Acknowledgements. We thank Max Wornowizki (TU Dortmund) for providing results for second evaluation. DK, SR are supported by the Collaborative Research Center (Sonderforschungsbereich, SFB) 876 "Providing Information by Resource-Constrained Data Analysis" within project TB1, see http://sfb876.tu-dortmund.de.

References

1. Bessa, V., Darwiche, K., Teschler, H., Sommerwerck, U., Rabis, T., Baumbach, J.I., Freitag, L.: Detection of volatile organic compounds (VOCs) in exhaled breath of patients with chronic obstructive pulmonary disease (COPD) by ion mobility spectrometry. International Journal for Ion Mobility Spectrometry 14, 7–13 (2011)
2. Bödeker, B., Vautz, W., Baumbach, J.I.: Peak finding and referencing in MCC/IMS-data. International Journal for Ion Mobility Spectrometry 11(1), 83–87 (2008)

3. Bunkowski, A.: MCC-IMS data analysis using automated spectra processing and explorative visualisation methods. Ph.D. thesis, University Bielefeld: Bielefeld, Germany (2011)
4. Bunkowski, A., Bödeker, B., Bader, S., Westhoff, M., Litterst, P., Baumbach, J.I.: MCC/IMS signals in human breath related to sarcoidosis – results of a feasibility study using an automated peak finding procedure. Journal of Breath Research 3(4), 046001 (2009)
5. D'Addario, M., Kopczynski, D., Baumbach, J.I., Rahmann, S.: A modular computational framework for automated peak extraction from ion mobility spectra. BMC Bioinformatics 15(1), 25 (2014)
6. Dempster, A.P., Laird, N.M., Rubin, D.B.: Maximum likelihood from incomplete data via the EM algorithm. Journal of the Royal Statistical Society. Series B (Methodological), 1–38 (1977)
7. Eiceman, G.A., Karpas, Z.: Ion Mobility Spectrometry, 2 ed. Taylor & Francis (2005)
8. Einstein, A.: Über die von der molekularkinetischen Theorie der Wärme geforderte Bewegung von in ruhenden Flüssigkeiten suspendierten Teilchen. Annalen der Physik 322(8), 549–560 (1905)
9. Ewing, R.G., Atkinson, D.A., Eiceman, G.A., Ewing, G.J.: A critical review of ion mobility spectrometry for the detection of explosives and explosive related compounds. Talanta 54(3), 515–529 (2001)
10. Hauschild, A.C., Kopczynski, D., D'Addario, M., Baumbach, J.I., Rahmann, S., Baumbach, J.: Peak detection method evaluation for ion mobility spectrometry by using machine learning approaches. Metabolites 3(2), 277–293 (2013)
11. Keller, T., Schneider, A., Tutsch-Bauer, E., Jaspers, J., Aderjan, R., Skopp, G.: Ion mobility spectrometry for the detection of drugs in cases of forensic and criminalistic relevance. Int. J. Ion Mobility Spectrom 2(1), 22–34 (1999)
12. Kolehmainen, M., Rönkkö, P., Raatikainen, O.: Monitoring of yeast fermentation by ion mobility spectrometry measurement and data visualisation with self-organizing maps. Analytica Chimica Acta 484(1), 93–100 (2003)
13. Kopczynski, D., Baumbach, J., Rahmann, S.: Peak modeling for ion mobility spectrometry measurements. In: 2012 Proceedings of the 20th European Signal Processing Conference (EUSIPCO), pp. 1801–1805. IEEE (August 2012)
14. Kopczynski, D., Rahmann, S.: Using the expectation maximization algorithm with heterogeneous mixture components for the analysis of spectrometry data. Pre-print - CoRR abs/1405.5501 (2014)
15. Kreuder, A.E., Buchinger, H., Kreuer, S., Volk, T., Maddula, S., Baumbach, J.: Characterization of propofol in human breath of patients undergoing anesthesia. International Journal for Ion Mobility Spectrometry 14, 167–175 (2011)
16. Munteanu, A., Wornowizki, M.: Demixing empirical distribution functions. Tech. Rep. 2, TU Dortmund (2014)
17. Needleman, S.B., Wunsch, C.D.: A general method applicable to the search for similarities in the amino acid sequence of two proteins. Journal of Molecular Biology 48(3), 443–453 (1970)
18. Nocedal, J., Wright, S.J.: Numerical Optimization, 2nd edn. Springer, New York (2006)
19. Spangler, G.E., Collins, C.I.: Peak shape analysis and plate theory for plasma chromatography. Analytical Chemistry 47(3), 403–407 (1975)
20. Westhoff, M., Litterst, P., Freitag, L., Urfer, W., Bader, S., Baumbach, J.: Ion mobility spectrometry for the detection of volatile organic compounds in exhaled breath of lung cancer patients. Thorax 64, 744–748 (2009)

A   Peak Descriptors and Parameters of Shifted Inverse Gaussian

The shifted Inverse Gaussian distribution with parameters dimension with parameters o (shift), μ (mean minus shift) and λ (shape) is given by (1). There is a bijection between (μ, λ, o) and (μ', σ, m) that is given by

$$\mu' = \mu + o, \quad \sigma = \sqrt{\mu^3/\lambda}, \quad m = \mu\left(\sqrt{1 + (9\mu^2)/(4\lambda^2)} - (3\mu)/(2\lambda)\right) + o.$$

in the forward direction and in the backward direction as follows [13]:

$$o = -p/2 - \sqrt{p^2/4 - q}, \quad \mu = \mu' - o, \quad \lambda = \mu^3/\sigma^2, \quad \text{where}$$
$$p := \left(-m(2\mu' + m) + 3 \cdot (\mu'^2 - \sigma^2)\right)/(2(m - \mu')),$$
$$q := \left(m(3\sigma^2 + \mu' \cdot m) - \mu'^3\right)/(2(m - \mu')).$$

Best-Fit in Linear Time for Non-generative Population Simulation

(Extended Abstract)

Niina Haiminen[1], Claude Lebreton[2], and Laxmi Parida[1,⋆]

[1] Computational Biology Center, IBM T. J. Watson Research, USA
[2] Limagrain Europe, Centre de Recherche de Chappes, France
parida@us.ibm.com

Abstract. Constructing populations with pre-specified characteristics is a fundamental problem in population genetics and other applied areas. We present a novel non-generative approach that deconstructs the desired population into essential local constraints and then builds the output bottom-up. This is achieved using primarily best-fit techniques from discrete methods, which ensures accuracy of the output. Also, the algorithms are fast, i.e., linear, or even sublinear, in the size of the output. The non-generative approach also results in high sensitivity in the algotihms. Since the accuracy and sensitivity of the population simulation is critical to the quality of the output of the applications that use them, we believe that these algorithms will provide a strong foundation to the methods in these studies.

1 Introduction

In many studies, it is important to work with an artificial population to evaluate the efficacy of different methods or simply generate a founder population for an in-silico breeding regimen. The populations are usually specified by a set of characteristics such as minimum allele frequency (MAF) and linkage disequilibrium (LD) distributions. A generative model simulates the population by evolving a population over time [1, 2]. Such an approach uses different parameters such as ancestor population characteristics and their sizes, mutation and recombination rates, and breeding regimens, if any. The non-generative models [3–5], on the other hand, do not evolve the population and often start with an exemplar population and perturb it either by a regimen of recombinations between the samples or other local perturbations.

We present a novel non-generative approach that first breaks up the specified (global) constraints into a series of local constraints. We map the problem onto a discrete framework by identifying subproblems that use best-fit techniques to satisfy these local constraints. The subproblems are solved iteratively to give an integrated final solution using techniques from linear algebra, combinatorics, basic statistics and probability. Using techniques from discrete methods, the algorithms are optimized to run in time linear with the size of the output, thus extremely time-efficient. In fact, for one of the problems, the algorithm completes the task in sublinear time.

⋆ Corresponding author.

D. Brown and B. Morgenstern (Eds.): WABI 2014, LNBI 8701, pp. 247–262, 2014.
© Springer-Verlag Berlin Heidelberg 2014

The first problem we address is that of constructing a deme (population) with pre-specified characteristics [6] More precisely, the problem is defined as:

Problem 1. (**Deme Construction**). The task is to generate a population (deme) of n diploids (or $2n$ haploids) with m SNPs that satisfy the following characteristics: MAF distribution p, LD distribution r^2.

Our approach combines algebraic techniques, basic quantitative genetics and discrete algorithms to best fit the specified distributions. The second problem we address is that of simulating *crossovers with interference* in a population. Capturing the crossover events in an individual chromosome as it is transmitted to its offspring, is a fundamental component of a population evolution simulator where the population may be under selection or not (neutral). The expected number of crossovers is d, also called the the genetic map distance (in units of Morgans). Let the recombination fraction be be denoted by r. Then:

Problem 2. (**F$_1$ Population with Crossover Interference**). The task is to generate a F$_1$ hybrid population with the following crossover interference models for a pair of parents:

1. Complete interference (Morgan [7]) model defined by the relationship $d = r$.
2. Incomplete interference (Kosambi [8]) model defined by the relationship

$$d = 0.25 \ln \frac{1 + 2r}{1 - 2r} \text{ or } r = 0.5 \tanh 2d.$$

3. No interference (Haldane [7, 8]) model defined by the relationship

$$d = -0.5 \ln(1 - 2r) \text{ or } r = e^{-d} \sinh d.$$

Again, we use combinatorics and basic probability to design a sub-linear time algorithm to best fit the distribution of the three different crossover interference models.

2   Problem 1: Deme Construction

Background. We recall some basic definitions here. Let p_1 and p_2 be the MAF at locus 1 and locus 2 and let r^2 be the LD between the two loci. Then D is defined as follows ([9, 10]):

$$D = \pm r \sqrt{p_1(1 - p_1)p_2(1 - p_2)}. \tag{1}$$

Equivalently, the *LD table* of the pairwise patterns, 00, 01, 10, 11, of the two loci, is written as:

$$
\begin{array}{c|c|c|c}
 & 0 & 1 & \\
\hline
0 & (1 - p_1)(1 - p_2) + D & (1 - p_1)p_2 - D & 1 - p_1 \\
\hline
1 & p_1(1 - p_2) - D & p_1 p_2 + D & p_1 \\
\hline
 & 1 - p_2 & p_2 & 1
\end{array}
\tag{2}
$$

With a slight abuse of notation we call D the LD between two loci, with the obvious interpretation.

The output deme (population) is a matrix M where each row is a haplotype and each column is a (bi-allelic) marker. Recall that the input is the MAF and LD distributions. By convention, the MAF of marker j, p_j, is the proportion of 1's in column j of M. Our approach to constructing the deme is to work with the markers one at a time and without any backtracking. We identify the following subproblem, which is used iteratively to construct the population.

*Problem 3 (k-**Constrained Marker Problem** (k-**CMP**)). Given columns $j_0, j_1, .., j_{k-1}$* and target values $r_0, r_1, ..., r_{k-1}$ and p_k, the task is to generate column j_k with MAF p_k such that the pairwise LD with column j_l is r_l, $l = 0, 1, .., k - 1$.

Outline of our approach to solving k-CMP. The 1's in column j_k are assigned at random respecting MAF p_k. Let $D_l(j_l, j_k)$ denote the LD between markers j_l and j_k. Then let the expected value, in the output matrix M, be $\overline{D}_l(\cdot, \cdot)$. When both the columns fulfill the MAF constraints of p_l and p_k respectively, let the observed value be $D_l^{\text{obs}}(\cdot, \cdot)$. In other words, if Q_{10} is the number of times pattern 10 is seen in these two markers in M with n rows (after the random initialization),

$$D_l^{\text{obs}} = \frac{1}{n}\left(np_l(1 - p_{k+1}) - Q_{10\cdot}\right). \tag{3}$$

Next, we shuffle the 1's in column j_k, such that it *simultaneously* satisfies k conditions. Thus we get a best-fit of $D_l^{\text{obs}}(j_l, j_k)$ to $\overline{D}(j_l, j_k)$. To achieve this, we compare column j_k with columns j_l, $l = 0, 1, 2, .., k - 1$, that have already been assigned. Thus, first, for each pair of markers j_l, j_k, compute the target deviation, D_l^{target}, based on input p and r values. Then, shuffle the 1's in column j_k of the output matrix, to get a best-fit to the targets $D_0^{\text{target}}, D_1^{\text{target}}, ..., D_{k-1}^{\text{target}}$ simultaneously.

3 k-CMP: Linear Algebraic Method

Given p_l, p_k and r_l, the following 4 values can be computed, for each l:

$$
\begin{aligned}
P_{lk} &= n((1 - p_l)(1 - p_k) + D_l) &&\text{(number of rows with 0 in column } l \text{ and 0 in column } k\text{),}\\
Q_{lk} &= n(p_l(1 - p_k) - D_l) &&\text{(number of rows with 1 in column } l \text{ and 0 in column } k\text{),}\\
C_{lk} &= n((1 - p_l)p_k - D_l) &&\text{(number of rows with 0 in column } l \text{ and 1 in column } k\text{),}\\
B_{lk} &= n(p_l p_k + D_l) &&\text{(number of rows with 1 in column } l \text{ and 1 in column } k\text{),}
\end{aligned}
$$

where

$$D_l = +r_l\sqrt{p_l(1 - p_l)p_k(1 - p_k)}.$$

Strictly speaking, the number of unknowns is 2×2^k, written as $y_{i,0}, y_{i,1}$, where $1 \le i \le 2^k$. Let X_i denote the binary k-pattern corresponding to binary pattern of the number $i - 1$. For example when $k = 3$, $X_1 = 000$ and $X_8 = 111$. Then in the solution, the number of rows with binary $(k + 1)$-pattern $X_i 0$ is $y_{i,0}$ and the number of rows with binary $(k + 1)$-pattern $X_i 1$ is $y_{i,1}$. Thus

$y_{i,0} + y_{i,1} = \#X_i$ where $\#X_i$ is the number of rows in the input M with k-pattern X_i in the given k columns.

Since the right hand side of the above 2^k equations can be directly obtained from the existing input data, the effective number of unknowns are 2^k, re-written as $y_i = y_{i,0}$,

$1 \leq i \leq 2^k$ (since $y_{i,1} = \#X_i - y_i$). Hence we focus on the computing the non-trivial 2^k unknowns y_i. We set up $k + 1$ linear equations in 2^k unknowns using P_{1k}, P_{2k}, .., $P_{(k-1)k}$ and $Q_{(k-1)k}$. For this we define the following ($0 \leq l < k$):

$$L_l^k = \left\{ i \mid 1 \leq i \leq 2^k \text{ and the } (l+1)\text{th entry is 0 in } k\text{-pattern } X_i \right\}.$$

For example, $L_1^3 = \{1, 2, 3, 4\}$ and $L_3^3 = \{1, 3, 5, 6\}$ (see Fig 1 for the binary patterns). Then the $k + 1$ linear equations are:

$$\sum_{i \in L_l^k} y_i = P_{lk}, \text{ for } l = 0, 1, .., k - 1,$$

$$\sum_{i \notin L_{k-1}^k} y_i = Q_{(k-1)k}.$$

This under-constrained system is solved using Gauss-Jordan elimination and we look for non-negative (integer) solutions, so that the results can be translated back to the number of 0's (and 1's) in column j_k. It is possible to save the solution space for each k-CMP problem and when a non-negative solution is not found, to backtrack and pick an alternative solution from the solution space. However, we did not experiment with the backtracking approach (instead we iteratively reduced the number of constraints to fit a subproblem with fewer columns, when necessary).

3.1   A Concrete Example

Consider the following concrete example where $k = 3$. We use the following convention: the given (constraint) columns are 0, 1 2 and the column under construction is 3. We solve for the eight variables $y_1, .., y_8$ and the conditions are derived below. Let $p_3 = 0.26$ with the three pairwise LD tables as:

$$r_{03}^2 = 0.27 \qquad r_{13}^2 = 0.29 \qquad r_{23}^2 = 0.37$$
$$D_0 = 0.0714 \qquad D_1 = 0.1178 \qquad D_2 = 0.133$$

	0	1			0	1			0	1	
0	73	16	89	0	51	2	53	0	54	1	55
1	1	10	11	1	23	24	47	1	20	25	45
	74	26	100		74	26	100		74	26	100

(4)

Exact Solution with Gauss-Jordan Elimination. Based on Equation 4, four linear equations are captured as:

$$y_1 + y_2 + y_3 + y_4 = 73 \quad \text{(using } P_{03}), \tag{5}$$
$$y_1 + y_2 + y_5 + y_6 = 51 \quad \text{(using } P_{13}), \tag{6}$$
$$y_1 + y_3 + y_5 + y_7 = 54 \quad \text{(using } P_{23}), \tag{7}$$
$$y_2 + y_4 + y_6 + y_8 = 20 \quad \text{(using } Q_{23}). \tag{8}$$

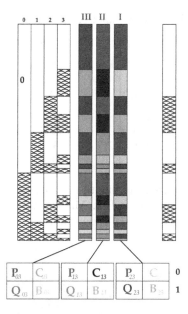

$$
\begin{array}{|c|c|l|}
\hline
000 & 27 & y_1 + y_{1,1} \\
001 & 21 & y_2 + y_{2,1} \\
010 & 22 & y_3 + y_{3,1} \\
011 & 19 & y_4 + y_{4,1} \\
100 & 4 & y_5 + y_{5,1} \\
101 & 1 & y_6 + y_{6,1} \\
110 & 2 & y_7 + y_{7,1} \\
111 & 4 & y_8 + y_{8,1} \\
\hline
 & 100 &
\end{array}
$$

Fig. 1. The problem set up for $k = 3$ showing effectively 8 unknowns in the rightmost column. The different numbers of the concrete example are shown on the right.

Recall from linear algebra that a column (or variable y_i) is *pivotal*, if it is the leftmost as well as the top-most non-zero element (which can be converted easily to 1). In the *reduced* row echelon form, a column is *pivotal*, if all elements to its left, all elements above it and all below are it zero. If the augmented column (the very last one with Ps and Qs) is also *pivotal*, then the system has no solution in $\mathbb{R}^8$. The different row transformations give the following:

input constraitns

y_1	y_2	y_3	y_4	y_5	y_6	y_7	y_8	
1	0	1	0	1	0	1	0	P_{23}
0	1	0	1	0	1	0	1	Q_{23}
1	1	1	1	0	0	0	0	P_{03}
1	1	0	0	1	1	0	0	P_{13}

$\Rightarrow$

row echelon form; pivotal columns are $y_1..y_4$

y_1	y_2	y_3	y_4	y_5	y_6	y_7	y_8	
1	0	1	0	1	0	1	0	P_{23}
0	1	0	1	0	1	0	1	Q_{23}
0	0	1	1	-1	-1	0	0	P_{03}-P_{13}
0	0	0	1	0	0	1	1	P_{23}-P_{13}+Q_{23}

$\Rightarrow$

reduced row echelon form; pivotal columns are $y_1..y_4$

y_1	y_2	y_3	y_4	y_5	y_6	y_7	y_8	
1	0	0	0	2	1	2	1	$2P_{23}$-P_{03}+Q_{23}
0	1	0	0	0	1	-1	0	P_{13}-P_{23}
0	0	1	0	-1	-1	-1	-1	P_{03}-P_{23}-Q_{23}
0	0	0	1	0	0	1	1	$P_{23} - P_{13} + Q_{23}$

All possible solutions are given by using constants c_1, c_2, c_3 and c_4:

$$
\begin{bmatrix} y_1 \\ y_2 \\ y_3 \\ y_4 \\ y_5 \\ y_6 \\ y_7 \\ y_8 \end{bmatrix}
=
\begin{bmatrix} 2P_{23} - P_{03} + Q_{23} \\ P_{13} - P_{23} \\ P_{03} - P_{23} - Q_{23} \\ P_{23} - P_{13} + Q_{23} \\ 0 \\ 0 \\ 0 \\ 0 \end{bmatrix}
+ c_1 \begin{bmatrix} -2 \\ 0 \\ 1 \\ 0 \\ 1 \\ 0 \\ 0 \\ 0 \end{bmatrix}
+ c_2 \begin{bmatrix} -1 \\ -1 \\ 1 \\ 0 \\ 0 \\ 1 \\ 0 \\ 0 \end{bmatrix}
+ c_3 \begin{bmatrix} -2 \\ 1 \\ 1 \\ -1 \\ 0 \\ 0 \\ 1 \\ 0 \end{bmatrix}
+ c_4 \begin{bmatrix} -1 \\ 0 \\ 1 \\ -1 \\ 0 \\ 0 \\ 0 \\ 1 \end{bmatrix}. \quad (9)
$$

Thus for this concrete example, the solution space is captured as:

$$
\begin{bmatrix} y_1 \le 27 \\ y_2 \le 21 \\ y_3 \le 22 \\ y_4 \le 19 \\ \hline y_5 \le 4 \\ y_6 \le 1 \\ y_7 \le 2 \\ y_8 \le 4 \end{bmatrix}
=
\begin{bmatrix} 55 \\ -3 \\ -1 \\ 23 \\ \hline 0 \\ 0 \\ 0 \\ 0 \end{bmatrix}
+ \begin{matrix} c_1 \\ \le 4 \end{matrix}
\begin{bmatrix} -2 \\ 0 \\ 1 \\ 0 \\ \hline 1 \\ 0 \\ 0 \\ 0 \end{bmatrix}
+ \begin{matrix} c_2 \\ \le 1 \end{matrix}
\begin{bmatrix} -1 \\ -1 \\ 1 \\ 0 \\ \hline 0 \\ 1 \\ 0 \\ 0 \end{bmatrix}
+ \begin{matrix} c_3 \\ \le 2 \end{matrix}
\begin{bmatrix} -2 \\ 1 \\ 1 \\ -1 \\ \hline 0 \\ 0 \\ 1 \\ 0 \end{bmatrix}
+ \begin{matrix} c_4 \\ \le 4 \end{matrix}
\begin{bmatrix} -1 \\ 0 \\ 1 \\ -1 \\ \hline 0 \\ 0 \\ 0 \\ 1 \end{bmatrix}.
$$

Fig 2 shows an example of applying the algebraic technique to a data set based on real human MAF and r^2 data provided by the International HapMap Project [11], from chr 22 in the LD data collection of Japanese in Tokyo (JPT) population.

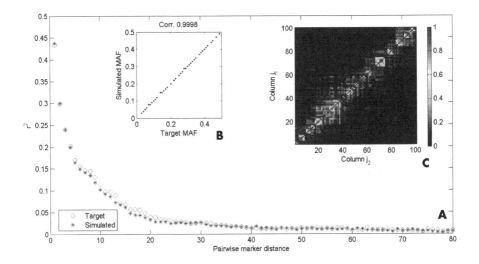

Fig. 2. Population construction using algebraic techniques, for HapMap JPT population. Here $k = 10$. (A) LD fit, (B) MAF fit, and (C) LD for each pair of columns, upper left triangle is the target and lower right triangle the constructed.

4 k-CMP: Single-Step Hill Climbing

One of the shortcomings of the linear algebraic approach is that, it is not obvious how to extract an approximate solution, when the exact solution does not exist. Recall that in the original problem the LD (r_{lk} values) are drawn from a distribution with non-zero variance. A single-step hill climbing algorithm is described here and the general hill-climbing (with compound-steps) will be presented in the full version of the paper.

Note that the cost function in the hill climbing process is crucial to the algorithm. Here we derive the cost function. To keep the exposition simple, we use $k = 3$ in the discussion. Let the column under construction be j_3 and in the pairwise comparison this column is being compared with, e.g., $j_l = j_0$. Then Q_{03} represents the number of rows with 1 in column j_0 and 0 in column j_3. A *flip* is defined as updating an entry of 0 to 1 (or 1 to 0). Further, in column j_3, we will exchange the position of 0 and 1, say at rows i_{from} and i_{to} respectively. This is equivalent to a flip at row i_{from} and at row i_{to}, in column j_3 (called Flip$_1$ and Flip$_2$ in the algorithm). Two flips are essential since the number of 1's (and 0's) in column j_3 must stay the same so as not to affect the allele frequency in column j_3. When i_{from} at column j_3 is 0 and i_{to} at column j_3 is 1, these two flips lead to a change in the LD between columns j_0 and j_3 as follows:

> Scenario I: The entry in row i_{from} of column j_0 is 0 and the entry in row i_{to} of column j_0 is 1. Then there is a negative change in the LD value D_0 since the count of the pattern 00 (and 11) went down by 1 and the count of pattern 01 (and 10) went up by 1.
>
> Scenario II: The entry in row i_{from} of column j_0 is 1 and the entry in row i_{to} of column j_0 is 0. Then there is a positive change in the LD value D_0 since the count of the pattern 00 (and 11) went up by 1 and the count of pattern 01 (and 10) went down by 1.
>
> Scenario III: The entry in row i_{from} of column j_0 is 0 and the entry in row i_{to} of column j_0 is 0. Then there is no change in the LD value D_0 since the count of the pattern 00 does not change and the count of pattern 01 does not change.
>
> Scenario IV: The entry in row i_{from} of column j_0 is 1 and the entry in row i_{to} of column j_0 is 1. Then there is no change in the LD value D_0 since the count of the pattern 11 does not change and the count of pattern 10 does not change.

The four scenarios and the effect on the LD is summarized below:

Cost function based on Q

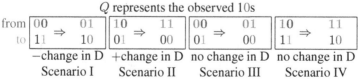

Q represents the observed 10s

Time complexity. The cost function is pre-computed for binary k-patterns and the data is sorted in a hash table (details in the full version of the paper). Based on this an entry of 1 in column j is processed no more than once. The number of such 1's is np_j. Thus for the k-CMP problem, the algorithm takes $\mathcal{O}(knp_j)$ time. Hence to to compute the entire matrix it takes $\mathcal{O}(knmp_j)$ time.

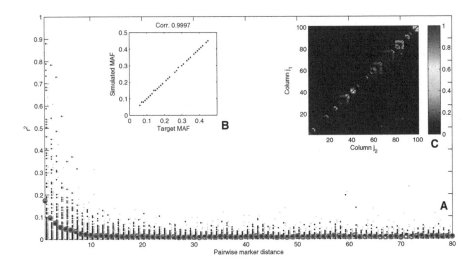

Fig. 3. Hill-climbing algorithm for ASW HapMap population with $k = 6$. (A) LD fit, (B) MAF fit, and (C) LD for each pair of columns, upper left triangle is the target and lower right triange is the constructed. The main figure (A) shows the target "o" and constructed "*" mean r^2 per distance, while the black dots show target and cyan dots constructed r^2 distribution per distance.

ALGORITHM: $(k = 3)$

1. Assign 1 to each column j_3 of M, written as M_{j_3}, with probability p_3 (the rest are assigned 0).
2. At column j_3 of matrix with the 3 constraints based on columns j_0, j_1, j_2:
   (a) Compute targets D_l^{target}, using r_{l3}, $G_l^{\text{target}} = n(D_l^{\text{target}} - D_l^{\text{obs}})$, for $l = 0, 1, 2$.
   (b) Initialize
       i. $G_0^t = G_1^t = G_2^t = 0$.
       ii. distance $= \sum_l (G_l^{\text{target}} - G_l^t)^2$.(goal of the LOOP is to minimize distance)
   (c) LOOP
       i. Move($Z_1 \rightarrow Z_2, 1$) in the following five steps:
           (1-from) Pick i_{from} in column j, with its k-neighbor pattern as Z_1, so that $M_{j_3}[i_{\text{from}}]$ is 0.
           (2-to) Pick i_{to} in column j, with its k-neighbor pattern as Z_2, such that $M_{j_3}[i_{\text{to}}]$ is 1 and the resulting distance decreases.
           (3-Update) For columns $l = 0, 1, 2$ (corresponding to G_0, G_1, G_2)
               IF $Z_{1l} = 1$ and $Z_{2l} = 0$ THEN G_l^t will go up by 1   (Scenario II)
               IF $Z_{1l} = 0$ and $Z_{2l} = 1$ THEN G_l^t will go down by 1 (Scenario I)
               (IF $(Z_{1l} = Z_{2l})$ then no change)   (Scenarios III & IV)
           (4-Flip$_1$) Flip $M_{j_3}[i_{\text{from}}]$.   (to maintain p_j)
           (5-Flip$_2$) Flip $M_{j_3}[i_{\text{to}}]$.   (to maintain p_j)
       ii. Update G_0^t, G_1^t, G_2^t and the distance
       WHILE the distance decreases

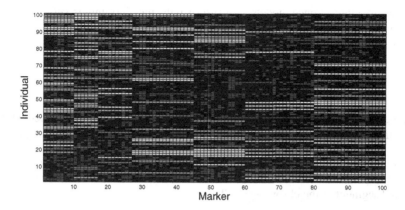

Fig. 4. Haplotype blocks in the simulated ASW HapMap population data, defined by HapBlock software. For each of the 7 identified blocks, three most frequent haplotype sequences are shown as white to gray horizontal lines. The remaining marker values are shown in darker gray/black.

Fig 3 shows an example of applying the hill climbing technique to a data set based on real human MAF and r^2 data provided by the International HapMap Project [11], from chr 22 in the LD data collection of the African ancestry in Southwest USA (ASW) population. Fig 4 shows the haplotype blocks in the simulated population, defined by the HapBlock [15] software. The results demonstrate reasonable haplotype block lengths in the simulated population.

5   Problem 2: F_1 Population with Crossover Interference

An individual of a diploid population draws its genetic material from its two parents and the interest is in studying this fragmentation and distribution of the parental material in the progeny. The difference in definition between recombinations and crossovers is subtle: the latter is what happens in reality, while the former is what is observable. Hence a simulator that reflects reality must simulate the crossover events. However, it is a well known fact that the there appears to be some interference between adjacent crossover locations. The reality of plants having large population of offsprings from the same pair of parents poses more stringent condition on the crossover patterns seen in the offspring population than is seen in humans or animals. Thus capturing the crossover events accurately in an individual chromosome as it is transmitted to its offspring, is a fundamental component of a population evolution simulator. Since the crossover event often dominates a population simulator, it determines both the accuracy as well as ultimately controls the execution speed of the simulator [16].

Various crossover models in terms of their overall statistical behavior have been proposed in literature. However, it is not obvious how they can be adapted to generating populations respecting these models since the distributions are some (indirect) nontrivial functions of the crossover frequencies. In Section 5.5, we present an utterly simple two-step probabilistic algorithm that is not only accurate but also runs in sublinear time for three interference models. Not surprisingly, the algorithm is very cryptic. The derivation of the two steps, which involves mapping the problem into a discrete framework, is discussed below. We believe that this general framework can be used to convert other such "statistical" problems into a discrete framework as well.

5.1 Background on Crossovers

Note that an even number of crossovers between a pair of loci goes unnoticed while an odd number (1, 3, 5, ..) is seen as a single crossover between the two loci.

Two point crosses. Let r_{ij} be the recombination fraction between loci i and j on the chromosome. Consider three loci 1, 2 and 3, occurring in that order in a closely placed or linked segment of the chromosome. Then:

$$r_{13} = r_{12} + r_{23} - 2r_{12}r_{23}. \tag{10}$$

Using combinatorics, and the usual interpretation of double frequency $r_{12}r_{23}$, we derive this as follows (1 is a crossover and $\cdot$ is an absence of crossover with the 4 possible distinct samples marked $i - iv$):

$$
\begin{array}{cccc}
& \begin{array}{cc} r_{12} & r_{23} \end{array} & r_{13} & r_{12}r_{23} \\
i & \begin{array}{|c|c|} \hline 1 & \cdot \\ \hline \end{array} & \begin{array}{|c|} \hline 1 \\ \hline \end{array} \text{ odd } x_1 & \cdot \\
ii & \begin{array}{|c|c|} \hline 1 & 1 \\ \hline \end{array} \Rightarrow & \begin{array}{|c|} \hline \cdot \\ \hline \end{array} \text{ even } x_2 & 1 \\
iii & \begin{array}{|c|c|} \hline \cdot & 1 \\ \hline \end{array} & \begin{array}{|c|} \hline 1 \\ \hline \end{array} \text{ odd } x_3 & \cdot \\
iv & \begin{array}{|c|c|} \hline \cdot & \cdot \\ \hline \end{array} & \begin{array}{|c|} \hline \cdot \\ \hline \end{array} \text{ even } x_4 & \cdot \\
\end{array}
$$

$$r_{13} = x_1 + x_3 = (x_1 + x_2) + (x_2 + x_3) - 2x_2 = r_{12} + r_{23} - 2r_{12}r_{23}.$$

If C is the interference factor [8, 10, 12, 13], then

$$r_{13} = r_{12} + r_{23} - 2Cr_{12}r_{23}, \tag{11}$$

5.2 Progeny as a 2D Matrix (Combinatorics)

The intent is to simulate N samples of expected length Z Morgans by generating a binary $N \times L$ matrix M where each row i corresponds to a sample and each column j corresponds to a position along the chromosomal segment (at the resolution of 1 cM). Each $M[i, j]$ of the matrix is also called a *cell*. For $1 \le i \le N$ and $1 \le j \le L$, let:

$$M[i,j] = \begin{cases} 0, & \text{if no crossover at } j \text{ cM distance from the left end of the chr in sample } i, \\ 1, & \text{if a crossover at } j \text{ cM distance from the left end of the chr in sample } i. \end{cases}$$

Then,

Definition 1 (Progeny Matrix M). M *represents a sample of N chr segments of expected length L cM if and only if*

(i)   *Expected number of 1's along a row* $= pL,$
(ii) *Expected number of 1's along a column* $= pN,$

where $p = 0.01$, based on the convention, there is a 1% chance of crossover in a chr segment of length 1 cM. Such a matrix M is called a progeny matrix.

The above also suggests an algorithm for simulating N chr segments of expected length L cM: M can be traversed (in any order) and at each cell a 1 is introduced with probability p (or 0 with probability $1 - p$). This results in a matrix M satisfying the conditions in Definition 1. In fact, when L is large and p is small, as is the case here, a Poisson distribution can be used with mean $\lambda = pL$, along a row to directly get the marker locations (the j's) with the crossovers.

Let $M[i_1, j_1] = x$ and $M[i_2, j_2] = y$ then $M[i_1, j_1] \bowtie M[i_2, j_2]$ denotes the exchange of the two values. In other words, after the $\bowtie$ operation, $M[i_1, j_1] = y$ and $M[i_2, j_2] = x$. The following is the central lemma of the section:

Lemma 1 (Exchange Lemma). *Let j be fixed. Let $I_1, I_2 \neq \phi$ be random subset of rows, and for random $i_1 \in I_1, i_2 \in I_2$, let M' be obtained after the following exchange operation:*

$$M[i_1, j] \bowtie M[i_2, j].$$

Then M' is also a progeny matrix.

This follows from the fact that each value in cell of M is independent of any other cell. □

Note that, given a progeny matrix M, if the values in some random cells are toggled (from 0 to 1 or vice-versa), the resulting matrix may not be a progeny matrix, but if a careful exchange is orchestrated respecting Lemma 1, then M continues to be a progeny matrix.

5.3 Approximations Based on Statistical Models

Recall Equation 11 that uses the two point cross model (with interference):

$$r_{13} = r_{12} + r_{23} - 2Cr_{12}r_{23}.$$

*Conjecture 1 ((ϵ, t)-**Interference Conjecture**).* For a (1) given interference function $C = f(r)$, (2) p and (3) a small small $\epsilon > 0$, there exists progeny matrix M satisfying the following conditions called the interference condition:

$$|r_{j_1 j_2} + r_{j_2 j_3} - 2f(r)r_{j_1 j_2}r_{j_2 j_3} - r_{j_1 j_3}| < \epsilon, \forall j_2, \tag{12}$$

$$\sum_{l=j_1}^{j_3} M[i, l] \leq 2, \tag{13}$$

where $j_1 = j_2 - t$ and $j_3 = j_2 + t$ and t is a small integer based on ϵ.

When $C = 1$ then $t = 0$ following the arguments presented in Section 5.1. Recall that $C = 1$ for the Haldane model and $C = 0$ for the Morgan model. For the Kosambi model, where $C = 2r$, we determine, t, the distance between the two points of the two point cross model empirically, for $\epsilon = 0.001$.

Definition 2 (M_C Respecting an Interference Model C). *For a given interference function $C = f(r)$, if a progeny matrix satisfies the interference conditions (of Equations 12 and 13) for some ϵ then it is said to respect the interference model $C = f(r)$ and is written as M_C.*

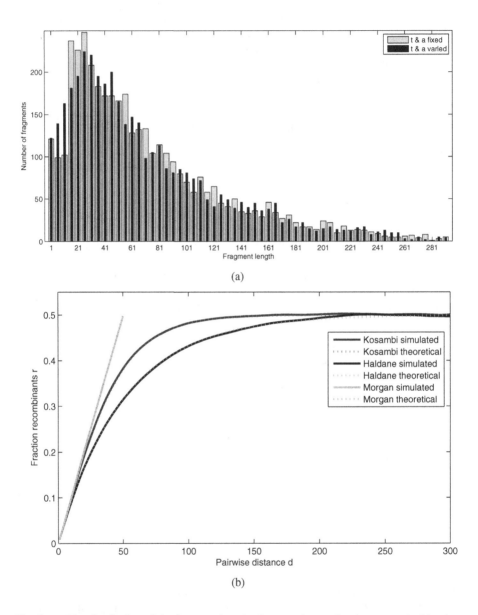

(a)

(b)

Fig. 5. (a) The distribution of the fragment length when t and a are fixed, compared with when they are picked at random from a small neighbourhood, in the Kosambi model. The latter gives a smoother distribution than the former (notice the jump at fragment length 20 in the fixed case). (b) Distance d (cM) versus recombination fraction r, for closed form solutions according to the Kosambi, Haldane, and Morgan models, and for observed data in the simulations. The results show average values from constructing a 5 Morgan chromosome $2,000$ times for each model.

5.4   Some Probability Calculations

Assume submatrix $M_{j_1 j_3}$ of M from columns j_1 to j_3 of Conjecture 1 and let $t = \mathfrak{t}$. Then,

Lemma 2

$$\textit{Fraction of samples with } (M[i, j_2] = 1) = p = \beta, \tag{14}$$

$$\textit{fraction of samples with } \left(\sum_{l=j_1, l \neq j_2}^{j_3} M[i, l] = 0 \right) \approx (1 - p)^{at} = \alpha, \tag{15}$$

where a is a correction factor. When the crossovers are assigned randomly and independently in M, due to the interference model C, the fraction with $M[\cdot, j_2] = 1$ that are in violation, are

$$(1 - \alpha) \times \beta \times (1 - C).$$

Also fraction that could potentially be exchanged with the violating i's while respecting the interference model are

$$\alpha \times (1 - \beta).$$

Thus for Kosambi model, $C = 2r \approx 2p$:

$$\textit{Violating fraction with } (M[\cdot, j_2] = 1) \approx (1 - (1 - p)^{at}) \times p \times (1 - 2p) = \gamma, \tag{16}$$

$$\textit{potential exchange fraction} \approx (1 - p)^{at} \times (1 - p) = \eta. \tag{17}$$

The correction factor a is empirically estimated. Thus an iterative procedure can exchange the violating fraction (γ) randomly with the values in the potential exchange fraction (η). Thus based on Lemma 1, M continues to be a progeny matrix. Since conditions of Conjecture 1 hold, the transformed matrix also respects the interference model. Thus the "interference adjustment" probability p' in Kosambi model is defined as:

$$p' \approx \frac{\gamma}{\eta} = p(1 - 2p) \frac{1 - (1 - p)^{at}}{(1 - p)^{at+1}}. \tag{18}$$

For Morgan model $C = 0$:

$$\gamma = (1 - \alpha)\beta, \eta = \alpha(1 - \beta), p' = p \frac{1 - (1 - p)^{at}}{(1 - p)^{at}(1 - p)}.$$

For Haldane model $C = 1$: $\eta = 0, p' = 0$. Now, we are ready for the details of the algorithm where each row of M_C can be computed independently using the two probabilities p and p'.

5.5   Algorithm

We encapsulate the above into a framework to generate crossovers based on the mathematical model of Eqn 11 and the generic interference function of the form $C = f(r)$. In this general framework a and t of Eqn 19 are estimated empirically, to match the expected r curves with $\epsilon = 0.001$ (of Conjecture 1). We present these estimated values for the $C = 2r$, $C = 1$ and $C = 0$ models in Eqn 19.

Crossover probability p. For historical reasons, the lengths of the chromosome may be specified in units of Morgan, which is the expected distance between two crossovers. Thus in a chromosomal segment of length 1 centiMorgan (cM), i.e., one-hundredth of M, there is only 1% chance of a crossover. Thus, in our representation each cM is a single cell in the matrix representation of the population leading to a crossover probability p at each position as $p = 0.01$ and this single cell may correspond to 1 nucleotide base [9, 10, 14].

INPUT: Length of chromosome: Z Morgans or $Z \times 100$ centiMorgans (cM).

ASSUMPTION: 1 cM is the resolution, i.e., is a single cell in the representation vector/matrix.

OUTPUT: Locations of crossover events in a chromosome.

ALGORITHM: Let

$$L = Z \times 100, \text{ (specified input length)}$$

$$p = 0.01,$$

$$(a, t) = \begin{cases} (-, 0) & \text{if } C = 1 \text{ (Haldane model)}, \\ (X_{1.1}, X_{16}) & \text{if } C = 2r \text{ (Kosambi model)}, \\ (X_{1.65}, X_{50}) & \text{if } C = 0 \text{ (Morgan model)}, \end{cases} \quad (19)$$

$$q = \begin{cases} 0 & \text{if } C = 1 \text{ (Haldane model)}, \\ 1 - 2p & \text{if } C = 2r \text{ (Kosambi model)}, \\ 1 & \text{if } C = 0 \text{ (Morgan model)}, \end{cases} \quad (20)$$

$$p' = pq \frac{1 - (1 - p)^{at}}{(1 - p)^{at+1}},$$

where X_c is a random variable drawn from a uniform distribution on $[b, d]$, for some $b < d$, where $c = (b + d)/2$. For example, uniform discrete distribution on $[1, 31]$ for t and uniform continuous distribution on $[1.0, 1.2]$ for a.

For each sampled chromosome:

Step 1. Draw the number of positions from a Poisson distribution with $\lambda = pL$. For each randomly picked position j, introduce a crossover. If crossovers in any of the previous t or next t positions (in cM) then the crossover at j is removed with probability q. [Interference]

Step 2. Draw the number of positions from a Poisson distribution with $\lambda' = p'L$. For each randomly picked position j', if no crossovers in the previous t and next t positions then a crossover is introduced at j'. [Interference adjustment]

Fragment lengths. Note that the careful formulation does not account for the following summary statistic of the population: the distribution of the length of the fragments, produced by the crossovers. It turns out that the fragment length is controlled by the choice of the empirical values of X_t and X_a in Eqn 19. In Fig 5 (a), we show the fragment length distribution where two values are fixed at $t = 16$ and $a = 1.1$ respectively in the Kosambi model. However, when the t and a are picked uniformly from a small neighborhood around these values, we obtain the distribution which is more acceptable by practitioners.

Running time analysis. The running time analysis is rather straightforward. Let c_p be the time associated with a Poisson draw and c_u with a uniform draw. Ignoring the initialization time and assuming t is negligible compared to L, the expected time taken by the above algorithm for each sample is $2c_p + (Z+1)c_u$. Since the time is proportional to the number of resulting crossovers, the running time is optimal.

6   Discussion and Open Problems

We achieved linearity in the method for simulating the demes by showing that a fairly small number of constraints, k, is sufficient for producing excellent results. In the full version of the paper, we report the comparison studies, where it not only gives a better fit than all the other methods, but additionally also is more sensitive to the input. Most methods are insensitive to the nuanced changes in the input while our non-generative method detects them, which is reflected in the output. We hypothesize that effectiveness of the small value of k is due to the fact that it avoids overfitting and is suitable for this problem setting since the input parameters are in terms of distributions, with non-zero variances.

With the availability of resequencing data, in plants as well as humans, the density of SNP markers continues to increase. However, this is not likely to affect the overall LD distributions already seen in the data with less dense markers. Under higher density markers, but similar LD distributions, a larger number of constraints (k) can be satisfied exactly by the algebraic method. But larger values of k would increase the running time. An option is to maintain the current density of markers and do a best-fit of the markers between a pair of adjacent low-density markers. Note that the hardest part of the deme problem is to fit high values of LD between a pair of markers. The theoretical upper limit on the LD between a pair of markers (as a function of the two MAFs) can be computed and we are currently already handling these values quite successfully. Based on this observation, we hypothesize that a simple best-fit interpolation between low density markers will be very effective. It is an interesting open question to devise such linear time methods that scale the core algorithms presented here.

A natural question arises regarding the specification of a deme: What are the other characteristics, independent of the MAF and LD? Further, can they be quantified? The latter question helps in measuring and evaluating the accuracy of a simulator. Although, somewhat non-intuitive, we find that MAF and LD distribution very effectively define the desired characteristics of a deme. We also studied the fragmentation by haplotype blocks of the population constructed by the non-generative approach. The problem of characterizing the fragmentation of the haplotypes in a deme and its exact relationship to the LD distribution is an interesting open problem.

Acknowledgments. We are very grateful to the anonymous reviewers for their excellent suggestions which we incorporated to substantially improve the paper.

References

1. Balloux, F.: EASYPOP (Version 1.7): A Computer Program for Population Genetics Simulations. Journal of Heredity 92, 3 (2001)
2. Peng, B., Amos, C.I.: Forward-time simulations of non-random mating populations using simuPOP. Bioinformatics 24, 11 (2008)
3. Montana, G.: HapSim: a simulation tool for generating haplotype data with pre-specified allele frequencies and LD coefficients. Bioinformatics 21, 23 (2005)
4. Yuan, X., Zhang, J., Wang, Y.: Simulating Linkage Disequilibrium Structures in a Human Population for SNP Association Studies. Biochemical Genetics 49, 5–6 (2011)
5. Shang, J., Zhang, J., Lei, X., Zhao, W., Dong, Y.: EpiSIM: simulation of multiple epistasis, linkage disequilibrium patterns and haplotype blocks for genome-wide interaction analysis. Genes & Genomes 35, 3 (2013)
6. Peng, B., Chen, H.-S., Mechanic, L.E., Racine, B., Clarke, J., Clarke, L., Gillanders, E., Feuer, E.J.: Genetic Simulation Resources: a website for the registration and discovery of genetic data simulators. Bioinformatics 29, 8 (2013)
7. Balding, D., Bishop, M., Cannings, C.: Handbook of Statistical Genetics, 3rd edn. Wiley J. and Sons Ltd. (2007)
8. Vinod, K.: Kosambi and the genetic mapping function. Resonance 16(6), 540–550 (2011)
9. Kearsey, M.J., Pooni, H.S.: The genetical analysis of quantitative traits. Chapman & Hall (1996)
10. Lynch, M., Walsh, B.: Genetics and Analysis of Quantitative Traits. Sinauer Associates (1998)
11. The International HapMap Consortium: The International HapMap Project. Nature 426, pp. 789–796 (2003)
12. Haldane, J.B.S.: The combination of linkage values, and the calculation of distance between linked factors. Journal of Genetics 8, 299–309 (1919)
13. Kosambi, D.D.: The estimation of map distance from recombination values. Journal of Genetics 12(3), 172–175 (1944)
14. Cheema, J., Dicks, J.: Computational approaches and software tools for genetic linkage map estimation in plants. Briefings in Bioinformatics 10(6), 595–608 (2009)
15. Zhang, K., Deng, M., Chen, T., Waterman, M.S., Sun, F.: A dynamic programming algorithm for haplotype block partitioning. Proceedings of the National Academy of Sciences 19(11), 7335–7339 (2002)
16. Podlich, D.W., Cooper, M.: QU-GENE: a simulation platform for quantitative analysis of genetic models. Bioinformatics 14(7), 632–653 (1998)

GDNorm: An Improved Poisson Regression Model for Reducing Biases in Hi-C Data

Ei-Wen Yang[1] and Tao Jiang[1,2,3]

[1] Department of Computer Science and Engineering, University of California, Riverside, CA
[2] Institute of Integrative Genome Biology, University of California, Riverside, CA
[3] School of Information Science and Technology, Tsinghua University, Beijing, China
{yyang027,jiang}@cs.ucr.edu

Abstract. As a revolutionary tool, the Hi-C technology can be used to capture genomic segments that have close spatial proximity in three dimensional space and enable the study of chromosome structures at an unprecedentedly high throughput and resolution. However, during the experimental steps of Hi-C, systematic biases from different sources are often introduced into the resultant data (*i.e.*, reads or read counts). Several bias reduction methods have been proposed recently. Although both systematic biases and spatial distance are known as key factors determining the number of observed chromatin interactions, the existing bias reduction methods in the literature do not include spatial distance explicitly in their computational models for estimating the interactions. In this work, we propose an improved Poisson regression model and an efficient gradient descent based algorithm, GDNorm, for reducing biases in Hi-C data that takes spatial distance into consideration. GDNorm has been tested on both simulated and real Hi-C data, and its performance compared with that of the state-of-the-art bias reduction methods. The experimental results show that our improved Poisson model is able to provide more accurate normalized contact frequencies (measured in read counts) between interacting genomic segments and thus a more accurate chromosome structure prediction when combined with a chromosome structure determination method such as ChromSDE. Moreover, assessed by recently published data from human lymphoblastoid and mouse embryonic stem cell lines, GDNorm achieves the highest reproducibility between the biological replicates of the cell lines. The normalized contact frequencies obtained by GDNorm is well correlated to the spatial distance measured by florescent in situ hybridization (FISH) experiments. In addition to accurate bias reduction, GDNorm has the highest time efficiency on the real data. GDNorm is implemented in C++ and available at http://www.cs.ucr.edu/~yyang027/gdnorm.htm

Keywords: chromosome conformation capture, Hi-C data, systematic bias reduction, Poisson regression, gradient descent.

1 Introduction

Three dimensional (3D) conformation of chromosomes in nuclei plays an important role in many chromosomal mechanisms such as gene regulation, DNA

D. Brown and B. Morgenstern (Eds.): WABI 2014, LNBI 8701, pp. 263–280, 2014.
© Springer-Verlag Berlin Heidelberg 2014

replication, maintenance of genome stability, and epigenetic modification [1]. Alterations of chromatin 3D conformations are also found to be related to many diseases including cancers [2]. Because of its importance, the spatial organization of chromosomes has been studied for decades using methods of varying scale and resolution. However, owing to the high complexity of chromosomal structures, understanding the spatial organization of chromosomes and its relation to transcriptional regulation is still coarse and fragmented [3].

An important approach for studying the spatial organization of chromosomes is florescent in situ hybridization (FISH) [4]. In FISH-based methods, florescent probes are hybridized to genomic regions of interests and then the inter-probe distance values on two dimensional fluorescence microscope images are used as the measurement for spatial proximity of the genomic regions. Because FISH-based methods rely on image analysis involving a few hundred cells under the microscope, they are generally considered to be of low throughput and resolution [2]. Recently, the limitation of throughput and resolution was alleviated by the introduction of the 3C technology that is able to capture the chromatin interaction of two given genomic regions in a population of cells by using PCR [5]. Combining this with microarray and next generation sequencing technologies has yielded more powerful variants of the 3C methods. For example, 4C methods [6,7] can simultaneously capture all possible interacting regions of a given genomic locus in 3D space while 5C methods can further identify complete pairwise interactions between two sets of genomic loci in a large genomic region of interests [8]. However, when it comes to genome-wide studies of chromatin interactions, 5C methods require a very large number of oligonucleotides to evaluate chromatin interactions for an entire genome. The cost of oligonucleotide synthesis makes the 5C methods unsuitable for genome-wide studies [2]. To overcome this issue, another NGS-based variant of the 3C technology, called Hi-C, was proposed to quantify the spatial proximity of the conformations of all the chromosomes [3]. By taking advantages of the NGS technology, Hi-C can quantify the spatial proximity between all pairs of chromosomal regions at an unprecedentedly high resolution. As a revolutionary tool, the introduction of Hi-C facilitates many downstream applications of chromosome spatial organization studies such as the discovery of the consensus conformation in mammalian genomes [9], the estimation of conformational variations of chromosomes within a cell population [10], and the discovery of a deeper relationship between genome spatial structures and functions [11].

The Hi-C technology involves the generation of DNA fragments spanning genomic regions that are close to each other in 3D space in a series of experimental steps, such as formaldehyde cross-linking in solution, restriction enzyme digestion, biotinylated junctions pull-down, and high throughput paired-end sequencing [3]. The number of DNA fragments spanning two regions is called the *contact frequency* of the two regions. The physical (spatial) distance between a pair of genomic regions is generally assumed to be inversely proportional to the contact frequency of the two regions and hence the chromosome structure can in principle be recovered from the contact frequencies between genomic regions [10,12].

However, during the experimental steps of Hi-C, systematic biases from different sources are often introduced into contact frequencies. Several systematic biases were shown to be related to genomic features such as number of restriction enzyme cutting sites, GC content and sequence uniqueness in the work of Yaffe and Tanay [13]. Without being carefully detected and eliminated, these systematic biases may distort many down-stream analyses of chromosome spatial organization studies. To remove such systematic biases, several bias reduction methods have been proposed recently. These bias reduction methods can be divided into two categories, the normalization methods and bias correction methods according to [2]. The normalization methods, such as ICE [14] and the method in [15], aims at reducing the joint effect of systematic biases without making any specific assumption on the relationships between systematic biases and related genomic features. Their applications are limited to the study of equal sized genomic loci [2]. In contrast, the bias correction methods, such as HiCNorm [16] and the method of Yaffe and Tanay (YT) [13], build explicit computational models to capture the relationships between systematic biases and related genomic features that can be used to eliminate the joint effect of the biases.

Although it is well known that observed contact frequencies are determined by both systematics biases and spatial distance between genomic segments, the existing bias correction methods do not take spatial distance into account explicitly. This incomplete characterization of causal relationships for contact frequencies is known to cause problems such as poor goodness-of-fitting to the observed contact frequency data [16]. In this paper, we build on the work in [16] and propose an improved Poisson regression model that corrects systematic biases while taking spatial distance (between genomic regions) into consideration. We also present an efficient algorithm for solving the model based on gradient descent. This new bias correction method, called GDNorm, provides more accurate normalized contact frequencies and can be combined with a distance-based chromosome structure determination method such as ChromSDE [12] to obtain more accurate spatial structures of chromosomes, as demonstrated in our simulation study. Moreover, two recently published Hi-C datasets from human lymphoblastoid and mouse embryonic stem cell lines are used to compare the performance of GDNorm with the other state-of-the-art bias reduction methods including HiCNorm, YT and ICE at 40kb and 1M resolutions. Our experiments on the real data show that GDNorm outperforms the existing bias reduction methods in terms of the reproducibility of normalized contact frequencies between biological replicates. The normalized contact frequencies by GDNorm are also found to be highly correlated to the corresponding FISH distance values in the literature. With regard to time efficiency, GDNorm achieves the shortest running time on the two real datasets and the running time of GDNorm increases linearly with the resolution of data. Since more and more high resolution ($e.g.$, 5 to 10kb) data are being used in the studies of chromosome structures [17], the time efficiency of GDNorm makes it a valuable bias reduction tool, especially for studies involving high resolution data.

The rest of this paper is organized as follows. Section 2.1 presents several genomic features that are used in our improved Poisson regression model. The details of the model as well as the gradient descent algorithm are described in Section 2.2. Several experimental results on simulated and real human and mouse data are presented in Section 3. Section 4 concludes the paper.

2   Methods

2.1   Genomic Features

A chromosome g can be binned into several disjoint and consecutive genomic segments. Given an ordering to concatenate the chromosomes, let $S = \{s_1, s_2, ..., s_n\}$ be a linked list representing all n genomic segments of interest such that the linear order of the segments in S is consistent with the sequential order in the concatenation of the chromosomes. For each genomic segment s_i, the number of restriction enzyme cutting sites (RECSs) within s_i is represented as R_i. The GC content G_i of segment s_i is the average GC content within the 200 bps region upstream of each RECS in the segment. The sequence uniqueness U_i of segment s_i is the average sequence uniqueness of 500 bps region upstream or downstream of each RECS. To calculate the sequence uniqueness for a 500 bps region, we use a sliding window of 36bps to synthesize 55 reads of 35 bps by taking steps of 10bps from 5′ to 3′ as done in [16]. After using the BWA algorithm [18] to align the 55 reads back to the genome, the percentage of the reads that is still uniquely mapped in the 500 bps region is considered as the sequence uniqueness for the 500 bps region. These three major genomic features have been shown to be either positively or negatively correlated to contact frequencies in the literature [13]. In the following, we will present a new bias correction method based on gradient search to eliminate the joint effect of the systematic biases correlated to the three genomic features, building on the Poisson regression model introduced in [16].

2.2   A Bias Correction Method Based on Gradient Descent

Let $F = \{f_{i,j} | 1 \leq i \leq n, 1 \leq j \leq n\}$ be the contact frequency matrix for the genomic segments in S such that each $f_{i,j}$ denotes the observed contact frequency between two segments s_i and s_j. HiCNorm [16] assumes that the observed contact frequency $f_{i,j}$ follows a Poisson distribution with rate determined by the joint effect of systematic biases and represents the joint effect as a log-linear model of the three genomic features mentioned above (*i.e.*, the number of RECSs, GC content and sequence uniqueness). In other words, if the Poisson distribution rate of $f_{i,j}$ is $\theta_{i,j}$, then

$$\log(\theta_{i,j}) = \beta_0 + \beta_{recs}\log(R_iR_j) + \beta_{gcc}\log(G_iG_j) + \beta_{seq}\log(U_iU_j), \quad (1)$$

where β_0 is a global constant, β_{recs}, β_{gcc} and β_{seq} are coefficients for the systematic biases correlated to RECS, GC content and sequence uniqueness, and

R_i, G_i and U_i are the number of RECSs, GC content and sequence uniqueness in segment s_i, respectively. The coefficient β_{seq} was fixed at 1 in [16] so the term $\log(U_iU_j)$ acts as the Poisson regression offset when estimating $\theta_{i,j}$.

However, this log-linear model does not capture all known causal relationships that affect the observed contact frequency $f_{i,j}$, because the spatial distance $d_{i,j}$ is not included in the model. To characterize more comprehensive causal relationships for observed contact frequencies, in a recently published chromosome structure determination method BACH [10], the spatial distance was modeled explicitly such that

$$\log(\theta_{i,j}) = \beta_0 + \beta_{dist}\log(d_{i,j}) + \beta_{recs}\log(R_iR_j) + \beta_{gcc}\log(G_iG_j) + \beta_{seq}\log(U_iU_j), \quad (2)$$

where $\beta = \{\beta_{recs}, \beta_{gcc}, \beta_{seq}\}$ again represents the systematic biases, β_{dist} represents the *conversion factor* and $D = \{d_{i,j}|1 \leq i \leq n, i < j\}$ are variables representing the spatial distance values to be estimated. However, without any constraint or assumption on spatial distance, the model represented by Eq. 2 is non-identifiable, because for any constant k, $\beta_{dist}\log(d_{i,j}) = k \times \beta_{dist}\log(d_{i,j}^{1/k})$. BACH solved this issue by introducing some spatial constraints from previously predicted chromosome structures. (Eq. 2 was used by BACH to iteratively refine the predicted chromosome structure.) Hence, Eq. 2 is infeasible for bias correction methods that do not rely on any spatial constraint. To get around this, we introduce a new variable $z_{i,j} = \beta_0 + \beta_{dist}\log(d_{i,j})$ and rewrite Eq. 2 as follows:

$$\log(\theta_{i,j}) = z_{i,j} + \beta_{recs}\log(R_iR_j) + \beta_{gcc}\log(G_iG_j) + \beta_{seq}\log(U_iU_j), \quad (3)$$

where the systematic biases β and $Z = \{z_{i,j}|1 \leq i \leq n, i < j\}$ are the variables to be estimated. Note that applying a Poisson distribution on read count data sometimes leads to the overdispersion problem, *i.e.*, underestimation of the variance [19], which is generally solved by using a negative binomial distribution instead. However, the results in [16] suggest that there is usually no significant difference in the performance of bias correction methods when a negative binomial distribution or a Poisson distribution is applied to Hi-C data. For the mathematical simplicity of our model, we use Poisson distributions.

Let θ denote the set of $\theta_{i,j}, 1 \leq i \leq n, 1 \leq j \leq n$. Given the observed contact frequency matrix F and genomic features of S, the log-likelihood function of the observed contact frequencies over the Poisson distribution rates can be written as:

$$\log(Pr(F|\beta,Z)) = \log(Pr(F|\theta)) = \log(\prod_{i=1,i<j}^{n} Pr(f_{i,j}|\theta_{i,j})) = \log(\prod_{i=1,i<j}^{n} \frac{e^{-\theta_{i,j}}\theta^{f_{i,j}}}{f_{i,j}!})$$

$$= \sum_{i=1,i<j}^{n} -\theta_{i,j} + f_{i,j}\log(\theta_{i,j}) - \log(f_{i,j}!). \quad (4)$$

We can estimate the variables Z and systematic biases β by finding parameters $x^* = \{\beta^*, Z^*\}$ to maximize the log-likelihood function in Eq. (4), which is equivalent to solving the following multivariate optimization problem:

$$x^* = \arg\min_{x} -\log(Pr(F|\beta,Z)) = \arg\min_{x} -\log(Pr(F|\theta)) = \arg\min_{x} \sum_{i=1,i<j}^{n} \theta_{i,j} - f_{i,j}\log(\theta_{i,j}) \quad (5)$$

However, without any constraint on the variables Z, the above model is still generally non-identifiable since for any β, we can always choose a $z_{i,j}$ such that $f_{i,j} = \theta_{i,j}$ and the likelihood function is maximized. Therefore, we require that for any i, j, $|z_{i,i+1} - z_{j,j+1}| \leq \epsilon$ for some threshold ϵ, since we expect that the distance between neighboring segments is roughly the same across a chromosome.

Observe that Eq. 5 cannot be solved by using the same Poisson regression fitting method as in HiCNorm, because Eq. 5 is no longer a standard log-linear model like Eq. 1. A popular technique for solving multivariate optimization problems is gradient descent. Gradient descent searches the optimum of a minimization problem with an objective function $g(x)$ from a given initial point x_1 at the first iteration and then iteratively moves toward a local minimum by following the negative of the gradient function $-\nabla g(x)$. In other words, at every iteration i, we compute $x_i \leftarrow x_{i-1} - \alpha \nabla g(x)$, where α is a constant. In our case, the objective function to be minimized is the negative of the above log-likelihood function $g(x) = g(\beta, Z) = -\log(Pr(F|\beta, Z))$. By taking partial derivatives of the objective function with respect to the variables β and Z, we have the gradient function $-\nabla g(x) = \{\frac{\partial g(x)}{\partial \beta}, \frac{\partial g(x)}{\partial Z}\}$ as

$$\frac{\partial g(\beta, Z)}{\partial z_{i,j}} = \theta_{i,j} - f_{i,j}$$

$$\frac{\partial g(\beta, Z)}{\partial \beta_{recs}} = \sum_{i=1, i<j}^{n} \log(R_i R_j)(\theta_{i,j} - f_{i,j})$$

$$\frac{\partial g(\beta, D)}{\partial \beta_{gcc}} = \sum_{i=1, i<j}^{n} \log(G_i G_j)(\theta_{i,j} - f_{i,j})$$

$$\frac{\partial g(\beta, D)}{\partial \beta_{seq}} = \sum_{i=1, i<j}^{n} \log(U_i U_j)(\theta_{i,j} - f_{i,j})$$

To initialize $x_1 = \{\beta^1, Z^1\}$, we first set the variable $z_{i,i+1}$ as a uniform constant z for every two neighboring segments, s_i and s_{i+1}, because we assume that the distance between every pair of neighboring segments is similar. The systematic biases are then initialized as β^1 by solving Eq. 1, with $z = \beta_0$, on neighboring segments only. To obtain initial variables $z_{i,j}$, where $j - i > 1$, $\theta_{i,j}$ is sampled from the conjugate prior of Poisson distribution $\Gamma(1, f_{i,j} + 1)$ and then $z_{i,j}$ is calculated by using Eq. 3 with the fixed parameter β^1. After the convergence of the gradient descent search, the normalized contact frequency $\hat{f}_{i,j}$ is computed by $\hat{f}_{i,j} = f_{i,j}/\{(R_i R_j)^{\beta_{recs}}(G_i G_j)^{\beta_{gcc}}(U_i U_j)^{\beta_{seq}}\}$. Our complete algorithm for GDNorm is summarized in Algorithm 1. Here, N_{max} denotes the maximum number of iterations allowed and its default is set to be 10 based on our empirical observation that the gradient descent search usually converges in no more than 10 iterations.

Input : Contact frequency matrix F and genomic features R, G and U
Output: Normalized contact frequency $\hat{F}$
begin

    Spatial Distance and Systematic Bias Estimation:
    Initialize $x_1 = \{\beta^1, Z^1\}$;
    for i *from 2 to* N_{max} **do**
        $x_i \leftarrow x_{i-1} - \alpha \nabla g(x)$;
        if$(g(x_i) > g(x_{i-1}))$
        Go to Contact Frequency Normalization;
    Contact Frequency Normalization:
    for $i < j$ **do**
        $\hat{f}_{i,j} = f_{i,j}/\{(R_iR_j)^{\beta_{recs}}(G_iG_j)^{\beta_{gcc}}(U_iU_j)^{\beta_{seq}}\}$
    return $\hat{F}$;

Algorithm 1. Bias Reduction Based on Gradient Descent

3 Experimental Results

We assess the performance of GDNorm in terms of (i) the accuracy of its normalized contact frequencies and (ii) the accuracy of structure determination using the normalized contact frequencies. The latter will be done by simulating biased Hi-C read count data from some simple reference chromosome structures and then trying to recover the reference structures from normalized contact frequencies in combination with the most recent chromosome structure determination algorithm, ChromSDE [12]. In other words, we will consider the impact of normalized contact frequencies on the chromosome structures predicted by ChromSDE. To measure the quality of bias correction, we consider the reproducibility of normalized contact frequencies between biological replicates of an mESC line [9] and the correlation between normalized contact frequencies and FISH distance values in the literature. The performance of GDNorm will be compared with the state-of-the-art bias reduction algorithms HiCNorm [16], YT [13] and ICE [14].

3.1 Simulation Studies

To evaluate the accuracy of chromosome structure prediction, two reference 3D structures, a helix and an arbitrary random walk, are constructed as shown in Fig. 1. In order to be close to the real chromosome structure prediction practice, each of the reference 3D structures consists of 44 segments, where the number 44 was determined by the average size of the chromosomal structure units studied in [10] (*i.e.*, conserved domains). Let B_i denote the systematic bias in a segment s_i and $T_{i,j}$ the true (unbiased) contact frequency between segments s_i and s_j. To synthesize observed contact frequencies $f_{i,j}$, we follow the assumption $f_{i,j} = T_{i,j}B_iB_j$ as in [14]. Here, $T_{i,j}$ is assumed to be inversely proportional to the spatial distance $d_{i,j}$. That is, $T_{i,j} = d_{i,j}^{\rho}$, where $\rho < 0$ is called the conversion factor between the unbiased contact frequency and its corresponding

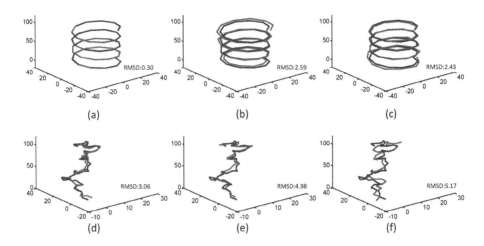

Fig. 1. Alignment between the reference chromosome 3D structures and structures predicted by GDNorm$_{sde}$, HiCNorm$_{sde}$ and BACH on simulated data. The red curves indicate the predicted structures and blue curves the reference structures. The results of GDNorm$_{sde}$, HiCNorm$_{sde}$ and BACH are shown from left to right. The top row is for the helix and bottom for the random walk. The quality of each structural alignment is evaluated by an RMSD value.

spatial distance. The value of $B_i B_j$ is estimated by using the log-linear function $\log(B_i B_j) = \beta_0 + \beta_{recs}\log(R_i R_j) + \beta_{gcc}\log(G_i G_j) + \beta_{seq}\log(U_i U_j)$ introduced in [16]. The coefficient β_{seq} is set to 1 as in [16] while ρ is set to -1.2 as estimated from a mouse cell line by ChromSDE [12]. To determine the coefficients β_0, β_{recs} and β_{gcc}, HiCNorm is run on the mm9 mESC data to form a pool of coefficients. A set of coefficients β_0, β_{recs} and β_{gcc} are then randomly drawn from the pool and used throughout the simulation study.

Because currently there is no tool to synthesize Hi-C reads reasonably from a given 3D structure and the methods YT and ICE require actual Hi-C reads as input, they are excluded from this simulation study but will be discussed in the real data experiments in the section 3.2. The method GDNorm and HiC-Norm are run on the simulated contact frequencies and their normalized contact frequencies are then used to predict chromosome 3D structures. Two structure prediction software, MCMC5C and ChromSDE, in the literature use normalized contact frequencies to predict chromosome 3D structures [20,12]. Here, we choose ChromSDE, instead of MCMC5C, as the structure prediction method because MCMC5C is not specific to Hi-C data and ChromSDE significantly outperformed MCMC5C in the most recent study [12]. The combination of HiCNorm and ChromSDE is denoted as HiCNorm$_{sde}$ while the combination of GDNorm and ChromSDE is called as GDNorm$_{sde}$ in the following discussion. To further study the performance of GDNorm$_{sde}$ and HiCNorm$_{sde}$ as chromosome structure prediction tools on biased Hi-C data, another independent prediction method, BACH [10], is also included in our comparisons. Note that BACH always normalizes the size of its predicted structure by fixing the distance between the first and the last segments to be 1 while ChromSDE does not perform this normalization. To obtain a fair comparison, we calibrate the predicted structure sizes in GDNorm$_{sde}$ and HiCNorm$_{sde}$ such that the distance between the first and last segment is fixed at 100. Finally, the accuracy of structure prediction is assessed using the root mean square difference (RMSD) measure after optimally aligning a predicted structure to the reference structure by Kabsch's algorithm [21].

GDNorm Provides the Most Accurate Chromosome Structure Prediction on Noise-Free Data. The optimal alignments of the predicted and reference chromosome structures are shown together with their RMSD values in Figure 1. In the structure predictions for both the helix and random walk, GDNorm$_{sde}$ predicted the chromosome structures with the minimum RMSDs. In the structure prediction for the helix, GDNorm$_{sde}$ obtained a structure that can be almost perfectly aligned with the reference structure with a very small RMSD value of 0.3. This is because GDNorm was able to significantly reduce the effect of systematic bias and the semi-definite programming method employed by ChromSDE can guarantee perfect recovery of a chromosome structure when the given distance values between segments are noise-free.

GDNorm Reduces Systematic Biases Significantly in Noise-Free Data. To examine how much the effect of systematic biases can be reduced by the selected bias reduction methods, we further analyze the predicted spatial distance values between neighboring segments in the structure prediction for the helix. Because the spatial distance between neighboring segments s_i and s_{i+1} in the reference structure of the helix is the same for all i, the difference in the observed contact frequency between s_i and s_{i+1}, for different i, is mainly a result of the systematic biases. If the systematic biases are correctly estimated and eliminated, the distance between any two consecutive segments in the predicted structure is expected to be the same. The spatial distance values between 10 pairs of consecutive segments with the greatest systematic biases are compared with the distance values between 10 pairs with the smallest systematic biases for each of the chromosome structures predicted by GDNorm$_{sde}$, HiCNorm$_{sde}$ and BACH. The box plots in Figure 2 summarizes the comparison results. The absolute differences between the means of the two sets of 10 distance values obtained by GDNorm$_{sde}$, HiCNorm$_{sde}$ and BACH are 0.045, 3.47 and 2.61, respectively. The statistical significance of the difference between two sets of 10 distance values obtained by each method is also examined by a two-tailed t-Test [22], which yielded a non-significant p-value of 0.42 for GDNorm$_{sde}$ and significant p-values of 1.3×10^{-12} and 1.56×10^{-6} for HiCNorm$_{sde}$ and BACH, respectively.

GDNorm Provides the Most Accurate Chromosome Prediction on Noisy Data. We have demonstrated the superior performance of GDNorm$_{sde}$ on Hi-C data without noise (but with systematic biases). To test its performance on noisy data, a uniformly random noise $\delta_{i,j}$ is injected into every contact frequency $f_{i,j}$ such that the noisy frequency $\tilde{f}^{ij} = f_{i,j}(1+\delta_{i,j})$. In this test, we consider two noise levels, 30% and 50%. Table 1 summarizes the RMSD values of the optimal alignments between the predicted structures and the reference structures. The

Table 1. RMSD values of the predicted structures on noisy data

Reference Structure	Noise Level	GDNorm$_{sde}$	HiCNorm$_{sde}$	BACH
Helix	30%	2.65	3.33	14.9
	50%	4.19	4.26	20.0
Random Walk	30%	4.26	6.40	5.26
	50%	5.17	7.11	6.43

results show GDNorm$_{sde}$ still outperforms the other two methods by achieving the overall smallest RMSD values at both noise levels. Note that BACH failed to predict the helix structure at both noise levels in this test, perhaps because its MCMC algorithm could sometimes be trapped in a local optimum when the input data contains a significant level of noise.

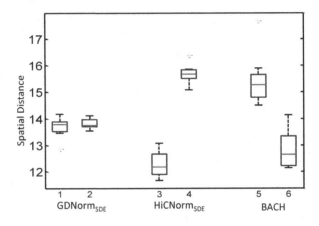

Fig. 2. Comparison of the predicted spatial distance values with the 10 greatest and 10 smallest systematic biases. For each structure prediction method studied, two sets of 10 distance values form the two boxes in a comparison group. The left box depicts the distribution of the distance values for contacts with the greatest systematic biases while the right shows the distribution of the distance values for contacts with the smallest systematic biases. Clearly, GDNorm$_{sde}$ produced the most consistent distance values and HiCNorm$_{sde}$ the least.

3.2   Performance on Real Hi-C Data

In addition to the simulation study, several experiments on real Hi-C data are conducted to evaluate the bias reduction capability of GDNorm, in comparison with other state-of-the-art bias reduction methods, HiCNorm, YT and ICE. Unlike the assessment in the previous simulation study, the reference structures for real Hi-C datasets are hardly obtainable because of the complexity of chromosome structures. To compare the performance of the studied bias reduction methods on real Hi-C data, a commonly used evaluation criterion is the similarity (or reproducibility) between normalized contact frequency matrices from biological replicates using different enzymes. Since these replicates are derived from the same chromosomal structures in the cell line, the contact frequencies normalized by a robust bias reduction algorithm using one enzyme are expected to be similar to those using another enzyme. However, a high reproducibility is a necessary but not sufficient condition for robust bias reduction algorithms. As suggested in [2], we further compare the correlation between normalized contact frequencies and the corresponding spatial distance values measured by FISH experiments. Both the similarity between the normalized contact frequency matrices and the correlation to FISH data will be measured in terms of Spearman's rank correlation coefficient that is independent to the conversion between normalized contact frequencies and spatial distance values.

To prepare benchmark datasets for the performance assessment, we use two recently published Hi-C data from human lymphoblastoid cells (GM06990) [3] and mouse stem cells (mESC) [9]. For the GM06990 dataset, the Hi-C raw reads, SRR027956 and SRR027960, of two biological replicates using restriction enzymes HindIII and NcoI, respectively, were downloaded from NCBI (GSE18199). Each of the chromosomes in the GM06990 cell line is binned into 1M bps segments and the pre-computed observed frequency matrices at 1M resolution were obtained from the publication website of [13]. For the mESC dataset, the mapped reads, uniquely aligned by the BWA algorithm [18], were downloaded from NCBI (GSE35156). Because of the enhanced sequencing depth in the mESC dataset, the Hi-C data can be analyzed at a higher resolution, $i.e.$, 40kb. In other words, the 20 chromosomes in the mESC cell line are binned into 40kb bps segments. To calculate observed contact frequencies from the mapped reads, the preprocessing protocols used in the literature [3,13] are followed. For every paired-end read, its total distance to the two closest RECSs is calculated. Any read with a total distance greater than 500 bps is defined as a non-specific ligation and thus removed to prevent reads from random ligation being used, as suggested in [13]. Reads from RECSs with low sequence uniqueness (smaller than 0.5) are also discarded. The remaining paired-end reads over the 20 chromosome, chr1 to chr20 (chrX), are used for calculating the observed contact frequencies.

The contact frequencies are derived from a cell population that may consist of several subpopulations of different chromosome structures. Without fully understanding the structural variations in a cell population, any structural inference from the Hi-C data can be distorted [10]. A recent single-cell sequencing study found that interchromosome (or trans) contacts have much higher variability among cells of the same cell line than intra-chromosome (or cis) contacts [23]. To avoid potential uncertainty that may be caused by significant variations in a cell line, we follow suggestions in the literature [9,2] and focus on cis contacts within a chromosome.

To obtain normalized frequencies of the bias reduction methods, we run both GDNorm and HiCNorm on the contact frequencies and ICE on the raw Hi-C reads. The normalized frequencies by the YT method are downloaded from the publication websites of the literature [9,13]. Note that although the primary objective of BACH is to predict chromosome structures, it also estimates systematic biases in the prediction of chromosome structures, using the log-linear regression model given in Eq. 2.

Hence, BACH can be regarded as a bias reduction method if we divide each observed contact frequency by its estimated systematic biases and use the quotient as the normalized frequency. To study the accuracy of bias estimation by BACH, we also include BACH in the comparison of bias correction methods. The reproducibility between the two biological replicates and correlation to FISH data achieved by the compared methods are discussed below.

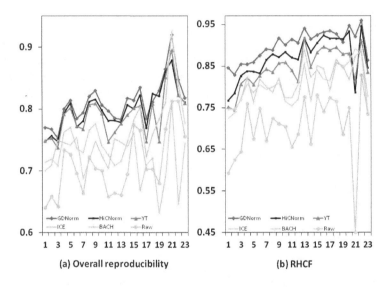

(a) Overall reproducibility   (b) RHCF

Fig. 3. Comparison of the reproducibility between two biological replicates achieved by GDNorm, HiCNorm, YT, ICE, and BACH on the 23 chromosomes, chr1 to chr23 (chrX), in the GM06990 cell line at 1M resolution. The distribution of Spearman's correlation coefficients achieved by a bias reduction method is represented as a solid curve over the 23 chromosomes. Plot (a) illustrates the overall reproducibility and plot (b) shows the reproducibility of high contact frequencies (RHCF).

GDNorm Achieves the Best Reproducibility on the Two Real Datasets.
The reproducibility between biological replicates is measured by Spearman's correlation coefficient. To prevent the assessment biased by background noise, when calculating Spearman's correlation coefficient, 2% of bins with lowest read counts in the matrices are deleted as done in [14]. The reproducibility over the remaining 98% of the bins is referred to as the overall reproducibility. Some recent studies in the literature using Hi-C data focused on high contact frequencies, *e.g.*,

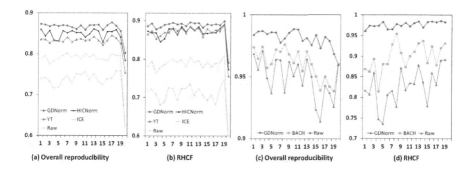

Fig. 4. Comparison of the reproducibility in the mESC dataset. Plots (a) and (b) illustrate the overall reproducibility and RHCF of GDNorm, HiCNorm, YT, and ICE on the 20 chromosomes, chr1 to chr20 (chrX), in the mESC cell line at 40kb resolution, respectively. Here, the distribution of Spearman's correlation coefficients achieved by each bias reduction method is represented as a solid curve over the 20 chromosomes. Plots (c) and (d) show the overall reproducibility and RHCF of GDNorm and BACH at 1M resolution, respectively.

studies concerning gene promoter-enhancer contacts [17] and spatial gene-gene interaction networks [24]. To assess the capability of reducing systematic biases in high contact frequencies, we calculate another Spearman's correlation coefficient, called the reproducibility of high contact frequencies (RHCF), by using only the top 20% of bins with the highest observed contact frequencies.

The Spearman's correlation coefficients over the 23 chromosomes in the GM06990 dataset are summarized in Figure 3. The average overall reproducibility of the observed (*i.e.*, raw) contact frequencies is 0.711 and GDNorm achieves the best overall reproducibility 0.811 on average while HiCNorm, YT, BACH, and ICE obtain 0.799, 0.789, 0,761, and 0.721, respectively. GDNorm improves the average overall reproducibility by up to 0.04 on an individual chromosome, over the second best method, HiCNorm. In terms of RHCF, the improvement by GDNorm over the second best method (HiCNorm) is more striking, 0.02 on average and up to 0.13 on an individual chromosome.

In the experiments on the mESC dataset, all the selected methods are run on the data at 40kb resolution except for BACH. The running time of BACH is prohibitive for performing chromosome-wide bias correction on the mESC dataset at the 40kb resolution, because it requires 5000 iterations to refine the predicted structure by default and each iteration takes about 30 minutes on average on our computer. So, we excluded BACH from the experiments at 40kb resolution, but will compare it with GDNorm at 1M resolution separately. The comparisons over the 20 chromosomes in the mESC dataset at 40kb resolution are summarized in Figure 4 (a) and (b). The average overall reproducibility of the observed (raw) contact frequencies is 0.734. The average overall reproducibility provided by GDNorm is 0.865, which is about 0.02 higher than the average overall reproducibility (0.846) obtained by HiCNorm and 0.03 higher

Table 2. Correlation between normalized contact frequencies at 40kb resolution and spatial distance measured by FISH experiments in the two biological replicates of the mESC data

Replicates	Raw	GDNorm	HiCNorm	YT	ICE
HindIII	-0.49	-0.66	-0.60	-0.66	-0.25
NcoI	-0.25	-0.37	-0.14	-0.37	0.31

than the third best (0.83) obtained by YT. Although ICE can eliminate systematic biases without assuming their specific sources, it achieves the lowest average overall reproducibility, 0.783, which is significantly lower than the average reproducibilities obtained by the other three methods. GDNorm achieves similar improvements in terms of RHCF, which is also 0.02 higher than the second best by HiCNorm on average and up to 0.04 on an individual chromosome. The comparisons between BACH and GDNorm at 1M resolution are shown in Figure 4 (c) and (d). GDNorm significantly outperforms BACH on both average overall reproducibility (0.02) and average RHCF (0.07). In the tests on individual chromosomes, the maximum improvement on RHCF by GDNorm is up to 0.15. This result shows that, although GDNorm and BACH both include spatial distance explicitly in their models, the gradient descent method of GDNorm can estimate the systematic biases more accurately than the MCMC based optimization procedure of BACH. These experimental results demonstrate that GDNorm is able to consistently improve on the reproducibility between biological replicates at both high (40kb) and low (1M) resolutions.

The Normalized Contact Frequencies Obtained by GDNorm are well Correlated to the FISH Data. To further validate the quality of normalized contact frequencies, we use an mESC 2d-FISH dataset that contains distance measurement for six pairs of genomic loci as our benchmark data. The six pairs of genomic loci are distributed on chromosomes 2 and 11 of the mESC genome, with three pairs on chromosome 2 and the other three on chromosome 11. The distance between each pair of the genomic loci is measured by inter-probe distance on 100 cell images from 2d-FISH experiments and normalized by the size of cell nucleus such that any change in the distance measurement is attributed solely to altered nucleus size on the images as described in the literature [4]. The average of the 100 normalized distance values for each pair of the genomic segments is used to correlate with the normalized contact frequency corresponding to the pair. The normalized frequencies are expected to be inversely correlated to the corresponding spatial distance values. Table 2 compares Spearman's correlation coefficients obtained by all four methods. The correlation coefficient between the 2d-FISH distance values and observed contact frequencies is low, -0.45 and -0.25 in the HindIII and NcoI replicates, respectively. YT and GDNorm are able to improve both correlation coefficients and achieve a strong correlation (smaller than -0.6) in the HindIII replicate while HiCNorm and ICE fail to deliver strongly correlated normalized frequencies in either replicate.

Table 3. The running time on the GM06990 and mESC datasets

Datasets	GDNorm	HiCNorm	BACH	ICE
GM06990	0.8 s	2.0 s	2 hr 17 m	5 hr 45 m
mESC	37 s	15 m 58 s	-	8 hr 36 m

The Time Efficiency of GDNorm. We evaluate the time efficiency of the selected methods by comparing their running time on the two real datasets. Our computing platform is a high-end compute server with eight 2.6GHz CPUs and 256GB of memory, but a single thread is used for each method. Because the normalized frequencies of YT were downloaded from the publication website, we did not run YT (in fact, we were unable to make YT run on our server) and will exclude YT from the comparison. The running time of the other four methods is summarized in Table 3. Due to the intensive computation requirement of the MCMC algorithm for refining chromosome structures, BACH is more than 10 time slower than HiCNorm and GDNorm on the 1M dataset (*i.e.*, GM06990). As mentioned before, the running time of BACH increases drastically with the number of genomic segments and becomes prohibitive when BACH is applied to the 40kb dataset (*i.e.*, mESC). ICE is significantly slower HiCNorm and GDNorm because it starts from raw Hi-C reads (instead of read counts) and requires additional time for iteratively mapping and processing the raw reads. Note that YT also uses raw Hi-C reads as its input and was found to be more than 1000 times slower than HiCNorm on the 1M dataset [16]. On both real datasets, GDNorm runs faster than HiCNorm. The standard iteratively reweighted least squares (IRIS) algorithm [25] was implemented in the software of HiCNorm to solve its log-linear regression model. In every iteration, the running time of the IRIS algorithm is quadratic in the number of genomic segment pairs.

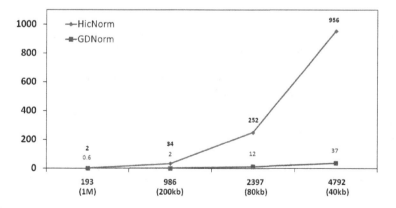

Fig. 5. The running time of GDNorm and HiCNorm on the mESC data at four different resolutions. The Y-axis shows the running time in seconds and the X-axis indicates the number of genomic segments at each resolution.

However, in our gradient descent method, the execution time of each iteration is only linear in the number of segment pairs, which makes GDNorm faster than HiCNorm. As illustrated in Figure 5, a simple experiment on the mESC data with resolutions at 40kb, 80kb, 200kb, and 1M shows that, when the number of genomic segments increases, the running time of HiCNorm grows much faster than that of GDNorm.

4   Conclusion

The reduction of systematic biases in Hi-C data is a challenging computational biology problem. In this paper, we proposed an accurate bias reduction method that takes advantage of a more comprehensive model of causal relationships among observed contact frequency, systematic biases and spatial distance. In our simulation study, GDNorm was able to provide more accurate normalized contact frequencies that resulted in improved chromosome structure prediction. Our experiments on two real Hi-C datasets demonstrated that GDNorm achieved a better reproducibility between biological replicates consistently at both high and low resolutions than the other state-of-the-art bias reduction methods and provided stronger correlation to published 2d-FISH data. The experiments also showed GDNorm's high time efficiency. With the rapid accumulation of high throughput genome-wide chromatin interaction data, the method could become a valuable tool for understanding the higher order architecture of chromosome structures.

Acknowledgement. We are grateful to Dr. Wendy Bickmore for sharing the FISH data of the mESC cell line with us. The research was partially supported by National Science Foundation grant DBI-1262107.

References

1. Dekker, J., Marti-Renom, M.A., Mirny, L.A.: Exploring the three-dimensional organization of genomes: interpreting chromatin interaction data. Nature Reviews. Genetics 14(6), 390–403 (2013)
2. Hu, M., Deng, K., Qin, Z., Liu, J.S.: Understanding spatial organizations of chromosomes via statistical analysis of Hi-C data. Quantitative Biology 1(2), 156–174 (2013)
3. Lieberman-Aiden, E., van Berkum, N.L., Williams, L., Imakaev, M., Ragoczy, T., Telling, A., Amit, I., Lajoie, B.R., Sabo, P.J., Dorschner, M.O., Sandstrom, R., Bernstein, B., Bender, M.A., Groudine, M., Gnirke, A., Stamatoyannopoulos, J., Mirny, L.A., Lander, E.S., Dekker, J.: Comprehensive mapping of long-range interactions reveals folding principles of the human genome. Science 326(5950), 289–293 (2009)
4. Eskeland, R., Leeb, M., Grimes, G.R., Kress, C., Boyle, S., Sproul, D., Gilbert, N., Fan, Y., Skoultchi, A.I., Wutz, A., Bickmore, W.A.: Ring1B compacts chromatin structure and represses gene expression independent of histone ubiquitination. Molecular Cell 38(3), 452–464 (2010)

5. Dekker, J., Rippe, K., Dekker, M., Kleckner, N.: Capturing chromosome conformation. Science 295(5558), 1306–1311 (2002)
6. Simonis, M., Klous, P., Splinter, E., Moshkin, Y., Willemsen, R., de Wit, E., van Steensel, B., de Laat, W.: Nuclear organization of active and inactive chromatin domains uncovered by chromosome conformation capture-on-chip (4C). Nature Genetics 38(11), 1348–1354 (2006)
7. Zhao, Z., Tavoosidana, G., Sjölinder, M., Göndör, A., Mariano, P., Wang, S., Kanduri, C., Lezcano, M., Sandhu, K.S., Singh, U., Pant, V., Tiwari, V., Kurukuti, S., Ohlsson, R.: Circular chromosome conformation capture (4C) uncovers extensive networks of epigenetically regulated intra- and interchromosomal interactions. Nature Genetics 38(11), 1341–1347 (2006)
8. Dostie, J., Richmond, T.A., Arnaout, R.A., Selzer, R.R., Lee, W.L., Honan, T.A., Rubio, E.D., Krumm, A., Lamb, J., Nusbaum, C., Green, R.D., Dekker, J.: Chromosome Conformation Capture Carbon Copy (5C): a massively parallel solution for mapping interactions between genomic elements. Genome Research 16(10), 1299–1309 (2006)
9. Dixon, J.R., Selvaraj, S., Yue, F., Kim, A., Li, Y., Shen, Y., Hu, M., Liu, J.S., Ren, B.: Topological domains in mammalian genomes identified by analysis of chromatin interactions. Nature 485(7398), 376–380 (2012)
10. Hu, M., Deng, K., Qin, Z., Dixon, J., Selvaraj, S., Fang, J., Ren, B., Liu, J.S.: Bayesian inference of spatial organizations of chromosomes. PLoS Computational Biology 9(1), e1002893 (2013)
11. Marti-Renom, M.A., Mirny, L.A.: Bridging the resolution gap in structural modeling of 3D genome organization. PLoS Computational Biology 7(7), e1002125 (2011)
12. Zhang, Z., Li, G., Toh, K.-C., Sung, W.-K.: Inference of spatial organizations of chromosomes using semi-definite embedding approach and hi-C data. In: Deng, M., Jiang, R., Sun, F., Zhang, X. (eds.) RECOMB 2013. LNCS, vol. 7821, pp. 317–332. Springer, Heidelberg (2013)
13. Yaffe, E., Tanay, A.: Probabilistic modeling of Hi-C contact maps eliminates systematic biases to characterize global chromosomal architecture. Nature Genetics 43(11), 1059–1065 (2011)
14. Imakaev, M., Fudenberg, G., Mccord, R.P., Naumova, N., Goloborodko, A., Lajoie, B.R., Dekker, J., Mirny, L.A.: Iterative correction of Hi-C data reveals hallmarks of chromosome organization. Nature Methods (September) (2012)
15. Cournac, A., Marie-Nelly, H., Marbouty, M., Koszul, R., Mozziconacci, J.: Normalization of a chromosomal contact map. BMC Genomics 13, 436 (2012)
16. Hu, M., Deng, K., Selvaraj, S., Qin, Z., Ren, B., Liu, J.S.: HiCNorm: removing biases in Hi-C data via Poisson regression. Bioinformatics 28(23), 3131–3133 (2012)
17. Jin, F., Li, Y., Dixon, J.R., Selvaraj, S., Ye, Z., Lee, A.Y., Yen, C.A., Schmitt, A.D., Espinoza, C.A., Ren, B.: A high-resolution map of the three-dimensional chromatin interactome in human cells. Nature 503(7475), 290–294 (2013)
18. Li, H., Durbin, R.: Fast and accurate short read alignment with Burrows-Wheeler transform. Bioinformatics (Oxford, England) 25(14), 1754–1760 (2009)
19. Lindsey, J.K., Altham, P.M.E.: Analysis of the human sex ratio by using overdispersion models. Journal of the Royal Statistical Society. Series C (Applied Statistics) 47(1), 149–157 (1998)
20. Rousseau, M., Fraser, J., Ferraiuolo, M.A., Dostie, J., Blanchette, M.: Three-dimensional modeling of chromatin structure from interaction frequency data using Markov chain Monte Carlo sampling. BMC Bioinformatics 12(1), 414 (2011)
21. Kabsch, W.: A solution for the best rotation to relate two sets of vectors. Acta Crystallographica Section A 32(5), 922–923 (1976)

22. Goulden, C.H.: Methods of Statistical Analysis, 2nd edn. Wiley, New York (1956)
23. Nagano, T., Lubling, Y., Stevens, T.J., Schoenfelder, S., Yaffe, E., Dean, W., Laue, E.D., Tanay, A., Fraser, P.: Single-cell Hi-C reveals cell-to-cell variability in chromosome structure. Nature 502(7469), 59–64 (2013)
24. Wang, Z., Cao, R., Taylor, K., Briley, A., Caldwell, C., Cheng, J.: The properties of genome conformation and spatial gene interaction and regulation networks of normal and malignant human cell types. PLoS ONE 8(3), e58793 (2013)
25. Dobson, A.J.: An Introduction to Generalized Linear Models. Chapman and Hall, London (1990)

Pacemaker Partition Identification

Sagi Snir*

Dept. of Evolutionary Biology, University of Haifa, Haifa 31905, Israel

Abstract. The universally observed conservation of the distribution of evolution rates across the complete sets of orthologous genes in pairs of related genomes can be explained by the model of the Universal Pacemaker (UPM) of genome evolution. Under UPM, the relative evolutionary rates of all genes remain nearly constant whereas the absolute rates can change arbitrarily. It was shown on several taxa groups spanning the entire tree of life that the UPM model describes the evolutionary process better than the traditional molecular clock model [26,25]. Here we extend this analysis and ask: how many pacemakers are there and which genes are affected by which pacemakers? The answer to this question induces a partition of the gene set such that all the genes in one part are affected by the same pacemaker. The input to the problem comes with arbitrary amount of statistical noise, hindering the solution even more. In this work we devise a novel heuristic procedure, relying on statistical and geometrical tools, to solve the pacemaker partition identification problem and demonstrate by simulation that this approach can cope satisfactorily with considerable noise and realistic problem sizes. We applied this procedure to a set of over 2000 genes in 100 prokaryotes and demonstrated the significant existence of two pacemakers.

Keywords: Molecular Evolution, Genome Evolution Pacemaker, Deming regression, Partition Distance, Gap Statistics.

1 Introduction

Comparative analysis of the rapidly growing collection of genomes of diverse organisms shows that the distribution of the evolutionary distances between orthologous genes remains remarkably constant across the entire history of life. All such distributions, produced for pairs of closely related genomes from different taxa, from bacteria to mammals, are approximately lognormal, span a range of three to four order of magnitude and are nearly identical in shape, up to a scaling factor [12,8,31]. Apparently, the simplest model of evolution that would imply the conservation of the shape of distance is that all genes evolve at approximately constant rates relative to each other. In other words, the changes in the gene-specific rates of evolution can be arbitrarily large (at least in principle) but are strongly correlated genome-wide. We denote this model of evolution the Universal PaceMaker (UPM) of genome evolution. Under the UPM, all genes in

---

* Research was supported in part by the USA-Israel Binational Science Foundation.

D. Brown and B. Morgenstern (Eds.): WABI 2014, LNBI 8701, pp. 281–295, 2014.

each evolutionary lineage adhere to the pace of a pacemaker (PM), and change their evolutionary rate (approximately) in unison although the pacemaker's pace at different lineages may differ. The UPM model is compatible with the large amount of data on fast-evolving and slow-evolving organismal lineages, primarily different groups of mammals [5]. An obvious alternative to the UPM is the Molecular Clock (MC) model of evolution under which genes evolved at roughly constant albeit different (gene-specific) rates [32] that implies the constancy of gene-specific relative evolution rates.

In a line of works [26,30,25] we established the superiority of the UPM model over the MC by explaining a larger fraction of the variance in the branch lengths of thousands of gene trees spanning the entire tree of life. Although highly statistically significant, in absolute terms however, the advantage of UPM over MC was small, and both models exhibited considerable evolution rate overdispersion. A plausible explanation to the latter is that instead of a single, apparently weak (overdispersed) PM, there are independent *multiple pacemakers* that each affect a (different) subset of genes and are less dispersed than the single pacemaker. Throughout, we use the notation UPM to refer to the model and the PM term for the pacemaker as an object.

Primarily, we investigate the requirements for the identification of distinct PMs and assignment of each gene to the appropriate PM. Such an assignment forms a partition over the set of genes and hence we denote this task as the *PM partition identification* (PMPI) problem. PM identification depends on the number of analyzed genes, the number of target PMs, the intrinsic variability of the evolutionary rate for each gene and the intrinsic variability of each PM. The PMPI problem is theoretically and practically hard as it concerns dealing with a lot of data obscured by a massive amount of noise. A possible direction to pursue is to exploit the signal from the data themselves in order to reduce the search space and focus only on relevant partitions.

In this work, a first attempt in this direction is made by devising and employing a novel technique using a series of analytic tools to solve the PMPI problem, and assess the quality of the derived solution. We tackle theoretical computational and statistical issues, as well as challenging engineering obstacles that arise along the way. These include guarantying *homoschedasticity* [29] by working in the log space, removing gene order dependency [1] by employing the Deming regression [6,10], and graph completion through most reliable paths. The result is the partial *gene correlation graph* where edge lengths represent (inversely) correlation, that we subsequently embed into the Euclidean space while preserving the distances. We apply standard clustering tools to this data and assess the significance of the result. We next formulate the PMPI problem as a recoloring problem [22,21] where a gene's PM is perceived as its color and the (set of) genes associated with a certain PM form a color class. To measure the quality of partition reconstruction, one may look for the minimum number of genes that need to be recolored in order that every part in the reconstructed partition is monochromatic. This number (the recolored genes) is denoted the *partition distance* [14] and can be solved by a matching algorithm. We however use a greedy maximum weighted matching algorithm, that is practically simpler for implementation and provided very good results empirically. Although theoretically this algorithm provides a 1/2-approximation guarantee for any input, under some statistical conditions the we note, with high probability the correct partition distance is returned.

The simulation results obtained using this approach are highly significant under a random model that we devise. The latter is significant as it implies that we were successful in both extracting the signal from the (noisy) data, and our technique is plausible. Finally, using insights from the simulation analysis, we analyzed the large set of phylogenetic trees of prokaryotic genes that was previously studied in [26]. Because the actual PM partition is unknown, we used the gap statistics criterion of Tibshirani et al. [27] to determine clustering significance and resulting in identification of two distinct genome evolution PMs.

2   The Evolutionary Model

Evolutionary history is described by a tree $T = (V, E)$ which is a combinatorial object composed of nodes representing (extant and extinct) species, and edges connecting these nodes such that there are no cycles in T. The edges are directed from an *ancestor* to its *descendant* nodes and also correspond to the time period between the respective nodes. There is one node with no ingoing edges, the *root*, and nodes with no outgoing edges are the *leaves* that are labeled by the *species* (or *taxa*) set. Therefore, the topology of T indicates the history of speciation events that led to the extant species at the leaves of T. Internal nodes correspond to ancestral forms existed at speciation events, and edges indicate ancestral relationships. A node (or a species) is a set of genes $G = \{g_i\}$ where a gene is a sequence of *nucleotides* of some given length. A gene evolves through a process in which mutations change its nucleotides from one state to another. In our model, all extant and extinct species possess the same set of genes $G = \{g_i\}$ and all genes g_i evolve along T according to an evolutionary model that is assumed to follow a continuous time Markov process. This process is represented by a given rate of mutations r per unit of time. In particular, every gene g_i evolves at an intrinsic rate $r_i \in \mathbf{r}$ that is constant along time but deviates randomly along the time periods (i.e. tree edges). Let $r_{i,j}$ be the *actual* (or *observed*) rate of gene i at period j. Then $r_{i,j} = r_i e^{\alpha_{i,j}}$ where $0 < e^{\alpha_{i,j}}$ is a multiplicative *error factor*. The number of mutations in gene g_i along period t_j is hence $\ell_{i,j} = r_{i,j} t_j$, commonly denoted as the *branch length* of gene g_i at period j. Throughout, we will use i to identify genes g_i and j for time periods t_j. As the topology of T is constant and assumed to be known, we will not make any reference to the tree and regard the edges only as independent time periods t_j for $1 \leq j \leq \tau$ where $\tau = |E|$.

We now extend this model to include a pacemaker that accelerates or decelerates a gene g_i, relative to its intrinsic rate r_i. Formally, a *pacemaker PM_k* is a set of τ *paces* $\beta_{k,j}$, $1 \leq j \leq \tau$ where $\beta_{k,j}$ is the relative pace of PM k during time period t_j and $-\infty < \beta < \infty$. Under the UPM model, a gene g_i that is *associated* with PM P_k has actual rate at time t_j: $r_{i,j} = r_i e^{\alpha_{i,j}} e^{\beta_{k,j}}$. Hence, for $\beta < 0$ the PM slows down its associated genes, for $\beta > 0$ genes are accelerated by their PM, and for $\beta = 0$, the PM is neutral. Assume every gene is associated with some PM and let $PM(g_i)$ be the PM of gene g_i. Then the latter defines a partition over the set of genes G, where genes g_i and $g_{i'}$ are in the same part if $PM(g_i) = PM(g_{i'})$.

Comment 1. *It is important to note that gene rates, as well as pace makers paces, are hidden and that we only see for each gene g_i, its set of edge lengths $\ell_{i,j}$.*

Comment 2. *The presence of two genes in the same part (PM) does not imply anything about their magnitude of rates, rather on their unison of rate divergence.*

The above gives rise to the *PM Partition identification Problem:*

Problem 1 (Pacemaker Partition Identification). Given a set of n genes g_i, each with τ branch lengths $\{\ell_{i,j}\}$, the *Pacemaker Partition Identification* (PMPI) problem is to find for each gene g_i, its pace maker $PM(g_i)$.

We first observe the following:

Observation 1. *Assume gene g_i has error factor $\alpha_{i,j} = 0$ for all time periods t_j, $1 \leq j \leq \tau$ and let $P' = PM(g_i)$ be the pace maker of gene g_i with relative paces e^{β_j}. Then at all periods t_j, $r_{i,j} = r_i e^{\beta_j}$.*

Observation 1 implies that if genes g_i and $g_{i'}$ belong to the same pace maker, and both genes have zero error factor at all periods, then at all periods, the ratio between the edge lengths at each period is constant and equals to $r_i/r_{i'}$. This however is not necessarily true if one of the error factor is not zero or genes g_i and $g_{i'}$ do not belong to the same pace maker. Recall that we do not see the gene intrinsic rates (and hence also the ratio between them). However if we see the same ratio between edge lengths across all time periods, we can conclude about the error factors and possibly their belonging to the same PM.

In order to tackle the PM identification problem, we impose some statistical structure (as observed in real data [12]) on the given setting. The goal is to assume that the error factor of each gene is small enough at every period, so that all genes belonging to the same PM, change their actual rate in unison.

Similarly, we assume that β_k varies so that genes from different PMs (parts) can be distinguished (otherwise, no difference except their random error factor exists)

Assumption 1.

1. *For all genes g_i and periods t_j, the gene error factors $\alpha_{i,j}$ follow a normal distribution $\alpha_{i,j} \sim N(0, \sigma_G^2)$,*
2. *For all PMs P_k and periods t_j, the PM paces $\beta_{k,j}$ follow a normal distribution $\beta_{k,j} \sim N(0, \sigma_P^2)$,*

3   The Pacemaker Partition Identification Procedure

Here we devise a procedure to solve the PMPI problem that entails a technique to infer distances between genes, constructing the *gene correlation graph*, embed reliably this graph in the plain and apply partitioning algorithms to this embedding. We now describe each of these steps.

3.1   Inferring Gene Distance

As outlined above, our first task is to infer gene pairwise distances from the raw data, which is gene edge lengths $\ell_{i,j}$ for every time period (edge) j. In particular, as the relevant information is encompassed in the random component of that value, the task of extracting that component is even more challenging.

We now proceed as follows: Given two sets of edge lengths $\ell_{i,j}$ and $\ell_{i',j}$ corresponding to genes g_i and $g_{i'}$, and time periods t_j for $1 \leq j \leq \tau$, we draw τ points on a plain $(\ell_{i,j}, \ell_{i',j})$. Now, if the error factors, $\alpha_{i,j} = \alpha_{i',j} = 0$ for all $1 \leq j \leq \tau$ and we connected all these points, we would obtain a straight line. The slope of that line is the multiplicative factor representing the ratio between the rates of evolution of the corresponding genes - $r_{g_i}/r_{g_{i'}}$; we denote it $\rho_{i,i'}$. Obviously, the above description refers to an idealized case. With real data, we never expect to find such a perfect correlation because the characteristic variance σ_G^2 is always non zero. Thus, we expect to find the points scattered around a trend line representing the rate ratio. The density of points around the trend line represents the level of correlation. Our goal is to obtain both the rate ratio $\rho_{i,i'}$ and the level of correlation where the latter will be used to classify between the genes. The method of choice to pursue here is to apply linear regression [29] between the points representing the two edge lengths. There are several outstanding issues that need to be addressed in such a task.

1. **Zero Intercept Requirement:** Linear regression, when applied to a set of points on a plane, finds a line $y = ax + b$ minimizing the sum of square distances of that line to all the points. As we deal with a multiplicative factor, the trend line has to cross the origin, i.e. $b = 0$. Hence we need to modify the standard procedure for regression.

2. **Homoschedasticity Requirement:** *homoschedasticity* is the property that the error in the dependent variable (y) is identically and independently distributed (IID) along the trend line. However, by our formulation $\ell_{i,j} = t_j r_{i,j} = t_j r_i e^{\alpha_{i,j}}$ and the expected value (the value on the trend line) is $t_j r_i$. The deviation then is $t_j r_i (e^{\alpha_{i,j}} - 1)$. As r_i is constant for all time period, we see that the longer the time period t_j, the larger the influence of $\alpha_{i,j}$. That is, assume two time periods j and j' with the same error factor $\alpha_{i,j} = \alpha_{i,j'}$ but different period lengths, WLOG $\ell_j < \ell_{j'}$. We obtain different deviations $t_j r_i (e^{\alpha_{i,j}} - 1) < t_{j'} r_i (e^{\alpha_{i,j'}} - 1)$, creating a bias toward longer periods. The following observation follows immediately from the definition of $\alpha_{i,j}$

   Observation 2. *If we take the* $\log \ell_{i',j} = \log t_j r_{i'} + \alpha_{i',j}$ *we arrive at Homoschedasticity.*

We denote this as the *log transformation* and also observe the following:

Observation 3. *Under the log transformation the trend line* $\log \ell_{i',j} = a \log \ell_{i,j} + b$ *has slope one* $(a = 1)$ *and intercept* $b = \log \rho_{i,i'}$.

We will use these properties in our calculations.

3. **Gene Order Independence:** The final problem with the linear regression has to do with the basic assumptions in least squares analysis. In standard least squares, the assumption is that the independent variable x is error-free while only the dependent variable y deviates from its expected values. In our case, however, the choice between the variables is arbitrary and both are subjected to deviation, according to their characteristic variance σ_G^2. Handling this case with standard least squares would cause arbitrary bias due to the selection of the variables [1]. To handle this case, we apply Deming Regression [6,10]. This approach assumes an explicit probabilistic model for the variables and extracts closed forms expressions (in the observed variables) for the sought expected values. To adjust to our specific case, we will use the observations drawn above. The linear model assumed is of type $\eta = \alpha \xi + \beta$ where the observations of both η and ξ , $(x_1, \ldots, x_n)$ and $(y_1, \ldots, y_n)$, respectively, have normally distributed errors: (i) $x_i = \xi_i + \varepsilon_{x_i}$, and (ii) $y_i = \eta_i + \varepsilon_{y_i} = \alpha + \beta \xi_i + \varepsilon_{y_i}$. As can be seen, this is exactly our case. The likelihood function of this model is:

$$f = \Pi_1^n (2\pi\sigma^2)^{-1/2} \exp\left(-\frac{(x_i - \xi_i)^2}{2\sigma^2}\right) (2\pi\sigma^2)^{-1/2} \exp\left(-\frac{(y_i - \alpha - \beta\xi_i)^2}{2\sigma^2}\right) \tag{1}$$

Under the general formulation, the ML value for α is: $\alpha = \bar{x} + \bar{y}\beta$ where $\bar{x}$ and $\bar{y}$ are the average values for x_i and y_i. However, in our formulation we have $\beta = 1$ and hence $\alpha = \bar{x} + \bar{y}$. Having α at hand, we can reconstruct the trend line and obtain the deviation of every point from it. Finally, by our formulation, $\rho_{i,i'}$ is given by $exp(\alpha)$ and the correlation between the rates is the standard *sample Pearson correlation coefficient* $r(X, Y)$ [29]:

$$r = \frac{\sum_{i=1}(X_i - \bar{X})(Y_i - \bar{Y})}{\sqrt{\sum_{i=1}(X_i - \bar{X})^2}\sqrt{\sum_{i=1}(Y_i - \bar{Y})^2}}. \tag{2}$$

3.2   The Gene Correlation Graph

After we inferred all pair-wise correlations, we can build the *Gene Correlation Graph* $G = (V, E, w)$ aiming at representing the correlation between the pairs of genes. $V = \{g_i\}$ and an edge $(i, i') \in E$ if $r(i, i')$ from Eq (2) is greater than some threshold δ_r, maintaining a minimal level of correlation in the graph. Hence we set $w(i, i') = r(i, i')$ and as $-1 \leq r \leq 1$ we are guaranteed no negative weighted edges exist. Note that we are not interested in r^2 which may reflect high *negative* correlation, rather only in high positive correlation.

Recall that our initial goal was to partition genes into clusters (PMs) according to correlation. Perhaps the most commonly used technique is *k-means* [15,19] that aims at minimizing the within-cluster sum of squares (WCSS). These techniques operate in the Euclidean space and hence some distance preserving technique is required to embed the correlation graph G in the space. Multidimensional Scaling [17] (also Euclidean embedding) is a family of approaches for this task. Kruskal's iterative algorithm [16] for non-metric multidimensional scaling (MDS) receives as input a (possibly partial) set of distances and the desired embedding should preserve the *order* of the original distances. It requires however a full matrix as a starting guess.

Our approach here is to join every two nodes by the most reliable connection and with the highest correlation. This translates to finding the path with the minimum number of nodes (hops) and that the multiplication of the corresponding weights is minimal. This distance measure, min hop min weight (MHMW), is also useful in communication networks, where hop distance corresponds to reliability [13]. While the naive algorithm for the latter runs in time $O(n^3)$ it can be easily seen that we can solve the problem in time $O(n^2 \log diam(G))$ where $diam(G)$ is the diameter of G. The completed graph $\hat{G}$ serves as input to the *Classical multidimensional scaling* (CMDS) [4] whose output serves as the initial guess to the Kruskal's non-metric MDS. Once we have the embedding, we can apply k-means and obtain the desired clustering.

Below is the complete formal procedure *PMPI*:

Procedure *PMPI(G, δ_r)*:

1. Set the correlation graph $G = (V, E)$ with $V = \emptyset$, $E = \emptyset$
2. $V = \{g | g$ is a gene in $\mathcal{G}\}$
3. for all $g_i, g_j \in \mathcal{G}$
   - apply the Deming regression between g_i and g_j to determine $r(g_i, g_j)$
   - if $r(g_i, g_j) \geq \delta_r$, then add $\{(g_i, g_j)\}$ to E and set $w(g_i, g_j) \leftarrow r(g_i, g_j)$
4. $\hat{G} \leftarrow MHMW(G)$
5. apply Classical Multidimensional Scaling (cmdscale) to the full graph $\hat{G}$
6. apply Kruskal's iterative algorithm (isoMDS) to the original distance matrix, starting from cmdscale output
7. apply *kmeans* to the resulted embedding

---

4 Simulation Analysis

In order to evaluate the PMPI procedure described in Section 3 and derive practical intuition over our model, we performed simulation according to the basic lines described above.

In a simulation study, a crucial part involves the assessment of the reconstruction quality with respect to the model on which the input was generated. As the PMPI is targeted at reconstruction of the original PM partition, we chose to use the partition distance measure.

4.1   Partition Distance

Once we obtain the reconstructed clustering, it should be compared to the original, model clustering. The task of comparing two clusterings can be casted as a partition distance where every clustering is a partition over the element set. We now define it formally. For two sets s_i and s_j, the distance $d(s_i, s_j)$ is the size of their symmetric difference set $s_i \triangle s_j = (s_i \setminus s_j) \cup (s_j \setminus s_i)$. Analogously, the *similarity* $s(s_i, s_j)$ is the size of their intersection set $s_i \cap s_j$ and it is easy to see that given the sizes of the two sets, one is derived from the other. A *partition* $\mathcal{P}$ over a ground element set N is a set of parts $\{p_i\}$ where every part is a subset of N, $\{p_i\}$ are pairwise disjoint (i.e. $p_i \cap p_j = \emptyset$ for every $i \neq j$), and their union is N. The cardinality of $\mathcal{P}$, denoted as $|\mathcal{P}|$ is the number of parts. A partition can also be perceived as a *coloring* function C from N to a set of colors $\mathcal{C}$ (the *color classes*) where $C(x)$ is the part of element $x \in N$ under partition P (or equivalently C). henceforth we will use the notions of PM identity and a color interchangeably. Given two partitions $\mathcal{P}$ and $\mathcal{P}'$ over the same element set N (or equivalently C and C'), denoted as the *source* and *target* partitions, we are interested in their *partition distance* $d(\mathcal{P}, \mathcal{P}')$ as some measure of similarity. The simplest approach is naturally the number of elements with different colors at the two partitions, i.e., $x \in N$ s.t. $C(x) \neq C'(x)$, and we call it the *identity similarity*. Under this approach, the partition distance between $\mathcal{P}$ and $\mathcal{P}'$, $d(\mathcal{P}, \mathcal{P}')$, is defined as: $d(\mathcal{P}, \mathcal{P}') = \sum_{x \in N} \bar{\delta}(C(x), C'(x))$ where $\hat{\delta}$ is the *inverse Kronecker delta*: $\bar{\delta}(i, j) = 1$ if $i \neq j$ and 0 otherwise.

This of course is simple and is an upper bound on a more accurate approach: colors can be permuted between the two partitions, in the sense that a color is mapped by a function f to another color in $\mathcal{C}$ and now $d(\mathcal{P}, \mathcal{P}')$ is defined as $d(\mathcal{P}, \mathcal{P}') = \sum_{x \in N} \bar{\delta}(f(C(x)), C'(x))$. It is easy to see that under this definition, f in the first approach is simply the identity function $f(c) = c$ for every $c \in \mathcal{C}$. This essentially defines a *recoloring problem*[22] where the goal is to recolor the least number of elements in $\mathcal{P}'$ (or C') such that $f(C(x)) = C'(x)$ for every element. Hence the cost of f is the number of elements x s.t. $f(C(x)) \neq C'(x)$.

Now, since the mapping is from $\mathcal{C}$ to $\mathcal{C}$, f is a bijection or simply a *matching* between the set of colors. In [14] Gusfield noted that the partition distance problem can be casted as an *assignment problem* [18] and hence be solved by a maximum flow in a bipartite graph in time $O(mn + n^2 \log n)$ [2]. Matching problems are among the most classical and well investigated in theoretical, as well as in practical, computer science [28]. Although it has a polynomial time exact algorithms with many flavors [2], a host of works on approximated solutions were introduced. For its very simple implementation and empirically accurate results that are based on theoretical properties we show below, we chose to use a very simple greedy algorithm, named *Greedy PartDist*. The algorithm works recursively and, at each recursion, chooses the heaviest edge (u, v) in the graph, adds it to the matching M and removes from the graph all other edges (u, v') and (u', v) for $u', v' \in V$. It is easy to see that the algorithm runs in time $O(m \log n)$ where the complexity of the sorting operation dominates. This algorithm provides a 1/2-approximation guarantee [23] for a general input and the same approximation guarantee can be obtain by the generic recursive analysis of the *local ratio* technique[3].

The Greedy Algorithm under the Stochastic Models. It is interesting to analyze the performance of the greedy algorithm under our stochastic model. It is easy to see (even simply for symmetry arguments) that under our model assumption, every gene remains in its part with probability α (that depends on the two variances σ_P and σ_G) and with probability $1 - \alpha$ chooses uniformly a partition (including its own partition). The expected identity similarity here is the sum over the elements maintaining their part plus those randomly chose that same (original) part.

Definition 1. *We say that a PM P is* correctly clustered *if most of the genes associated with P as a source PM, choose P as their target PM.*

Definition 1 implies that under a correctly clustered PM, a significant core set of genes stay together is the target PM (part). It is easy to see that, under our stochastic model, if enough genes are associated with every source PM, then all PMs are correctly clustered.

Claim. Assume every color is correctly clustered. Then Algorithm Greedy Part-Dist returns the correct result.

Proof. The proof follows by induction on the number of PMs $|\mathcal{P}|$. For a single PM, there is a single edge in the bipartite graph and this edge is chosen. For $|\mathcal{P}| > 1$, note that by the assumption, the heaviest edge emanating from each PM (node) in $\mathcal{P}$ to its corresponding color in the partition $\mathcal{P}'$. In particular, this is true for the heaviest edge in the bipartite graph, linking between the nodes corresponding to some PM P. Then the algorithm chooses that edge and remove all edges adjacent to it. Therefore, PM P was correctly chosen and by the induction hypothesis the algorithm returns the correct result.

4.2   Simulation Results

To asses the effectiveness of our PM partitioning identification procedure *PMPI*, we conducted the following simulation study. Number of genes n was held constant $n = 100$ giving rise to $\binom{100}{2} = 4950$ pairs of correlation tests. The number of edges per a gene tree was set to 25, reflecting the average size of the agreement tree among our real data trees. To simulate low agreement similarly to our real data (low MAST value) we discarded every pair with probability $2/3$ maintaining approximately $1/3$ of the pairs (see more details in Section 5). Every PM P_k was associated with an intrinsic variance σ_P^2 that sets its relative pace to $e^{\beta_{k,j}}$ where $\beta_{k,j} \sim N(0, \sigma_P^2)$. Similarly, every gene sets its rate at period j to $r_{i,j} = r_i e^{\alpha_{i,j}} e^{\beta_{k,j}}$ where $\alpha_{i,j} \sim N(0, \sigma_G^2)$ (See Model Section 2 for full details). Every gene was associated with a source PM, same number of genes for each PM. Number of PMs k varied from 2 to 10 (i.e. 10 to 50 genes per PM). Distance between genes was set as $1 - r$ from the regression line where the latter was derived by the Deming regression. This has defined our correlation graph described above.

In order to apply clustering algorithms on the elements, the elements need to be embedded in some Euclidean space. Multidimensional scaling takes a set of dissimilarities (over a set of elements) and returns a set of points in a Euclidean space, such that the distances between the points are approximately equal to the dissimilarities. A set of Euclidean distances on n points can be represented exactly in at most $n-1$ dimensions. The procedure *cmdscale* follows the analysis of Mardia [20], and returns the best-fitting k-dimensional representation, where k may be less than the argument k (and by definition smaller than n). In our implementation, in order to avoid any distortion, we set k to the maximum value as determined by the data (and is found and returned by the method). We used a version of *cmdscale* that is implemented in R. As *cmdscale* requires a complete graph, we used the min-hop-min-weight (MHMW) algorithm. The output of the MHMW is a complete graph where the weight between any two points is the lightest (min weight) path among all min hop reliable paths (paths between trees for which correlation was derived). At this point we can use *cmdscale* to map this graph to the Euclidean space. Note however, that this mapping corresponds *not* to the original graph, rather to some approximation of it derived by the output of the MHMW algorithm. This mapping however serves as an initial guess to the iterative mapping of the original, partial, distance matrix. This iterative

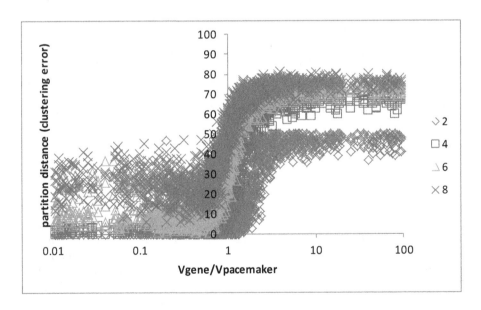

Fig. 1. Partition distance obtained by applying the PMPI technique on simulated data versus the gene/pacemaker variance ratio; the plots are shown for 2, 4, 6 and 8 clusters (PMs)

process is done by the function *isoMDS* implemented in R. This mapping will serve us for the clustering operation. Now, as opposed to real data, here we know the original number of clusters, we can just set this as the number of clusters required. We used *kmeans* implemented by R to obtain the optimal clustering. Our results appear in Figure 1. The measured quantity is (normalized) partition distance as measured by our *greedy PartDist*. The independent variable is the ratio between σ_G and σ_P. The larger σ_P the more dispersed are the PMs and hence farther from one another. Equivalently, the smaller σ_G, the more concentrated around their PM are the genes. Therefore, we expect that the smaller the ratio σ_G/σ_P is, i.e. PMs are spaced away from each other while their associated genes are more concentrated, we get better results in the sense that more genes remain in their original cluster and successfully identified. Also, we expect that the larger the number of PMs, the greater the mixing between them with genes end up in PMs that are neighboring to their original PMs. Indeed it can be seen that for two and four PMs, for any ratio of $\sigma_G/\sigma_P \leq 1$ a very accurate reconstruction is achieved and so as to six clusters, but for ratio a little less than 1. It is also shown that for every number of PMs, at some critical σ_G/σ_P ratio (that depends on #PMs) the reconstruction curve reaches a saturation that tends to the random similarity as we computed above.

5   Results on Real Data

Working with real data poses some other serious problems requiring solution. The first, is that we don't have here exactly τ periods with edge length $\ell_{i,j}$ for every gene g_i rather a set of trees with loose pairwise agreement. This loose agreement is due to vast discordance between the histories of the various genes as a result of phenomena such as horizontal gene transfer (HGT) or incomplete lineage sorting (ILS, see more details below). However, discordance can arise even from the simple fact that some gene is missing in some specific species, resulting in a contraction of internal nodes.

To cope with this problem, we employ the idea of Maximum Agreement Subtrees (MAST) [9], that seeks for the largest subset of species under which the two trees are the same. Under MAST (or in general, any subset of the leaf set), edges not connecting any species to the induced tree, are removed, and internal nodes with degree two are contracted, while maintaining the original length of the path. Hence for every pair of genes (trees) we need to find the MAST and compare lengths of corresponding edges.

Additionally, here as opposed to a simulation study, we do not know the "real" partition and cannot compare the resultant clustering to it. Therefore, another method for assessing the results should be employed. Here we need to compare the result to the probability of being obtained under a random model. Recall that at the final stage of the *PMPI* procedure we employ the *kmeans* algorithm which seeks to minimize some *error measure* W_K. This error measure holds the sum of all pairwise distances between members of the same cluster, across all clusters in the partition. It is clear that the more clusters, the smaller W_K is.

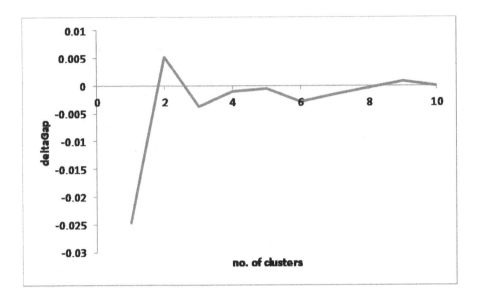

Fig. 2. The deltaGap function for 2755 analyzed genes, k from 1 to 10. According to Tibshirani et al [27], the smallest k producing a non-negative value of deltaGap[k] = Gap[k]-Gap[k+1]+sigma[k+1] indicates the optimal number of clusters.

However, the decrease in W_K is the largest near the real value of the number of clusters $k = K$, and vanishes slowly for $k > K$. Therefore, a threshold for the improvement (decrease) in W_K must be defined as a stopping condition, above which we don't increase the number of clusters k. The *gap statistics analysis* [27] compares the improvement in W_K under the real data, to that of a random model. The gap (between the improvements) forms an "elbow" at the optimal (real) K and this is the stopping condition.

The real data we chose to analyze is the one used by us [26] previously, of a set of gene trees that covers 2755 orthologous families from 100 prokaryotic genomes [24]. Prokaryotic evolution is characterized by the pervasive phenomena of horizontal gene transfer (HGT) [7,11], resulting in different topologies for almost any two gene trees. To account for this we employed the MAST procedure for every gene pair and considered this pair only if the MAST contained at least 10 leaves (species). Branch lengths of the original trees were used to compute the branch lengths of the corresponding MAST components (by computing path lengths). The variant of Deming regression in the log space as described in Section 4 was performed on the logarithms of the lengths of equivalent branches in both MAST components. The standard sample Pearson correlation coefficient was used as the measure of correlation between the branch lengths. The graph of correlations between the gene trees contained a giant connected component containing 2755 genes and 1,250,972 edges, 33% of the maximum possible number (an edge in the graph exists only when the MAST for the corresponding pair of trees consists of at least 10 species). To cluster these genes according to the correlation between their branch lengths, the data were projected using isoMDS into a 30-dimensional space based on the sparse matrix where $1 - r$ (correlation coefficient) was used as a distance. We ran *k-means* for k spanning the range from 2 to 30. The random model we chose to consider is the fully random uniform model (i.e., $\alpha = 0$, no advantage to source PM) and we compared the results to this model. Grouping these 2755 genes in two clusters containing 1550 and 1205 members, respectively, yields the optimal partitioning according to the gap function statistics (Figure 2). We see the typical "elbow" at the value of $k = 2$. The absolute results were $5,587,960$ for the total graph weight, $2,686,914$ and $2,285,921$ weight within each of the clusters, and $615,125$ between them. Analysis of the cluster membership reveals small albeit significant differences in the representation of functional categories of genes but no outstanding biologically relevant trends were detected. Therefore, we can hypothesize that if indeed the data gives rise to multiple PMs, this signal is completely obscured by the amount of noise produced by the genes themselves (i.e. loose adherence to the associated PM), and noise introduced by artificial factors such as MAST, multiple sequence alignment, and phylogenetic reconstruction.

6 Conclusions

The universal pacemaker (UPM) model provides a more general framework to analyze genome evolution than the MC model as it makes no assumptions of the absolute evolutionary rates of gene, only on the relative rates. This provides a better explanation to the data observed at extant species. However, similarly to the MC, the UPM is extremely over-dispersed, with the noise complicating detailed analysis. The difficulty in PM analysis is caused both by the weak informative signal and by the large volume of the data.

A natural expectation, however, is for different gene groups, to adhere to different PMs, characterized by different functions. This classification imposes a partition over the gene set where each gene is associated with its own PM. The inference of such a partition is challenging twofold; first from information perspective, as it needs to overcome a high level of "noise", both biological, as well as artificial. Next, the computational task of solving the PMPI problem requires investigating all possible partitions over the gene set.

In this work we provide the first heuristic procedure for detecting such a partitioning that is based on theoretical ground. We use the Deming regression to infer correlation between pairs of genes, and represent this correlation relationship in a graph. Subsequently, we embed this graph in the Euclidean space and apply a clustering procedure to it.

We also provide simulation and empirical results of the application of this procedure. In the simulation study, we have shown that the proposed procedure is sound and is capable of detecting the original partition with high accuracy for a fairly small (up to 6) number of PMs as long as the intrinsic gene rate variance is at the size of the PM variance. In the real data realm, we succeed in showing that the analyzed genome-wide set of gene trees is optimally partitioned between two PMs, and the improvement in the statistical explanation is small albeit highly significant. The partition of different functional gene groups between the two PMs is also statistically significant (WRT random partitioning of each group) however the biological interpretation of this partitioning is challenging and remained for future research.

Acknowledgments. We thank Eugene Koonin and Yuri Wolf for helpful discussions, in particular in interpretation of the biological significance of the resulted clustering of the real data in Section 5.

References

1. Adcock, R.J.: A problem in least squares. Annals of Mathematics 5, 53–54 (1878)
2. Ahuja, R.K., Magnanti, T.L., Orlin, J.B.: Network Flows. Prentice-Hall, Englewood Cliffs (1993)
3. Bar-Yehuda, R.: One for the price of two: A unified approach for approximating covering problems. Algorithmica 27, 131–144 (2000)
4. Borg, I., Groenen, P.: Modern multidimensional scaling, theory and applications. Springer, New York (1997)
5. Bromham, L.: Why do species vary in their rate of molecular evolution? Biology Letters 5(3), 401–404 (2009)
6. Deming, W.E.: Tatistical adjustment of data. J. Wiley & Sons (1943)
7. Doolittle, W.F.: Phylogenetic classification and the universal tree. Science 284(5423), 2124–2129 (1999)
8. Drummond, D.A., Wilke, C.O.: Mistranslation-induced protein misfolding as a dominant constraint on coding-sequence evolution. Cell 134(2), 341–352 (2008)
9. Finden, C.R., Gordon, A.D.: Obtaining common pruned trees. Journal of Classification 2, 225–276 (1985)
10. Fuller, W.A.: Measurement error models. John Wiley & Sons, Chichester (1987)
11. Gogarten, J.P., Doolittle, W.F., Lawrence, J.G.: Prokaryotic evolution in light of gene transfer. Mol. Biol. Evol. 19, 2226–2238 (2002)
12. Grishin, N.V., Wolf, Y.I., Koonin, E.V.: From complete genomes to measures of substitution rate variability within and between proteins. Genome Research 10(7), 991–1000 (2000), doi:10.1101/gr.10.7.991
13. Guérin, R., Orda, A.: Computing shortest paths for any number of hops. IEEE/ACM Trans. Netw. 10(5), 613–620 (2002)
14. Gusfield, D.: Partition-distance: A problem and class of perfect graphs arising in clustering. Information Processing Letters 82(3), 159 (2002)
15. Hartigan, J.A., Wong, M.A.: A k-means clustering algorithm. Applied Statistics 28, 100–108 (1979)
16. Kruskal, J.B.: Nonmetric multidimensional scaling: a numerical method. Psychometrika 29, 115–130 (1964)
17. Kruskal, J.B., Wish, M.: Multidimensional Scaling. Sage Publications (1978)
18. Lawler, E.L.: Combinatorial optimization: networks and matroids. The University of Michigan (1976)
19. Lloyd, S.P.: Least squares quantization in pcm. IEEE Transactions on Information Theory 28, 129–137 (1982)

20. Mardia, K.V.: Some properties of classical multidimensional scaling. Communications on Statistics – Theory and Methods A7 (1978)
21. Moran, S., Snir, S.: Efficient approximation of convex recolorings. J. Comput. Syst. Sci. 73(7), 1078–1089 (2007)
22. Moran, S., Snir, S.: Convex recolorings of strings and trees: Definitions, hardness results and algorithms. J. Comput. Syst. Sci. 74(5), 850–869 (2008)
23. Preis, R.: Linear time $\frac{1}{2}$-approximation algorithm for maximum weighted matching in general graphs. In: Meinel, C., Tison, S. (eds.) STACS 1999. LNCS, vol. 1563, p. 259. Springer, Heidelberg (1999)
24. Puigbo, P., Wolf, Y., Koonin, E.: Search for a 'tree of life' in the thicket of the phylogenetic forest. Journal of Biology 8(6), 59 (2009)
25. Snir, S., Wolf, Y.I., Koonin, E.V.: Universal pacemaker of genome evolution in animals and fungi and variation of evolutionary rates in diverse organisms. In: Genome Biology and Evolution (2014)
26. Snir, S., Wolf, Y.I., Koonin, E.V.: Universal pacemaker of genome evolution. PLoS Comput Biol. 8, e1002785 (2012)
27. Tibshirani, R., Walther, G., Hastie, T.: Estimating the number of clusters in a data set via the gap statistic. Journal of the Royal Statistical Society: Series B (Statistical Methodology) 63(2), 411–423 (2001)
28. Tutte, W.T.: Connectivity in graphs. Mathematical expositions. University of Toronto Press (1966)
29. Wasserman, L.: All of Statistics, ch. 4. Springer, New York (2004)
30. Wolf, Y.I., Snir, S., Koonin, E.V.: Stability along with extreme variability in core genome evolution. Genome Biology and Evolution 5(7), 1393–1402 (2013)
31. Wolf, Y.I., Novichkov, P.S., Karev, G.P., Koonin, E.V., Lipman, D.J.: The universal distribution of evolutionary rates of genes and distinct characteristics of eukaryotic genes of different apparent ages. Proceedings of the National Academy of Sciences 106(18), 7273–7280 (2009)
32. Zuckerkandl, E.: On the molecular evolutionary clock. Journal of Mol. Evol. 26(1), 34–46 (1987)

Manifold de Bruijn Graphs

Yu Lin and Pavel A. Pevzner

Department of Computer Science and Engineering,
University of California, San Diego, La Jolla, California
{yul280,ppevzner}@ucsd.edu

Abstract. Genome assembly is usually abstracted as the problem of reconstructing a string from a set of its k-mers. This abstraction naturally leads to the classical de Bruijn graph approach—the key algorithmic technique in genome assembly. While each vertex in this approach is labeled by a string of the fixed length k, the recent genome assembly studies suggest that it would be useful to generalize the notion of the de Bruijn graph to the case when vertices are labeled by strings of variable lengths. Ideally, we would like to choose larger values of k in high-coverage regions to reduce repeat collapsing and smaller values of k in the low-coverage regions to avoid fragmentation of the de Bruijn graph. To address this challenge, the *iterative de Bruijn graph assembly* (IDBA) approach allows one to increase k at each iterations of the graph construction. We introduce the *Manifold de Bruijn (M-Bruijn) graph* (that generalizes the concept of the de Bruijn graph) and show that it can provide benefits similar to the IDBA approach in a single iteration that considers the entire range of possible k-mer sizes rather than varies k from one iteration to another.

1   Introduction

The de Bruijn graphs are the key algorithmic technique in genome assembly [1–3] that resulted in dozens of software tools [4–10]. In addition, the de Bruijn graphs have been used for repeat classification [11], de novo protein sequencing [12], synteny block construction [13], multiple sequence alignment [14], and other applications in genomics and proteomics. In fact, the de Bruijn graphs have become so ubiquitous in bioinformatics that one rarely questions what are the intrinsic limitation of this approach.

We argue that the original definition of the de Bruijn graph is far from being optimal for the challenges posed by the assembly problem. We further propose a new notion of the *Manifold de Bruijn (M-Bruijn) graph* (that generalizes the concept of the de Bruijn graph) and show that it has advantages over the classical de Bruijn graph in assembly applications.

The disadvantages of the de Bruijn graphs became apparent when bioinformaticians moved from assembling cultivated bacterial genomes (with rather uniform read coverage) to assembling genomes from single cells (with 4 orders of magnitude variations in coverage [15]). In such projects, selecting a fixed k-mer size is detrimental since k should be small in low-coverage regions (otherwise the

D. Brown and B. Morgenstern (Eds.): WABI 2014, LNBI 8701, pp. 296–310, 2014.

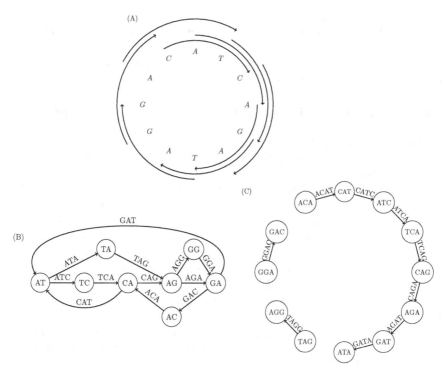

Fig. 1. (A) A circular string *String* =CATCAGATAGGA. The de Bruijn graphs (B) *DB(Reads, 3)* and (C) *DB(Reads, 4)* on a set of *Reads* = {CATC, ATCA, TCAG, CAGA, AGAT, GATA, TAGG, GGAC, ACAT } drawn from that circular genome. Small value of $k = 3$ "glues" many repeats and makes *DB(Reads, 3)* *tangled*, while larger value of $k = 4$ fails to detect overlaps and makes *DB(Reads, 4)* *fragmented*.

graph becomes *fragmented*) and large in high-coverage regions (otherwise the graph becomes *tangled*). Figure 1 illustrates the tradeoff.

Since the standard de Bruijn graph does not allow one to vary k, the leading single cell assemblers SPAdes [10] and IDBA-UD [16] use a heuristic called an *Iterative De Bruijn Assembly (IDBA)* proposed by Peng *et al.* [6]. IDBA starts from small k (resulting in a *tangled* de Bruijn graph), uses contigs from the resulting graph as *pseudoreads*, and mixes *pseudoreads* with original reads to construct the de Bruijn graph for larger k. The recent benchmarking studies demonstrated that SPAdes and IDBA-UD improve on other assemblers not only in single cell but also in standard multi-cell projects [17, 18].

However, the IDBA approach, while valuable, remains a heuristic that requires manual parameter setup (no automated parameter learning approach for IDBA has been proposed yet). Moreover, the running time for t iterations increases by a factor of t, forcing researchers to jump between various values of k (e.g., $21 \rightarrow 33 \rightarrow 55 \rightarrow 77 \rightarrow 99 \rightarrow 127$ for reads of length $250bp$ in the default setting of SPAdes [10] for multicell data) rather than increasing k by 1 in each iteration, a more accurate but impractical strategy. The question thus arises whether one can provide benefits similar to the IDBA approach in a single iteration that considers the entire range of possible k-mer sizes rather than varies it from one iteration to another. The *Manifold de Bruijn (M-Bruijn) graph* achieves this goal by automatically varying k-mer sizes according to the input data.

2   From the de Bruijn Graph to the A-Bruijn Graph

To introduce the *Manifold de Bruijn (M-Bruijn) graph*, we first need to depart from the classical definition of the de Bruijn graph (edges coded by k-mers and vertices coded by $(k$-1$)$-mers). We will use the concept of the *A-Bruijn graph* [11, 19] to provide an equivalent definition of the de Bruijn graph. In the *A-Bruijn graph* framework, the classical de Bruijn graph $DB(String, k)$ of a string *String* is defined as follows. Let $Path(String, k)$ be a path consisting of $|String| - k + 1$ edges, where the i-th edge of this path is labeled by the i-th k-mer in *String* and the i-th vertex of the path is labeled by the i-th $(k$-1$)$-mer in *String*. The de Bruijn graph $DB(String, k)$ is formed by gluing identically labeled vertices in $Path(String, k)$ as described in [11] (Figure 2). Note that this somewhat unusual definition results in exactly the same de Bruijn graph as the standard definition. In the case when instead of a string *String*, we are given a set of reads *Reads*, the definition of $DB(String, k)$ naturally generalizes to $DB(Reads, k)$ by constructing a path for each read and further gluing all identically labeled vertices in all paths.

We have defined $Path(String, k)$ as a path through all $(k$-1$)$-mers occurring in $String = s_1 s_2 \ldots s_n$, i.e., all substrings $s_i s_{i+1} \ldots s_{i+k-1} \in \Sigma^{k-1}$, where Σ^{k-1} is the set of all $(k$-1$)$-mers in alphabet Σ. We will thus change the notation from $Path(String, k)$ to $Path(String, \Sigma^{k-1})$.

We now consider an arbitrary *substring-free* set V where no string in V is a substring of another string in V. V consists of words (of any length) in the alphabet Σ and the new concept $Path(String, V)$ is defined as a path through all words from V appearing in *String* (in order) as shown in Figure 3. We further assign integer i (called the *shift*) to the edge (v, w) if the start of v precedes the start of w by i symbols in *String*. If $i > |v|$, we also assign a *shift tag* (the characters between the end of v and the start of w in *String*) to the edge (v, w). Afterwards, we glue identically labeled vertices as before to derive the *A-Bruijn* graph $AB(String, V)$ as shown in Figure 3. Clearly, $DB(String, k)$ is $AB(String, \Sigma^{k-1})$ with shifts of all edges equal to 1.

The definitions of $AB(String, \Sigma^{k-1})$ and $AB(String, V)$ naturally generalize to $AB(Reads, \Sigma^{k-1})$ and $AB(Reads, V)$ by constructing a path for each read and further gluing all identically labeled vertices in all paths. Figure 4 illustrate the construction of $AB(Reads, V)$. Note that when strings from V do not cover the first (last) symbol of a read, we add an additional *prefix (suffix) tag* to the first/last node of the read. For example, the read TCAG in Figure 4 contains two strings from V (TC and CA) that do not cover the last symbol G of this read. We have thus added a suffix tag "G" to the last node of the corresponding tag.

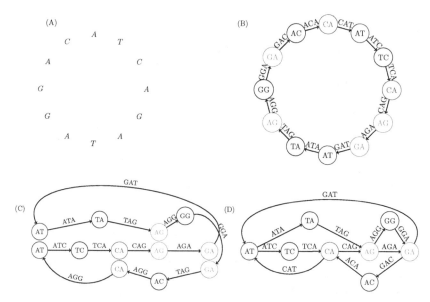

Fig. 2. A circular *String* =CATCAGATAGGA (A) and *Path*(*String*, 3) (B). Bringing identically labeled vertices (in (B)) closer to each other (in (C)) to eventually glue them into a single vertex in *DB*(*String*, 3) (in (D)).

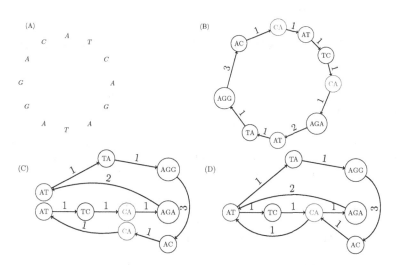

Fig. 3. A circular *String* =CATCAGATAGGA (A) and *Path*(*String*, *V*), where where *V* = {CA, AC, TC, AGA, AT, TA, AGG} (B). Bringing identically labeled vertices (in (B)) closer to each other (in (C)) to eventually glue them into a single vertex in *AB*(*String*, *V*) (in (D)).

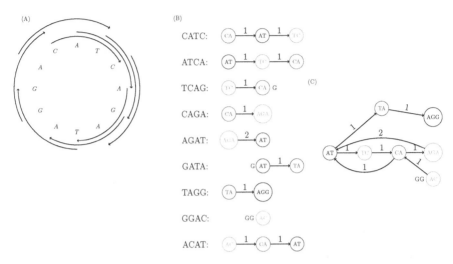

Fig. 4. (A) A set of *Reads* = {CATC, ATCA, TCAG, CAGA, AGAT, GATA, TAGG, GGAC, ACAT }. (B) All paths corresponding to reads from *Reads* and V = {CA, AC, TC, AGA, AT, TA, AGG } (C) $AB(Reads, V)$.

Below we address the question of how to choose V so that the resulting assembly $AB(Reads, V))$ improves on the classical assembly approach represented by $AB(Reads, \Sigma^{k-1}) = DB(Reads, k)$.

3   From the A-Bruijn Graph to the Manifold de Bruijn Graph

A word is called *irreducible* with respect to a string *String* if it appears once in *String* but all its substrings appear multiple times in *String*. The irreducible words with respect to a string *String* are also known as the *minimum unique substrings* [20]. Let *Irreducible(String)* be the set of all irreducible words with respect to *String*, e.g., *Irreducible(CAGGCA)* = {AG, GG, GC}. The set *Irreducible(String)* can be constructed in linear time [20] using the suffix arrays [21].

The *Manifold de Bruijn (M-Bruijn) graph* $MB(String)$ is defined as $AB(String, Irreducible(String))$. Please note that there is no parameter k in the definition of the M-Bruijn graph of a string. Obviously, no vertices are glued in the M-Bruijn graph (see Figure 5).

A word is called *right irreducible (r-irreducible)* with respect to *String* if it appears exactly once in *String* but all its prefixes appear multiple times in *String*. A word is called *left irreducible (l-irreducible)* with respect to *String* if it appears once in *String* but all its suffixes appear multiple times in *String*. For example {CAG, AG, GG, GC} and {AG, GG, GC, GCA} are the sets of *r*-irreducible and *l*-irreducible words for $CAGGCA$. Obviously, each irreducible

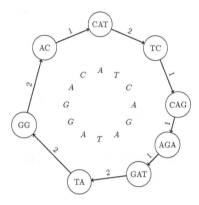

Fig. 5. A circular string *String* =CATCAGATAGGA with *Irreducible(String)* = { CAT, TC, CAG, AGA, GAT, TA, GG, AC}, and the M-Bruijn graph *MB(String)*

word is both *r*-irreducible and *l*-irreducible. Below we describe how to efficiently construct the sets of *r*-irreducible/*l*-irreducible/irreducible words.

While the linear time algorithm for constructing the set *Irreducible(String)* has been described by Ilie *et al.* [20], we describe a different approach that is better suited for the Manifold de Bruijn graph and its generalization to a set of reads described in the next section. Given a string $String = s_1 s_2 \ldots s_n$, we add a special termination character \$ to the end of *String*, define $Suf[i]$ as its suffix $s_i s_{i+1} \ldots s_n \$$, and define $T(String)$ as the *suffix tree* on $s_1 s_2 \ldots s_n \$$ [21]. An edge in $T(String)$ is called *trivial* if it is labeled by a single special character \$.

Given a string $String = s_1 s_2 \ldots s_n$, a word $w(i,j)$ represents the substring $s_i s_{i+1} \ldots s_j$. Consider a root-to-leaf path in $T(String)$ that corresponds to $Suf[i] = s_i s_{i+1} \ldots s_n \$$. If this path "ends" in a non-trivial edge labeled by $s_j s_{j+1} \ldots s_n \$$, we define the word $w(i,j) = s_i s_{i+1} \ldots s_j$ as an *outpost* with respect to *String* (Figure 6).

Proposition 1. *A word is r-irreducible if and only if it is an outpost with respect to String.*

Through a depth-first search of the suffix tree, we can derived the set of all *r*-irreducible words, represented by pairs of indices $\{(i_1,j_1),(i_2,j_2),\ldots,(i_m,j_m)\}$ with respect to *String*, e.g., each *r*-irreducible word $w(i,j)$ is denoted as a pair of indices (i,j). This set contains all the irreducible words with respect to *String*. To construct *Irreducible(String)*, we need to find all *r*-irreducible words that are also *l*-irreducible. Below we show how to do it in linear time.

Lemma 1. *If an r-irreducible word v is a substring of another r-irreducible word w then v is a suffix of w.*

Proof. If v is not a suffix of w, then it is contained in a prefix of w. Since w is *r*-irreducible, each prefix of w appears multiple times in *String*, and thus v also appears multiple times in *String*, implying that it can not be *r*-irreducible, a contradiction.   □

The corollary below reduces the search for irreducible words to the search for *r*-irreducible words:

Corollary 1. *An r-irreducible word $w(i, j)$ with respect to String is irreducible with respect to String if it is the shortest among all r-irreducible words ending at position j.*

Thus a simple linear time algorithm that scans the set of pairs of indices of all *r*-irreducible words reveals the set of irreducible words.

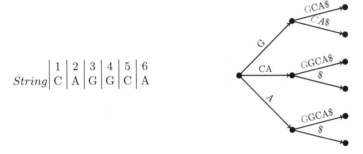

String	1	2	3	4	5	6
	C	A	G	G	C	A

Fig. 6. The suffix tree (right) for a string *String* =CAGGCA. The four outposts (GG, GC, CAG, and AG) end in symbols colored in red in the suffix tree. While CAG and AG are two *r*-irreducible words ending at position 3, only one of them (AG) is irreducible with respect to *String*.

Theorem 1. *The set Irreducible(String) can be constructed in linear time.*

Proof. The suffix tree for *String* can be constructed in linear time [21], the set of *r*-irreducible words (outposts, represented by pairs of indices) can be computed in linear time by a depth-first search of this suffix tree, and all irreducible words can be derived from the *r*-irreducible words in linear time (Corollary 1).   □

4   Manifold de Bruijn Graphs: From a Single String to a Set of Reads

We have defined the notion of an irreducible word with respect to a single string *String*. We now define the notion of an irreducible word with respect to a set of strings *Reads*.

4.1 Consistent and Irreducible Words with Respect to a Set of Reads

Let *Reads* be a substring-free set of reads $\{read_1, read_2, \ldots, read_m\}$ and $w(x, i, j)$ be the substring spanning positions from i to j in $read_x$. We refer to all substrings of reads as *words from Reads*. A string is *Reads-consistent* if it contains all strings from *Reads* as substrings.

A word from *Reads* is *irreducible* (with respect to *Reads*) if there exists a *Reads*-consistent string where this word is irreducible. Let *Irreducible(Reads)* be the set of all irreducible words with respect to *Reads*.

A word from *Reads* is *consistent* (with respect to *Reads*) if there exists a *Reads*-consistent string where this word appears exactly once.

To check whether a word w is consistent with respect to *Reads*, consider all reads containing w and represent each such read $read_t$ as a three-part concatenate of $affix_t(w)$, w, and $postfix_t(w)$, where $affix_t(w)$ ($postfix_t(w)$) are formed by symbols preceding (following) w in $read_t$. We further select the longest affix and postfix among all reads containing w and form the three-part concatenate of the longest affix, w, and the longest postfix (denoted by $Superstring(w)$). For example, for a set of reads {CAGCA, AGATT, ATTGC} and a word w=AG, the reads CAGCA and AGATT are represented as C-AG-CA and -AG-ATT in the affix-w-postfix notation. Therefore, $Superstring$(AG)=CAGATT.

The following proposition (illustrated in Figure 7) describes how to check if a word is consistent and implies that the word AG is not consistent because $Superstring$(AG) does not contain one of the reads (CAGCA) containing AG. Using this proposition, one can verify that for a set of reads *Reads* = {CAGCA, AGATT, ATTGC}, *Irreducible(Reads)* = {AGC, AT, CAG, GA, GCA, TG, TT}.

Fig. 7. Illustration of a consistent word (i) and an inconsistent word (ii),(iii) and (iv) with respect to *Reads*. Perfectly aligned symbols are shown by the same color (marked by dotted lines) while misalignments are shown by different colors (shown with $\neq$ sign).

Proposition 2. *A word w is consistent with respect to Reads if and only if w appears at most once in each read and each read containing w is a substring of $Superstring(w)$.*

4.2 Generalized Suffix Trees

Given the set *Reads* = $\{read_1, read_2, \ldots, read_m\}$, we add special termination character $\$_x$ to the end of each $read_x$, i.e. there will be m different termination characters overall. We define $T(Reads)$ as the *generalized suffix tree* [21] for *Reads* and define $Suf[x, i]$ as the suffix starting at position i of $read_x\$_x$. In $T(Reads)$, each $Suf[x, i]\$_x$ corresponds to a root-to-leaf path. An edge in $T(Reads)$ is called *trivial* if it is labeled by $\$_x$ for $1 \leq x \leq m$ (see Figure 8). A vertex in $T(Reads)$ is called a *branching* vertex if it has at least two non-trivial outgoing edges in $T(Reads)$. Given an edge from vertex u to vertex v, we say that all suffixes (leaves) in the subtree rooted at v are *after* v in $T(Reads)$.

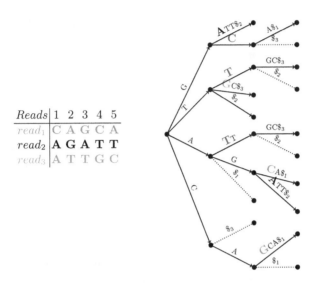

Reads	1	2	3	4	5
read₁	C	A	G	C	A
read₂	A	G	A	T	T
read₃	A	T	T	G	C

Fig. 8. The generalized suffix tree $T(Reads)$ (right) for $Reads = \{$CAGCA, AGATT, ATTGC $\}$ (left). Trivial edges are shown by dotted lines in $T(Reads)$. If an outpost appears only in one read, the outpost ends in a symbol colored in the color of that read; if an outpost appears in multiple reads, the outpost ends in a symbol colored in red. The three outposts (CAG, AGC, GC) in $Read_1$ end in symbols colored in brown (or red), the four outposts (AGA, GA, AT, TT) in $Read_2$ end in symbols colored in blue (or red), and the four outposts (AT, TT, TG, GC) in $Read_3$ end in symbols colored in green (or red) in $T(Reads)$.

Consider a path from the root to a leaf in $T(Reads)$ that corresponds to $Suf[x, i] = w(x, i, |read_x|)\$_x$. We find the first edge, denoted by (u, v), in the path such that all the suffixes after vertex v belong to distinct reads, and there is no branching vertices in the subtree rooted at v (including v). If such edge exists and is labeled by $w(x, j, l)$ (if $l < |read_x|$) or $w(x, j, |read_x|)\$_x$, we define the word $w(x, i, j)$ as an *outpost* in $Read_x$ (see Figure 8), and define $Right_x(i) = j$; otherwise $Right_x(i) = |read_x| + 1$. Note that if we ignore the differences in the indices and treat each outpost as a string, then the set of outposts are defined uniquely with respect to $T(Reads)$, although the same outpost may appear in multiple reads. Thus the positions of all outposts in $T(Reads)$ can be computed by the depth-first search of $T(Reads)$.

We define the *reverse* of string $String = s_1 s_2 \ldots s_n$ as the string $\overline{String} = s_n s_{n-1} \ldots s_1$. We further define the reverse of a set $Reads = \{read_1, read_2, \ldots, read_m\}$ as the set of reads $\overline{Reads} = \{\overline{read_1}, \overline{read_2}, \ldots, \overline{read_m}\}$ (see Figure 8 for $\overline{Reads}$ and $T(\overline{Reads})$).

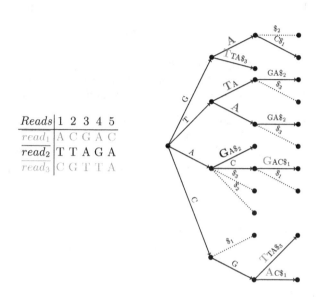

Reads	1 2 3 4 5
$read_1$	A C G A C
$read_2$	T T A G A
$read_3$	C G T T A

Fig. 9. The generalized suffix tree $T(\overline{Reads})$ (left) for $\overline{Reads}$ = {ACGAC, TTAGA, CGTTA } (top). Trivial edges are shown by dotted lines in $T(\overline{Reads})$. The three outposts (ACG, CGA, GA) in $\overline{read_1}$ end in symbols colored in brown (or red), the four outposts (TT, TA, AG, GA) in $\overline{read_2}$ end in symbols colored in blue (or red), and the four outposts (CGT, GT, TT, TA) in to $\overline{read_3}$ end in symbols colored in green (or red) in $T(\overline{Reads})$.

Consider a path from the root to a leaf in $T(\overline{Reads})$ that corresponds to $\overline{word(x, 1, j)}\$_x$. We find the first edge, denoted by (u, v), in the path such that all the suffixes after v belong to distinct reads, and there is no branching vertices in the subtree rooted at v (including v). If such edge exists and is labeled by $\overline{w(x, l, i)}$ (if $l > 1$) or $\overline{w(x, 1, i)}\$_x$, we define the word $\overline{w(x, i, j)}$ as an *outpost* in $\overline{Read_x}$ (see Figure 9), and define $Left_x(j) = i$; otherwise $Left_x(j) = 0$.

Given a word $w(x, i, j)$, if $j \geq Right_x(i)$, all postfixes of $w(x, i, j)$ in different reads are prefixes of the longest postfix among them and $w(x, i, j)$ appears at most once in each read; if $i \leq Left_x(j)$, all affixes of $w(x, i, j)$ in different reads are suffixes of the longest affix among them. Thus the two conditions, $(j \geq Right_x(i))$ and $(i \leq Left_x(j))$, imply that there exists $Superstring(w(x, i, j))$ such that each read containing $w(x, i, j)$ is a substring of $Superstring(w(x, i, j))$ and $w(x, i, j)$ appears at most once in each read. From Proposition 2 we have

Proposition 3. *A word $w(x, i, j)$ is consistent with respect to Reads if and only if $j \geq Right_x(i)$ and $i \leq Left_x(j)$.*

A word is *right irreducible (r-irreducible)* with respect to *Reads* if it is consistent but all its prefixes are inconsistent with respect to *Reads*. A word is *left irreducible (l-irreducible)* with respect to *Reads* if it is consistent but all its suffixes are inconsistent with respect to *Reads*. Obviously, a word is irreducible if and only if it is both *r*-irreducible and *l*-irreducible. This observation implies

Proposition 4. *A word $w(x, i, j)$ is irreducible with respect to Reads, if and only if*
(i) $j \geq Right_x(i)$ and $i \leq Left_x(j)$ (the consistent condition),
(ii) $j < Right_x(i + 1)$ or $i + 1 > Left_x(j)$ (the l-irreducible condition),
(iii) $i > Left_x(j - 1)$ or $j - 1 < Right_x(i)$ (the r-irreducible condition).

The proposition 4 reduces the construction of $Irreducible(Reads)$ to checking 3 conditions for each triple of indices (x, i, j). We further note that if both $w(x, i, j)$ and $w(x, i + 1, j')$ are irreducible then $j < j'$. This observation leads to the linear time Algorithm 1 for computing $Irreducible(Reads)$ (see Figure 10, Figure 11 and Figure 12).

Algorithm 1. Computing irreducible words in $read_x$

INPUT: Arrays $Right_x(i)$ $(1 \geq i \geq |read_x|)$ and $Left_x(j)$ $(1 \geq j \geq |read_x|)$
OUTPUT: Array $Irreducible_x$ represented as pairs of indices $\{(a, b) | w(x, a, b)$ is an irreducible word$\}$
INITIAL: $i = 1$, $j = Right_x(1)$, $Irreducible_x = \emptyset$
while $j \leq |read_x|$ **do**
    if (i, j) satisfy all three conditions in Proposition 4 **then**
        add (i, j) to $Irreducible_x$
        $i = i + 1$; $j = j + 1$;
    else
        if (i, j) violates condition (i) in Proposition 4 **then**
            $j = j + 1$
        else
            $i = i + 1$
        end if
    end if
end while

Theorem 2. *The set $Irreducible(Reads)$ can be constructed in linear time.*

Proof. The generalized suffix trees for *Reads* and $\overline{Reads}$ can be constructed in linear time [21]. We can then compute all $Right_x(i)$ and $Left_x(i)$ for any i $(1 \leq i \leq |read_x|)$ of $read_x$ in linear time through the depth-first search of each trees. Then Algorithm 1 derives all the irreducible words in linear time.   □

The *M-Bruijn graph* $MB(Reads)$ is defined as $AB(Reads, Irreducible(Reads))$. Similar to $MB(String)$, the construction of $MB(Reads)$ does not specify the k-mer size. Note that in the "gluing" process, if two vertices are connected by multiple

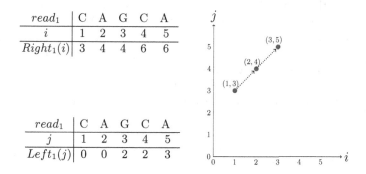

$read_1$	C	A	G	C	A
i	1	2	3	4	5
$Right_1(i)$	3	4	4	6	6

$read_1$	C	A	G	C	A
j	1	2	3	4	5
$Left_1(j)$	0	0	2	2	3

Fig. 10. Algorithm 1 identifies three irreducible words in $read_1$: $w(1,1,3) = $ CAG, $w(1,2,4) = $ AGC and $w(1,3,5) = $ GCA (shown as red points)

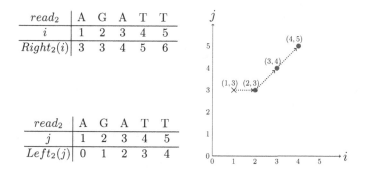

$read_2$	A	G	A	T	T
i	1	2	3	4	5
$Right_2(i)$	3	3	4	5	6

$read_2$	A	G	A	T	T
j	1	2	3	4	5
$Left_2(j)$	0	1	2	3	4

Fig. 11. Algorithm 1 identifies three irreducible words in $read_2$: $w(2,2,3) = $ GA, $w(2,3,4) = $ AT and $w(2,4,5) = $ TT (shown as red points)

$read_3$	A	T	T	G	C
i	1	2	3	4	5
$Right_3(i)$	2	3	4	5	6

$read_3$	A	T	T	G	C
j	1	2	3	4	5
$Left_3(j)$	0	1	2	3	3

Fig. 12. Algorithm 1 identifies three irreducible words in $read_3$: $w(3,1,2) = $ AT, $w(3,2,3) = $ TT and $w(3,3,4) = $ TG (shown as red points)

"parallel" edges, all these edges have the same shift and the same shift tag. We thus substitute all such edges by a single edge. It is easy to see that $MB(Reads)$ is a set of paths (including paths consisting of a single vertex) or cycles. Each cycle spells a sequence that we refer to as a *cyclic contig*. Each path spell a sequence that we refer to as a *linear contig* after concatenating it with the longest prefix tag of its first vertex and the longest suffix tag of its last vertex.

Figure 13 shows an example of an M-Bruijn graph on a set of reads.

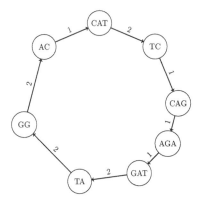

Fig. 13. The *M-Bruijn graph* $MB(Reads)$, where $Reads = \{$CATC, ATCA, TCAG, CAGA, AGAT, GATA, TAGG, GGAC, ACAT $\}$ drawn from a circular string $String =$CATCAGATAGGA, and $Irreducible(Reads) = \{$CAT, TC, CAG, AGA, GAT, TA, GG, AC$\}$. Compared to Figure 5, $MB(Reads)$ reconstructs $MB(String)$ and thus the circular string $String$. $MB(Reads)$ is neither tangled (like $DB(Reads, 3)$ in Figure 1(B)) nor fragmented (like $DB(Reads, 4)$ in Figure 1(C)).

5   Conclusion

The Iterative de Bruijn graph Assembly (IDBA) approach starts from small k, uses contigs from the de Bruijn graph on k-mers as pseudoreads, and mixes pseudoreads with original reads to construct the de Bruijn graph for larger k. The key step in IDBA is to maintain the *accumulated de Bruijn graph* (H_k) to carry the contigs forward as k increases [6].

We have proposed a notion of the Manifold de Bruijn (M-Bruijn) graph that does not require any parameter setup, e.g., it does not require one to specify the k-mer size. The M-Bruijn graph provides an alternative way to generate pseudoreads (as its contigs) that incorporate information for k-mers of varying sizes.

Our introduction of M-Bruijn graph is merely a preliminary theoretical concept that may seem impractical since we have not addressed various challenges posed by the real datasets in genome assembly. When Idury and Waterman [2] introduced the de Bruijn graph approach for genome assembly, the high error rates in Sanger reads also made that approach seem impractical. Pevzner *et al.* [3] later removed this obstacle by introducing an error correction procedure that made the vast majority of reads error-free. Thus, our ability to handle errors in reads is crucial for future applications of the M-Bruijn graph approach.

References

1. Pevzner, P.A.: l-tuple DNA sequencing: computer analysis. J. Biomol. Struct. Dyn. 7, 63–73 (1989)
2. Idury, R.M., Waterman, M.S.: A new algorithm for DNA sequence assembly. J. Comput. Biol. 2(2), 291–306 (1995)
3. Pevzner, P.A., Tang, H., Waterman, M.S.: An Eulerian path approach to DNA fragment assembly. Proc. Nat'l Acad. Sci. 98(17), 9748 (2001)
4. Zerbino, D.R., Birney, E.: Velvet: algorithms for de novo short read assembly using de Bruijn graphs. Genome Research 18(5), 821–829 (2008)
5. Chaisson, M.J., Pevzner, P.A.: Short read fragment assembly of bacterial genomes. Genome Research 18(2), 324–330 (2008)
6. Peng, Y., Leung, H.C.M., Yiu, S.M., Chin, F.Y.L.: IDBA – A practical iterative de bruijn graph de novo assembler. In: Berger, B. (ed.) RECOMB 2010. LNCS, vol. 6044, pp. 426–440. Springer, Heidelberg (2010)
7. Butler, J., MacCallum, I., Kleber, M., et al.: ALLPATHS: de novo assembly of whole-genome shotgun microreads. Genome Research 18(5), 810–820 (2008)
8. Li, R., Zhu, H., Ruan, J., et al.: De novo assembly of human genomes with massively parallel short read sequencing. Genome Research 20(2), 265–272 (2010)
9. Chitsaz, H., Yee-Greenbaum, J.L., Tesler, G., et al.: Efficient de novo assembly of single-cell bacterial genomes from short-read data sets. Nature biotechnology (2011)
10. Bankevich, A., Nurk, S., et al.: Spades: A new genome assembly algorithm and its applications to single-cell sequencing. J. Comput. Biol. 19(5), 455–477 (2012)
11. Pevzner, P.A., Tang, H., Tesler, G.: De novo repeat classification and fragment assembly. Genome Research 14(9), 1786–1796 (2004)
12. Böcker, S.: Sequencing from compomers: Using mass spectrometry for DNA de-novo sequencing of 200+ nt. In: Benson, G., Page, R.D.M. (eds.) WABI 2003. LNCS (LNBI), vol. 2812, pp. 476–497. Springer, Heidelberg (2003)
13. Pham, S.K., Pevzner, P.A.: DRIMM-Synteny: decomposing genomes into evolutionary conserved segments. Bioinformatics 26(20), 2509–2516 (2010)
14. Raphael, B., Zhi, D., Tang, H., Pevzner, P.A.: A novel method for multiple alignment of sequences with repeated and shuffled elements. Genome Research 14(11), 2336–2346 (2004)
15. Dean, F.B., Nelson, J.R., Giesler, T.L., Lasken, R.S.: Rapid amplification of plasmid and phage dna using phi29 dna polymerase and multiply-primed rolling circle amplification. Genome Research 11(6), 1095–1099 (2001)
16. Peng, Y., Leung, H., Yiu, S., Chin, F.: IDBA-UD: a de novo assembler for single-cell and metagenomic sequencing data with highly uneven depth. Bioinformatics 28(11), 1420–1428 (2012)
17. Gurevich, A., Saveliev, V., Vyahhi, N., Tesler, G.: QUAST: quality assessment tool for genome assemblies. Bioinformatics 29(8), 1072–1075 (2013)

18. Magoc, T., Pabinger, S., Canzar, S., et al.: GAGE-B: an evaluation of genome assemblers for bacterial organisms. Bioinformatics 29(14), 1718–1725 (2013)
19. Compeau, P.E.C., Pevzner, P.A.: Bioinformatics Algorithms: An Active-Learning Approach. Active Learning Publishers (2014)
20. Ilie, L., Smyth, W.F.: Minimum unique substrings and maximum repeats. Fundamenta Informaticae 110(1), 183–195 (2011)
21. Gusfield, D.: Algorithms on strings, trees and sequences: computer science and computational biology. Cambridge University Press (1997)

Constructing String Graphs in External Memory

Paola Bonizzoni, Gianluca Della Vedova, Yuri Pirola, Marco Previtali,
and Raffaella Rizzi

DISCo, Univ. Milano-Bicocca, Milan, Italy
{bonizzoni,dellavedova,pirola,marco.previtali,rizzi}@disco.unimib.it

Abstract. In this paper we present an efficient external memory algorithm to compute the string graph from a collection of reads, which is a fundamental data representation used for sequence assembly.

Our algorithm builds upon some recent results on lightweight Burrows-Wheeler Transform (BWT) and Longest Common Prefix (LCP) construction providing, as a by-product, an efficient procedure to extend intervals of the BWT that could be of independent interest.

We have implemented our algorithm and compared its efficiency against SGA—the most advanced assembly string graph construction program.

1 Introduction

De novo sequence assembly is a fundamental step in analyzing data from Next-Generation Sequencing (NGS) technologies. NGS technologies produce, from a given (genomic or transcriptomic) sequence, a huge amount of short sequences, called reads—the most widely used current technology can produce 10^9 reads with average length 150. The large majority of the available assemblers [1,10,15] are built upon the notion of de Bruijn graphs where each k-mer is a vertex and an arc connects two k-mers that have a $k-1$ overlap in some input read. Also in transcriptomics, assembling reads is a crucial task, especially when analyzing RNA-seq in absence of a reference genome.

Alternative approaches to assemblers based on de Bruijn graphs have been developed recently, mostly based on the idea of *string graph*, initially proposed by Myers [9] before the advent of NGS technologies and further developed [13,14] to incorporate some advances in text indexing, such as the FM-index [7]. This method builds an overlap graph whose vertices are the reads and where an arc connects two reads with a sufficiently large overlap. For the purpose of assembling a genome some arcs might be uninformative. In fact an arc (r_1, r_2) is called *reducible* if its removal does not change the strings that we can assemble from the graph, therefore reducible arcs can be discarded. The final graph, where all reducible arcs are removed, is called the *string graph*. More precisely, an arc (r_1, r_2) of the overlap graph is labeled by a suffix of r_2 so that traversing a path $r_1, \cdots, r_k$ and concatenating the first read r_1 with the labels of the arcs of the path gives the assembly of the reads along the path [9].

The naïve way of computing all overlaps consists of pairwise comparisons of all input reads, which is quadratic in the number of reads. A main contribution of [13]

D. Brown and B. Morgenstern (Eds.): WABI 2014, LNBI 8701, pp. 311–325, 2014.
© Springer-Verlag Berlin Heidelberg 2014

is the use of the notion of Q-interval to avoid such pairwise comparisons. More precisely, for each read r in the collection R, the portion of BWT (Q-interval), identifying all reads whose overlap with r is a string Q, is computed in time linear in the length of r. In a second step, Q-intervals are extended to discover irreducible arcs. Both steps require to keep the whole FM-index and BWT for R and for the collection of reversed reads in main memory since the Q-intervals considered cover different positions of the whole BWT. Notice that the algorithm of [13] requires to recompute Q-intervals a number of times that is equal to the number of different reads in R whose suffix is Q, therefore that approach cannot be immediately translated into an external memory algorithm. For this reason, an open problem of [13] is to reduce the space requirements by developing an external memory algorithm to compute the string graph.

Recently, an investigation of external memory construction of the Burrows-Wheeler Transform (BWT) and of related text indices (such as the FM-index) and data structures (such as LCP) has sprung [2,3,6] greatly reducing the amount of RAM necessary. In this paper, we show that two scans of the BWT, LCP and the generalized suffix array (GSA) for the collection of reads are sufficient to build a compact representation of the overlap graph, mainly consisting of the Q-intervals for each overlap Q.

Since each arc label is a prefix of some reads and a Q-interval can be used to represent any substring of a read, we exploit the above representation of arcs also for encoding labels. The construction of Q-intervals corresponding to labels is done by iterating the operation of backward σ-extension of a Q-interval, that is computing the σQ-interval on the BWT starting from a Q-interval. The idea of backward extension is loosely inspired by the pattern matching algorithm using the FM-index [7]. A secondary memory implementation of the operation of backward extension is a fundamental contribution of [5]. They give an algorithm that, with a single scan of the BWT, reads a lexicographically sorted set of disjoint Q-intervals and computes all possible σQ-intervals, for every symbol σ (the original algorithm extends all Q-intervals where all Qs have the same length, but it is immediate to generalize that algorithm to an input set of disjoint Q-intervals). Our approach requires to backward extend generic sets of Q-intervals. For this purpose, we develop a procedure (ExtendIntervals) that will be a crucial component of our algorithm to build the overlap and string graph.

Our main result is an efficient external memory algorithm to compute the string graph of a collection of reads. The algorithm consists of three different phases, where the second phase consists of some iterations. Each part will be described as linear scans and/or writes of the files containing the BWT, the GSA and the LCP array, as well as some other intermediate files. We strive to minimize the number of passes over those files, as a simpler adaptation of the algorithm of [13] would require a number of passes equal to the number of input reads in the worst case, which would clearly be inefficient.

After building the overlap graph, where each arc consists of two reads with a sufficiently large overlap, the second phase iteratively extends the Q-intervals found in the first phase, and the results of the previous iterations to compute an additional symbol of some arc labels (all labels are empty at the end of the first phase). At the end of the second phase, those labels allow to reconstruct the entire assembly (*i.e.* the genome/transcriptome from which the reads have been extracted). Finally, the third phase is devoted to testing whether an arc is reducible, in order to obtain the final string graph, using a new characterization of reducible arcs in terms of arc labels, *i.e.* prefixes of reads.

The algorithm has $O(d\ell^2 n)$ time complexity, where ℓ and n are the length and the number of input reads and d is the indegree of the string graph. We have developed an open source implementation of the algorithm, called LightString-Graph (*LSG*), available at http://lsg.algolab.eu/. We have compared LSG with SGA [13] on a dataset of 37M reads, showing that LSG is competitive (its running time is 5h 28min while SGA needed 2h 19min) even if disk accesses are much slower than those in main memory (SGA is an in-memory algorithm).

2 Preliminaries

We briefly recall the standard definitions of Generalized Suffix Array and Burrows-Wheeler Transform on a set of strings. Let Σ be an ordered finite alphabet and let S a string over Σ. We denote by $S[i]$ the i-th symbol of S, by $\ell = |S|$ the length of S, and by $S[i : j]$ the substring $S[i]S[i + 1] \cdots S[j]$ of S. The *reverse* of S is the string $S^{rev} = S[\ell]S[\ell - 1] \cdots S[1]$. The *suffix* and *prefix* of S of length k are the substrings $S[\ell - k + 1 : \ell]$ and $S[1 : k]$, respectively. The k-suffix of S is the suffix of length k. Given two strings (S_i, S_j), we say that S_i *overlaps* S_j iff a nonempty suffix Z of S_i is also a prefix of S_j, that is $S_i = XZ$ and $S_j = ZY$. In that case we say that S_j *extends* S_i by $|Y|$ symbols, that Z is the overlap of S_i and S_j, denoted as $ov_{i,j}$, that Y is the *extension* of S_i with S_j, denoted as $ex_{i,j}$, and X is the *prefix-extension* of S_i with S_j, denoted as $pe_{i,j}$.

In the following of the paper we will consider a collection $R = \{r_1, \ldots, r_n\}$ of n reads (*i.e.*, strings) over Σ. As usual, we append a sentinel symbol \$ $\notin \Sigma$ to the end of each string (\$ lexicographically precedes all symbols in Σ). Then, let $R = \{r_1\$, \ldots, r_n\$\}$ be a collection of n strings (or *reads*), where each r_i is a string over Σ; we denote by $\Sigma^\$$ the extended alphabet $\Sigma \cup \{\$\}$. Moreover, we assume that the sentinel symbol \$ is not taken into account when computing overlaps between two strings.

The *Generalized Suffix Array (GSA)* [12] of R is the array SA where each element $SA[i]$ is equal to (k, j) if and only if the k-suffix of string r_j is the i-th smallest element in the lexicographic order of the set of all the suffixes of the strings in R. In the literature (as in [2]), the relative order of two elements (k, i) and (k, j) of the GSA such that reads r_i and r_j share their k-suffix is usually determined by the order in which the two reads appear in the collection R (*i.e.*, their indices). However, starting from the usual definition of the order of the elements of the GSA, it is possible to compute the GSA with the order of their elements determined by the lexicographic order of the reads with two sequential scans of the GSA itself. The first scan extracts the sequence of pairs (k, j) where k is equal to the length of r_j, hence obtaining the reads of R sorted lexicographically. The second scan uses the sorted R to reorder consecutive entries of the

GSA sharing the same suffix. This ordering will be essential in the following since a particular operation (namely, the backward $-extension, as defined below) is possible only if this particular order is assumed. The *Longest Common Prefix* of R, denoted by LCP, is an array of size equal to the total length of the strings in R and such that $LCP[i]$ is equal to the length of the longest prefix shared by the suffixes pointed to by $GSA[i]$ and $GSA[i-1]$ (excluding the sentinel $). For convenience, we assume that $LCP[1] = 0$. Notice that no element of LCP is larger than the maximum length of a read of R.

The *Burrows-Wheeler Transform (BWT)* of R is the sequence B such that $B[i] = r_j[|r_j| - k]$, if $SA[i] = (k, j)$ and $k < |r_j|$, or $B[i] = \$$, otherwise. Informally, $B[i]$ is the symbol that precedes the k-suffix of string r_j where such suffix is the i-th smallest suffix in the ordering given by SA. Given a string Q, all suffixes of the GSA whose prefix is Q appear consecutively in GSA, therefore they induce an interval $[b, e)$ which is called Q-*interval* [2] and denoted by $q(Q)$. We define the *length* and *width* of the Q-interval $[b, e)$ as $|Q|$ and the difference $(e - b)$, respectively. Notice that the width of the Q-interval is equal to the number of occurrences of Q as a substring of some string $r \in R$. Whenever the string Q is not specified, we will use the term *string-interval* to point out that it is the interval on the GSA of all suffixes having a common prefix. Since the BWT and the GSA are closely related, we also say that $[b, e)$ is a string-interval (or Q-interval for some string Q) on the BWT. Let B^{rev} be the BWT of the set $R^{rev} = \{r^{rev} \mid r \in R\}$, let $[b, e)$ be the Q-interval on B for some string Q, and let $[b', e')$ be the Q^{rev}-interval on B^{rev}. Then, $[b, e)$ and $[b', e')$ are called *linked*. The linking relation is a 1-to-1 correspondence and two linked intervals have same width and length, hence $(e - b) = (e' - b')$.

Given a Q-interval and a symbol $\sigma \in \Sigma$, the *backward σ-extension* of the Q-interval is the σQ-interval (that is, the interval on the GSA of the suffixes sharing the common prefix σQ). We say that a Q-interval has a *nonempty* (*empty*, respectively) backward σ-extension if the resulting interval has width greater than 0 (equal to 0, respectively). Conversely, the *forward σ-extension* of a Q-interval is the $Q\sigma$-interval. Given the BWT B, the FM-index [7] is essentially composed of two functions C and Occ: $C(\sigma)$, with $\sigma \in \Sigma$, is the number of occurrences in B of symbols that are alphabetically smaller than σ, while $Occ(\sigma, i)$ is the number of occurrences of σ in the prefix $B[1 : i - 1]$ (hence $Occ(\cdot, 1) = 0$). These two functions can be used to efficiently compute a backward σ-extension on B of any Q-interval [7] and the corresponding forward σ-extension of the linked Q^{rev}-interval on B^{rev} [8]. The same procedure can be used also for computing backward σ-extensions only thanks to the property that the first $|R|$ elements of the GSA corresponds to R in lexicographical order. Notice that the order we assumed on the elements of the GSA allows us to compute also the backward $-extension of a Q-interval (hence determining the set of reads sharing a common prefix Q), whereas this operation is not possible according to the usual order of the elements of the GSA. The backward $-extension will be used in several parts of our algorithms in order to compute and represent such a set of reads. Moreover, for the purpose of computing σ-extensions, notice that the BWT can be obtained assuming any order for equal suffixes in different reads, since there not exists any string-interval including only some of them.

3   The Algorithm

Since short overlaps are likely to appear by chance, they are not meaningful for assembling the original sequence. Hence, we will consider only overlaps at least τ long, where τ is a positive constant. For simplicity, we assume that the set R of the reads is *substring-free*, that is, there are no two reads $r_1, r_2 \in R$ such that r_1 is a substring of r_2. The *overlap graph* of R is the directed graph $G_O = (R, A)$ whose vertices are the strings in R, and two reads r_i, r_j form the arc (r_i, r_j) if they overlap. Moreover, each arc (r_i, r_j) of G_O is labeled by the extension $ex_{i,j}$ of r_i with r_j. Each path $(r_1, \cdots, r_k)$ in G_O represents a string that is obtained by assembling the reads of the path. More precisely, such string is the concatenation $r_1 ex_{1,2} ex_{2,3} \cdots ex_{k-1,k}$ [9, 14]. An arc (r_i, r_j) of G_O is called *reducible* if there exists another path from r_i to r_j representing the same string of the path (r_i, r_j) (*i.e.*, the string $r_i ex_{i,j}$). Notice that reducible arcs are not helpful in assembling reads, therefore we are interested in removing (or in avoiding computing) them. The resulting graph is called *string graph* [9].

Let us denote by $R^s(Q)$ and $R^p(Q)$ the set of reads whose suffix (prefix, resp.) is a given string Q. If $|Q| \geq \tau$, then each pair of reads $r_s \in R^s(Q)$, $r_p \in R^p(Q)$ forms an arc (r_s, r_p) of G_O. Conversely, given an arc (r_s, r_p) of G_O, then $r_s \in R^s(ov_{s,p})$ and $r_p \in R^p(ov_{s,p})$. Therefore, the arc set of the overlap graph is the union of $R^s(Q) \times R^p(Q)$ for each Q at least τ characters long. Observe that a $\$Q$-interval represents the set $R^p(Q)$ of the reads with prefix Q, while a $Q\$$-interval represents the set $R^s(Q)$ of the reads with suffix Q. As a consequence, we can represent the sets $R^s(Q)$ and $R^p(Q)$ as two string-intervals.

Our algorithm for building the string graph is composed of three steps. The first step computes a compact representation of the overlap graph in secondary memory, the second step computes the prefix-extensions of each arc of the overlap graph that will be used in the third step for removing the reducible arcs from the compact representation of the overlap graph (hence obtaining the string graph). In the first step, since the cartesian product $R^s(S) \times R^p(S)$ represents all arcs whose overlap is S, we compute the (unlabeled) arcs of the overlap graph by computing all S-intervals ($|S| \geq \tau$) such that the two sets $R^s(S)$, $R^p(S)$ are both nonempty. We compactly represent the set of arcs whose overlap is S as a tuple $(q(S\$), q(\$S), 0, |S\$|)$, that we call *basic arc-interval*. We will use S for denoting a string that is an overlap among some reads.

The three steps of the algorithm work on the three files—$\mathcal{B}$, $\mathcal{SA}$ and $\mathcal{L}$— containing the BWT, the GSA, and the LCP of the set R, respectively. We first discuss the ideas used to compute the overlap graph, while we will present the other steps in the following parts of the section. Observe that the arcs of the overlap graph correspond to nonempty $S\$$-intervals and $\$S$-intervals for every overlap S of length at least τ. As a consequence, the computation of the overlap graph reduces to the task of computing the set of S-intervals that have a nonempty backward and forward \$-extension (along with the extensions themselves). We first show how to compute in secondary memory all such S-intervals and their nonempty \$-extensions with a single sequential scan of $\mathcal{L}$ and $\mathcal{SA}$. Then, we will describe the procedure **ExtendIntervals** that computes, in secondary memory and with a single scan of files $\mathcal{B}$ and $\mathcal{L}$, the backward σ-extensions of a collection of string-intervals (in particular, those computed before). Such a collection is not necessarily composed of pairwise-disjoint string-intervals, hence the procedure of [5] cannot be applied since it stores only a couple of Occ entry, called Π and π, while extending multiple nested intervals requires to store multiple values of Π. We point out that **ExtendIntervals** is of more general interest and, in fact, it will be also used in the second step of the algorithm.

An S-interval $[b, e)$ corresponds to a maximal portion $LCP[b + 1 : e - 1]$ of values greater than or equal to $|S|$, that we call $|S|$-*superblock*. Moreover, if S is an overlap between at least two reads, the width of such superblock is greater than 1. Notice that for each position i of the LCP and for each integer j, there exists at most one j-superblock containing i. During a single scan of the LCP, for each position i, we can maintain the list of j-superblocks for all possible j (*i.e.*, all the string-intervals for some string S such that $|S| = j$) that contain i. Such a list of superblocks represents the list of possible string-intervals that need to be forward and backward \$-extended to compute $R^s(S)$ and $R^p(S)$. Since the GSA contains all suffixes in lexicographic order, the S\$-interval (if it exists) is the initial portion $[b, e_1)$ of the S-interval $[b, e)$ such that, for each $b \leq i < e_1$, we have $SA[i] = (|S|, \cdot)$. Thus, by a single scan of the LCP file and of the GSA file, we complete the computation of all the S\$-intervals. This first scan can also maintain the corresponding S-intervals. Then, a backward \$-extension of this collection of S-intervals determines if the \$$S$-interval is nonempty. As noted before, the S-intervals might not be disjoint, therefore the procedure of [5] cannot be applied. However, we produce this collection of S-intervals ordered by their end boundary. We developed the procedure **ExtendIntervals** (illustrated below) that, given a list of string-intervals ordered by their end boundary on the BWT, with a single scan of the files $\mathcal{B}$ and $\mathcal{L}$, outputs the backward σ-extensions of all the string-intervals given in input. Moreover, if pairs of linked intervals (*i.e.*, pairs composed of an S-interval on B and the linked S^{rev}-interval on B^{rev}) are provided as input of **ExtendIntervals**, then it simultaneously computes the backward extensions of the intervals on B and the forward extensions of the intervals on B^{rev}. Consequently, if we give as input of **ExtendIntervals** the collection of all S-intervals that have a nonempty forward \$-extension, then we will obtain the collection of \$$S$-intervals, that, coupled with the S\$-intervals computed before, provide the desired compact representation of the overlap graph. Finally, we remark that the same procedure **ExtendIntervals** will be also crucial for computing the prefix-extensions in the second step of our algorithm.

Backward Extending Q-Intervals. In this section, we will describe a procedure for computing the backward extensions of a generic set I of string-intervals. Differently from the procedure in [5], which is only able to backward extend sets of pairwise disjoint string-intervals, we exploit the LCP array in order to efficiently deal with the inclusion between string-intervals (in fact, any two string-intervals are either nested or disjoint). Each Q-interval $[b, e)$ in I is associated to a record $(Q, [b, e), [b', e'))$ such that $[b, e)$ is the Q-interval on B and $[b', e')$ is the Q^{rev}-interval on B^{rev}, that is, the intervals in each record are linked. Moreover, a set $x([b, e))$ of symbols are associated to each string-interval $[b, e)$ in I, and $x([b, e))$ contains the symbols that must be used to extend the record. For each string-interval and for each character σ in the associated set of symbols, the result must contain a record $(\sigma Q, [b_\sigma, e_\sigma), [b'_\sigma, e'_\sigma))$ where $[b_\sigma, e_\sigma)$ is the backward σ-extension of $[b, e)$ on B and $[b'_\sigma, e'_\sigma)$ is the forward σ-extension of $[b', e')$ on B^{rev}. Notice that also the intervals in the output records are linked.

The algorithm **ExtendIntervals** performs only a single pass over the BWT B and the LCP $\mathcal{L}$, and maintains an array $\Pi[\cdot]$ which stores for each symbol in $\Sigma^\$$ the number of its occurrences in the prefix of the BWT preceding the current position. In other words, when the first p symbols of B have been read, the array Π gives the number of occurrences of each symbol in $\Sigma^\$$ in the first $p - 1$ characters of B. The procedure also maintains some arrays $\mathcal{E}\Pi_j[\cdot]$ so that, for each symbol σ and each integer j, $\mathcal{E}\Pi_j[\sigma] = Occ(\sigma, p_j)$ where p_j is the starting position of the Q-interval containing the current position of the BWT such that (1) $|Q| = j$ and (2) the width of the Q-interval is larger than 1. Notice that, for each position p and integer j, at most one such Q-interval exists. If no such Q-interval exists then the value of $\mathcal{E}\Pi_j$ is undefined. We recall that $Occ(\sigma, p)$ is the number of occurrences of σ in $B[0 : p - 1]$ [7]. Since **ExtendIntervals** accesses sequentially the arrays B and LCP, it is immediate to view the procedure as an external memory algorithm where B and LCP are two files. Notice also that line 3, that is finding all Q-intervals whose end boundary is p, can be executed most efficiently if the intervals are already ordered by end boundary. Lemmas 1 and 2 show the correctness of Alg. 1.

Lemma 1. *At line 3 of Algorithm 1, for each $c \in \Sigma$ (1) $\Pi[c]$ is equal to the number of occurrences of c in $B[1 : p - 1]$ and (2) $\mathcal{E}\Pi_k[c] = Occ(c, p_k)$ for each Q-interval $[p_k, e_k)$ of width larger than 1 which contains p and such that $|Q| = k$.*

Proof. We prove the lemma by induction on p. When $p = 1$, there is no symbol before position p, therefore Π must be made of zeroes, and the initialization of line 1 is correct. Moreover all string-intervals containing the position 1 must start at 1 (as no position precedes 1), therefore line 1 sets the correct values of $\mathcal{E}\Pi_k$.

Assume now that the property holds up to step $p - 1$ and consider step p. The array Π is updated only at line 16, hence its correctness is immediate. Let $[p_k, e_k)$ be the generic Q-interval $[p_k, e_k)$ containing p and such that (1) $|Q| = k$, and (2) the width of the Q-interval is larger than 1, that is $e_k - p_k \geq 2$. Since all suffixes in the interval $[p_k, e_k)$ of the GSA have Q as a common prefix and $|Q| = k$, $LCP[i] \geq k$ for $p_k < i \leq e_k$.

If $p_k < p$, then $[p_k, e_k)$ contains also $p - 1$, that is Q is a prefix of the suffix pointed to by $SA[p - 1]$. Hence $LCP[p] \geq k$ and the value of $\mathcal{E}\Pi_k$ at iteration p is the same as at iteration $p - 1$. By inductive hypothesis $\mathcal{E}\Pi_k = Occ(c, p_k)$. The value of $\mathcal{E}\Pi_k$ is correct, since the line 15 of the algorithm is not executed.

Algorithm 1. ExtendIntervals

Input : The BWT B and the LCP array L of a set R of strings. A set I of Q-intervals, each one associated with a record and with a set $x(\cdot)$ of characters driving the extension.

Output : The set of extended Q-intervals.

1 Initialize Π and each $\mathcal{E}\Pi_j$ (for $1 \leq j \leq \max_i\{L[i]\}$) to be $|\Sigma|$-long vectors $\bar{0}$;
2 **for** $p \leftarrow 1$ **to** $|B|$ **do**
3   **foreach** Q-*interval* $[b,e)$ *in* I *such that* $e = p$ **do**
4          $(Q,[b,e),[b',e')) \leftarrow$ the record associated to $[b,e)$ // $p = e$
5   **foreach** *character* $c \in x([b,e))$ **do**
6   **if** $b = e - 1$ **then**
7   **if** $B[p-1] = c$ **then**
8                      $t \leftarrow 0$
9   **else**
10                      $t \leftarrow 1$;
11   Output $(cQ, [C[c] + \Pi[c] + t, C[c] + \Pi[c] + 1),$ $[b' + \sum_{\sigma < c}\left(\Pi(\sigma) - \mathcal{E}\Pi_{|Q|}(\sigma)\right), b' + \sum_{\sigma < c}\left(\Pi(\sigma) - \mathcal{E}\Pi_{|Q|}(\sigma)\right) + (1-t)));$
12   **else**
13   Output $(cQ, [C[c] + \mathcal{E}\Pi_{|Q|}[c] + 1, C[c] + \Pi[c] + 1),$ $[b' + \sum_{\sigma < c}\left(\Pi(\sigma) - \mathcal{E}\Pi_{|Q|}(\sigma)\right), b' + \sum_{\sigma < c}\left(\Pi(\sigma) - \mathcal{E}\Pi_{|Q|}(\sigma)\right) + (\Pi[c] - \mathcal{E}\Pi_{|Q|}[c])));$
14   **foreach** j *such that* $L[p] \leq j < L[p+1]$ **do**
15          $\mathcal{E}\Pi_j \leftarrow \Pi$ // a Q-interval with $|Q| = j$ begins at position p
16      $\Pi[B[p]] \leftarrow \Pi[B[p]] + 1$;

Consider now the case $p_k = p$, that is p is the beginning of the Q-interval, for some Q with $|Q| = k$. In this case $LCP[p] < k$. Therefore $\mathcal{E}\Pi_k$ is updated at line 15 and, by the correctness of Π, is set to $Occ(\cdot, p)$. □

Lemma 2. *Let* $(Q, [b,e), [b', e'))$ *be a record and let* $c \in x([b,e))$ *be a character. Then Algorithm 1 outputs the correct c-extension of such record.*

Proof. When the algorithm reaches position $p = e$, it outputs a c-extension of the record $(Q, [b,e), [b',e'))$. Therefore we only have to show that the computed extension is correct. The backward c-extension of $[b,e)$ is $[C(c) + Occ(c,b) + 1, C(c) + Occ(c,e) + 1)$ [7], while the forward c-extension of its linked interval $[b', e')$ has starting point b' plus the number of occurrences in $B[b : e-1]$ of the symbols smaller than c [8]. Moreover two linked intervals have the same width [8].

These observations, together with Lemma 1 and the fact that Π and $\mathcal{E}\Pi$ are not modified in lines 5–13, establish the correctness for the case $b < e-1$.

Let us now consider the case $b = e - 1$. Notice that $Occ(c,p) = Occ(c,p-1)$ unless $B[p-1] = c$, and $Occ(c,p) = Occ(c,p-1)+1$ if $B[p-1] = c$. Therefore the assignment of t at lines 7–10 guarantees that, at line 11, $Occ(c,p) = Occ(c,p-1)+(1-t)$. Together with Lemma 1, we get $\Pi[c]+t = Occ(c,p)+t = Occ(c,p-1)$. □

Computing Arc Labels. In this part, we describe how the compact representation of the overlap graph computed in the first step can be further processed in order to easily remove reducible arcs without resorting to (computationally expensive) string comparisons. First, we give an easy-to-test characterization of reducible arcs of overlap graphs in terms of string-intervals (Lemmas 3 and 4). Then, we show how such string-intervals (that we call *arc-labels*) can be efficiently computed in external memory starting from the collection of basic arc-intervals computed in the first step.

Lemma 3. *Let G_O be the overlap graph for R and let $(r_{i_1}, r_{i_2}, \ldots, r_{i_k})$ be a path of G_O. Then, such a path represents the string $pe_{i_1, i_2} pe_{i_2, i_3} \cdots pe_{i_{k-1}, i_k} r_{i_k}$.*

Proof. We will prove the lemma by induction on k. Let (r_h, r_j) be an arc of G_O. Notice that the string represented by such arc is $pe_{h,j} ov_{h,j} ex_{h,j}$. Since $r_h = pe_{h,j} ov_{h,j}$ and $r_j = ov_{h,j} ex_{h,j}$, applying the property to the arc (r_{i_1}, r_{i_2}) settles the case $k = 2$. Assume now that the lemma holds for paths of length smaller than k and consider the path $(r_{i_1}, \ldots, r_{i_k})$. By definition, the string represented by such path is $r_{i_1} ex_{i_1, i_2} \cdots ex_{i_{k-1}, i_k}$ which, by inductive hypothesis on the path $(r_{i_1}, r_{i_2}, \ldots, r_{i_{k-1}})$, is equal to $pe_{i_1, i_2} \cdots pe_{i_{k-2}, i_{k-1}} r_{i_{k-1}} ex_{i_{k-1}, i_k}$. But $r_{i_{k-1}} ex_{i_{k-1}, i_k} = pe_{i_{k-1}, i_k} ov_{i_{k-1}, i_k} ex_{i_{k-1}, i_k}$ which can be rewritten as $pe_{i_{k-1}, i_k} r_{i_k}$. Hence $pe_{i_1, i_2} \cdots pe_{i_{k-2}, i_{k-1}} r_{i_{k-1}} ex_{i_{k-1}, i_k} = pe_{i_1, i_2} \cdots pe_{i_{k-2}, i_{k-1}} pe_{i_{k-1}, i_k} r_{i_k}$. □

Lemma 4. *Let G_O be the overlap graph for a substring-free set R of reads and let (r_i, r_j) be an arc of G_O. Then, (r_i, r_j) is reducible iff there exists another arc (r_h, r_j) such that $pe_{h,j}$ is a proper suffix of $pe_{i,j}$ (or, equivalently, that $pe_{h,j}^{rev}$ is a proper prefix of $pe_{i,j}^{rev}$).*

Proof. By definition, (r_i, r_j) is reducible if and only if there exists a second path $(r_i, r_{h_1}, \ldots, r_{h_k}, r_h, r_j)$ representing the string XYZ, where X, Y and Z are respectively the prefix-extension, the overlap and the extension of r_i with r_j (notice that $(r_{h_1}, \ldots, r_{h_k})$ might be empty). Assume that such a path $(r_i, r_{h_1}, \ldots, r_{h_k}, r_h, r_j)$ exists, hence r_{h_k} is a substring of XYZ. Since r_{h_k} overlaps with r_j, $r_{h_k} = X_1 Y Z_1$ where X_1 is a suffix of X and Z_1 is a proper prefix of Z. Notice that $X_1 = pe_{h_k, j}$ and R is substring free, hence X_1 is a proper suffix of X, otherwise r_i would be a substring of r_{h_k}, completing this part of the proof.

  Assume now that there exists an arc (r_h, r_j) such that $pe_{h,j}$ is a proper suffix of $pe_{i,j}$. Again, $r_h = X_1 Y_1 Z_1$ where X_1, Y_1 and Z_1 are respectively the prefix-extension, the overlap and the extension of r_{h_k} with r_j. By hypothesis, X_1 is a suffix of X. Since r_h is not a substring of r_i, the fact that X_1 is a suffix of X implies that Y is a substring of Y_1, therefore r_i and r_h overlap and $|ov_{i,h}| \geq |Y| \geq \tau$, hence (r_i, r_h) is an arc of G_O. The string associated to the path r_i, r_h, r_j is $r_i ex_{i,h} ex_{h,j}$. By Lemma 3, $r_i ex_{i,h} ex_{h,j} = pe_{i,h} pe_{h,j} r_j$. At the same time the string associated to the path r_i, r_j is $r_i ex_{i,j} = pe_{i,j} r_j$ by Lemma 3, hence it suffices to prove that $pe_{i,h} pe_{h,j} = pe_{i,j}$. Since $pe_{h,j}$ is a proper suffix of $pe_{i,j}$, by definition of prefix-extension, $pe_{i,h} pe_{h,j} = pe_{i,j}$, completing the proof. □

  A corollary of Lemma 4 is that an arc (r_i, r_j) is reducible iff there exists another arc (r_h, r_j) such that the $pi_{h,j}^{rev}$-interval strictly contains the $pi_{i,j}^{rev}$-interval.

As a consequence, it would be useful to compute and store the prefix-extensions of the arcs, obtaining a partition of each set $R^s(S) \times R^p(S)$ (*i.e.*, of each basic arc-interval) in classes with the same prefix-extension P. More precisely, for each of those classes, we need the $PS\$$-interval as well as the P-interval to represent all arcs (r_i, r_j) with $ov_{i,j} = S$ and label $pe_{i,j} = P$. However, in order to perform the reducibility test, the $pe_{i,j}$-interval alone is not sufficient, and we also need the $pe_{i,j}^{rev}$-interval. All these concepts are fundamental for describing our algorithm and are formally defined as follows.

Definition 5. *Let B be a BWT for a collection of reads R and let B^{rev} be the BWT for the reversed reads R^{rev}. Let S and P be two strings. Then, the arc-interval associated to (P, S) is the tuple $(q(PS\$), q(\$S), |P|, |PS\$|)$, where $q(PS\$)$ and $q(\$S)$ are the $PS\$$-interval and the $\$S$-interval on B. Moreover, $|PS\$|$, S and P are respectively called the* length, *the* overlap-string, *and the* prefix-extension *of the arc-interval.*

An arc-interval is terminal *if the $PS\$$-interval has a nonempty backward $\$$-extension. The triple $(q(P), q^{rev}(P^{rev}), |P|)$ is called the* arc-label *of the arc-interval associated to (P, S).*

To obtain the labels of the arcs we need to compute the terminal arc-intervals, that is the arc-intervals where P is the (complete) prefix-extension of an arc, since $q(PS\$)$ (in a terminal arc-interval) has a nonempty backward $\$$-extension. If R is a substring-free set of strings, $q(PS\$)$ represents a unique read $r = PS$. The associated arc-labels are used to test efficiently whether an arc is reducible.

Terminal arc-intervals are computed by procedure **ExtendArcIntervals** (Algorithm 2) that extends string-intervals $q(PS\$)$ of arc-intervals by increasing length $|PS|$. This step is done by modifying the approach in [5] to deal, at each iteration of backward extension, with string-intervals that can be disjoint or duplicated. In fact, we may have two arc-intervals, associated to the pairs (P_1, S_1) and (P_2, S_2), which correspond to the same string-interval $q(P_1 S_1 \$) = q(P_2 S_2 \$)$, where $P_1 S_1 = P_2 S_2$ but $S_1 \neq S_2$. Such duplicated arc-intervals will occur consecutively in the input list.

Arc-labels are computed by incrementally backward extending with **ExtendIntervals** the linked intervals $q(P)$ and $q^{rev}(P^{rev})$ at the same iteration where interval $q(PS\$)$ of the associated arc-interval has been extend. In fact, we maintain a link between an arc-interval and its arc-label. While at each iteration all string-intervals originating from arc-intervals have the same length, the string-intervals associated to arc-labels can have different lengths. However, in each file they are ordered by their end boundary, hence we can apply directly the **ExtendIntervals** procedure. By maintaining a suitable organization of the files, we are able to keep a 1-to-1 correspondence between arc-intervals and arc-labels.

When we compute an extension, we test if the arc-interval $(q(PS\$), q(S\$), |P|, |PS\$|)$ is terminal, that is if $q(PS\$)$ has a nonempty backward $\$$-extension. In that case we have found the set $\{r\} \times R^p(S)$ of arcs of the overlap graph outgoing from the read $r = PS$ and with overlap S.

Managing Q-Intervals Using Files. During the first step, the algorithm computes a file $\mathcal{BAI}(\sigma, \ell_1)$, for each symbol $\sigma \in \Sigma$ and for each integer ℓ_1, such

Algorithm 2. ExtendArcIntervals

Input : Two files $\mathcal{B}$ and $\mathcal{SA}$ containing the BWT and the GSA of the set R, respectively. A set of files $\mathcal{AI}(\cdot, \ell_1)$ containing the arc-intervals of length ℓ_1. A set of files $\mathcal{BAI}(\cdot, \ell_1)$ containing the basic arc-intervals of length ℓ_1.

Output: A set of files $\mathcal{AI}(\cdot, \ell_1 + 1)$ containing the arc-intervals of length $\ell_1 + 1$. The arcs of the overlap graph coming out from reads of length $\ell_1 - 1$.

1 $\Pi(\sigma) \leftarrow 0$, for each $\sigma \in \Sigma$;
2 $\pi(\sigma) \leftarrow 0$, for each $\sigma \in \Sigma$;
3 $[b_{prev}, e_{prev}) \leftarrow null$;
4 **foreach** $\sigma \in \Sigma$ **do**
5   **foreach** $([b, e), q(\$S), l_e, \ell_1) \in SortedMerge(\mathcal{AI}(\sigma, \ell_1), \mathcal{BAI}(\sigma, \ell_1))$ **do**
      // If the $PS\$$-interval $[b, e)$ is different from the one previously processed,
      then vectors Π and π must be updated, otherwise $[b, e)$ is extended using the
      values previously computed.
6   **if** $([b, e) \neq [b_{prev}, e_{prev})$ **then**
7          $\Pi(\sigma) \leftarrow \Pi[\sigma] + \pi[\sigma]$, for each $\sigma \in \Sigma$;
8   Update Π while reading $\mathcal{B}$ until the BWT position $b - 1$;
9          $\pi(\sigma) \leftarrow 0$, for each $\sigma \in \Sigma$;
10         $r \leftarrow null$;
11   **while** *reading $\mathcal{B}$ from the BWT position b to $e - 1$* **do**
12           $\sigma \leftarrow$ symbol of the BWT at the current position p;
13   **if** $\sigma \neq \$$ **then**
14             $\pi[\sigma] \leftarrow \pi[\sigma] + 1$;
15   **else**
            // The arc-interval is terminal and r is the read equal to PS
16             $r \leftarrow$ the read pointed to by GSA at position p;
17   **if** $r \neq null$ **then**
        // Update the file $\mathcal{A}$ of the output arcs, since the arc-interval is terminal
18   Append $\{r\} \times R^p(S)$ to $\mathcal{A}_{|\ell_e|}$;
19   **else**
20   **foreach** $\sigma \in \Sigma$ **do**
21   **if** $\pi[\sigma] > 0$ **then**
22                $b' \leftarrow C[\sigma] + \Pi[\sigma] + 1$;
23                $e' \leftarrow b' + \pi[\sigma]$;
24   Append $([b', e'), q(S\$), l_e + 1, \ell_1 + 1)$ to $\mathcal{AI}(\sigma, \ell_1 + 1)$;
25       $[b_{prev}, e_{prev}) \leftarrow [b, e)$;

that $\tau + 1 \leq \ell_1 \leq \ell_{max}$, where ℓ_{max} is the maximum length of a read in R. More precisely, the file $\mathcal{BAI}(\sigma, \ell_1)$ contains the basic arc-intervals of length ℓ_1, whose overlap-string begins with the symbol σ (observe that the overlap-string has length $\ell_1 - 1$). The basic arc-intervals are stored (in each file) by non-decreasing values of the start boundary e of the interval $q(S\$) = [b, e)$.

The algorithm also uses a file $\mathcal{AI}(\sigma, \ell_1)$ and a file $\mathcal{AL}(\sigma, \ell_1, \ell_2)$ for each symbol $\sigma \in \Sigma$, for each integer ℓ_1, such that $\tau + 1 < \ell_1 \leq \ell_{max}$, and for each integer ℓ_2 such that $1 \leq \ell_2 \leq \ell_{max} - \tau$. The file $\mathcal{AI}(\sigma, \ell_1)$ consists of the arc-intervals of length ℓ_1, whose prefix-extension P begins with the symbol σ, while the file $\mathcal{AL}(\sigma, \ell_1, \ell_2)$ contains the arc-labels related to arc-intervals of length ℓ_1 whose

ℓ_2-long prefix-extension P begins with the symbol σ. These files are tightly coupled, since there is a 1-to-1 correspondence between records of $\mathcal{AI}(\sigma, \ell_1)$ and records of $\mathcal{AL}(\sigma, \ell_1, \cdot)$, where those records refer to the same pair (P, S) of prefix-extension (of length ℓ_2) and overlap-string. Each of the $\mathcal{BAI}(\cdot, \cdot)$, $\mathcal{AI}(\cdot, \cdot)$ and $\mathcal{AL}(\cdot, \cdot, \cdot)$ files contains string-intervals of the same length and ordered by start boundary, hence those intervals are also sorted by end boundary.

Computing Terminal Arc-Intervals and Arc-Labels. After the first step, the algorithm computes terminal arc-intervals and arc-labels. The first fundamental observation is that an arc-interval of length $\ell_1 + 1$ (that is an arc-interval that will be stored in $\mathcal{BAI}(\cdot, \ell_1 + 1)$), and corresponding to a pair (P, S) with $|PS\$| = \ell_1 + 1$, can be obtained by extending a basic arc-interval of length ℓ_1 (taken from $\mathcal{BAI}(\cdot, \ell_1)$) or a non-basic arc-interval of length ℓ_1 (taken from $\mathcal{AI}(\cdot, \ell_1)$). Since all those files are sorted, we can assume to have a SortedMerge procedure which receives two sorted files and returns their sorted union. Notice that we do not actually need to write a new file, as SortedMerge basically consists in choosing the file from which to read the next record.

The algorithm performs some extension steps, each mainly backward σ-extending string-intervals. In fact, at each extension step i (to simplify some formulae, the first step is set to $i = \tau + 1$), the algorithm scans all files $\mathcal{BAI}(\cdot, i)$, $\mathcal{AI}(\cdot, i)$, $\mathcal{AL}(\cdot, i, j)$ and computes the files $\mathcal{AI}(\cdot, i + 1)$, $\mathcal{AL}(\cdot, i + 1)$. At iteration i, for each $\sigma_1 \in \Sigma$, all records in SortedMerge($\mathcal{BAI}(\sigma_1, i)$, $\mathcal{AI}(\sigma_1, i)$) are σ_2-extended, for each $\sigma_2 \in \Sigma$, via the procedure **ExtendArcIntervals**, outputting the results in the file $\mathcal{AI}(\cdot, i + 1)$. We recall that σ-extending a record means, in this case of the procedure ExtendArcIntervals, to backward σ-extend the $q(PS\$)$ of the arc-interval (or the $q(S\$)$ of the basic arc-interval). If the record to σ_2-extend is read from a file $\mathcal{BAI}(\cdot, i)$ (*i.e.*, it is a basic arc-interval), when the algorithm writes a record of $\mathcal{AI}(\cdot, i + 1)$ (*i.e.*, the σ_2-extension of those record), it also writes the corresponding record of $\mathcal{AL}(\cdot, i + 1)$, that is an arc-label where the prefix-extension is equal to the symbol σ_2. On the other hand, if the current record to σ_2-extend is read from a file $\mathcal{AI}(\cdot, i)$, we consider also the corresponding record of $\mathcal{AL}(\cdot, i)$ to write a record of $\mathcal{AI}(\cdot, i + 1)$ and the corresponding record of $\mathcal{AL}(\cdot, i + 1)$ which is the σ_2-extension of the record in $\mathcal{AL}(\cdot, i)$. Each time a terminal arc-interval associated to (P, S) is found, the arcs $\{r\} \times R^p(S)$, where $r = PS$, are written in the file $\mathcal{A}_{|P|}$.

Testing Irreducible Arcs. The algorithm reads the arcs of the overlap graph, stored in the files $\mathcal{A}_i$, for increasing values of i. Each arc a is added to the set A of the arcs of the string graph if there is no arc already in A reducing a. Notice that A is stored in main memory in the current implementation. Lemma 4 implies that an arc (of the overlap graph) associated to a pair (P_1, S_1) can be reduced only by an arc associated to a pair (P_2, S_2), such that $|P_1| > |P_2|$. Hence, an arc in $\mathcal{A}_i$ can be reduced by an arc in $\mathcal{A}_j$ only if $j < i$. Since we examine the files $\mathcal{A}_i$ by increasing values of i, either an arc a is reduced by an arc that is already in A, or no subsequently read arc of the overlap graph can reduce a. Notice also that, by the reducibility test of Lemma 4, an arc associated to a pair (P_1, S_1) is reduced by an arc associated to (P_2, S_2) if and only if P_2^{rev} is a proper prefix of P_1^{rev}. Thus, the test is equivalent to determine whether the P_2^{rev}-interval on B^{rev} properly contains the P_1^{rev}-interval. The latter test can be easily performed by outputting in the files $\mathcal{A}_j$ a representation of the prefix-interval of each arc.

On the Complexity. Notice that Algorithm 1 scans once $\mathcal{B}$ and $\mathcal{L}$ and recall the total length of $\mathcal{B}$ is ℓn where ℓ is the length of the reads and n is number of reads. Since the input Q-intervals are nested or disjoint, there are at most $O(\ell n)$ distinct Q-intervals, that is, block at lines 6–10 and line 12 are executed $O(\ell n)$ times. A stack-based data structure allows to store the distinct $\mathcal{E}\Pi$ arrays while requiring $O(1)$ time for each iteration, hence the time complexity of Algorithm 1 is $O(\ell n)$, while its space complexity is $O(\ell|\Sigma|)$ since it stores at most ℓ arrays of $|\Sigma|$ elements (plus a constant space). The second phase of our algorithm consists of $(\ell - \tau)$ iterations, each requiring a call to **ExtendIntervals** and **ExtendArcIntervals**, therefore the overall time complexity to compute the overlap graph is $O(\ell^2 n)$, which is also an upper bound on the number of arcs of the overlap graph. The time complexity of the third phase is $O(de)$, where e and d are respectively the number of arcs of the overlap graph and the maximum indegree of the resulting string graph, as each arc must be tested for reducibility against each adjacent vertex.

4   Experimental Analysis

We performed a preliminary experimental comparison of LSG with SGA, a state-of-the-art assembler based on string-graphs [13], on the dataset of the Human chromosome 14 used in the recent Genome Assembly Gold-standard Evaluation (GAGE) project [11]. We used a slightly modified version of BEETL 0.9.0 [2] to construct the BWT, the GSA, and the LCP of the reads needed by LSG. Note that BEETL is able to compute the GSA by setting a particular compilation flag. Since the current version of BEETL requires all input reads to have the same length, we harmonized the lengths of the reads ($\sim$36M) of the GAGE dataset to 90bp: we discarded shorter reads ($\sim$6M), whereas we split longer reads into overlapping substrings with a minimum overlap of 70bp. We further preprocessed and filtered the resulting $\sim$50M reads according to the workflow used for SGA in GAGE [11]: no reads were discarded by the preprocess step, while $\sim$13M reads were filtered out as duplicated. As a result, the final dataset was composed of $\sim$37M reads of length 90bp.

We generated the index of the dataset using `sga-index` and `beetl-bwt` and we gave them as input to SGA and LSG requiring a minimum overlap (τ) of 65. We performed the experimental analysis on a workstation with 12GB of RAM and standard mechanical hard drives, as our tool is designed to cope with a limited amount of main memory. The workstation has a quad-core Intel Xeon W3530 2.80GHz CPU running Ubuntu Linux 12.04. To perform a fair comparison, we slightly modified SGA to disable the computation of overlaps on different strands (*i.e.*, when one read is reversed and complemented w.r.t. the other).

For the comparison, we focused on *running times* and *main memory allocation*. During the evaluation of the tools we do not consider the index generation step because such part is outside the scope of this paper. Regarding the running times, SGA built the string graph in 2 hours and 19 minutes, whereas LSG built the string graph in a total time of 5 hours and 28 minutes (9min were required for computing the basic arc-intervals with 98% CPU time, 5h and 13min for arc labeling with 76% CPU time, 4min for graph reduction with 60% CPU time, and 2min for producing the final output with 59% CPU time). Regarding the main memory usage, SGA had a peak memory allocation of 3.2GB whereas LSG required less than 0.09GB for basic arc-interval computation and for arc labeling, and less than 0.25GB for graph reduction.

We chose to write the output in ASQG format (the format used by SGA) to allow processing the results obtained by LSG by the subsequent steps of the SGA workflow (such as the assembly and the alignment steps). In the current straightforward implementation of this part, we store the whole set of read IDs in main memory, which pushes the peak memory usage of this part to 2.5GB. However, more refined implementations (that, for example, store only part of the read IDs in main memory) could easily reduce the memory usage for this (non-essential) part. We also want to point out that the memory required by the reduction step can be arbitrarily reduced by reducing iteratively arcs incident to subsets of nodes with only a small penalty in running times.

Furthermore, we point out that this experimental part was performed on commodity hardware equipped with mechanical hard disks. As a consequence, the execution of LSG on systems equipped with faster disks (*e.g.*, SSDs) will significantly decrease its running time, especially when compared with SGA.

5   Conclusions and Future Work

We have proposed an external memory algorithm for building a string graph from a set R of reads and we have shown that our approach is efficient in theory (the time complexity is not much larger than that of the lightweight BWT construction, which is a necessary step) and in practice (the time required by LSG is less than 3 times that of SGA, while memory usage in all the most computationally expensive steps is less than 12 times that of SGA) on a regular PC.

Since LSG potentially scales on very large datasets, we expect to be able to use our approach to assemble RNA-seq reads even when the entire transcriptome is sequenced at high coverage. In fact, an important research direction is to face the problem of assembling RNA-seq data and building graph models of gene structures (such as splicing graphs [4]) in absence of a reference genome.

Acknowledgments. The authors thank the reviewers for their detailed and insightful comments. The authors acknowledge the support of the MIUR PRIN 2010-2011 grant 2010LYA9RH (Automi e Linguaggi Formali: Aspetti Matematici e Applicativi), of the Cariplo Foundation grant 2013-0955 (Modulation of anti cancer immune response by regulatory non-coding RNAs), of the FA 2013 grant (Metodi algoritmici e modelli: aspetti teorici e applicazioni in bioinformatica).

References

1. Bankevich, A., Nurk, S., Antipov, D., et al.: SPAdes: A new genome assembly algorithm and its applications to single-cell sequencing. J. Comput. Biol. 19(5), 455–477 (2012)
2. Bauer, M., Cox, A., Rosone, G.: Lightweight algorithms for constructing and inverting the BWT of string collections. Theor. Comput. Sci. 483, 134–148 (2013)
3. Bauer, M.J., Cox, A.J., Rosone, G., Sciortino, M.: Lightweight LCP construction for next-generation sequencing datasets. In: Raphael, B., Tang, J. (eds.) WABI 2012. LNCS, vol. 7534, pp. 326–337. Springer, Heidelberg (2012)
4. Beretta, S., Bonizzoni, P., Della Vedova, G., Pirola, Y., Rizzi, R.: Modeling alternative splicing variants from RNA-Seq data with isoform graphs. J. Comput. Biol. 16(1), 16–40 (2014)
5. Cox, A.J., Jakobi, T., Rosone, G., Schulz-Trieglaff, O.B.: Comparing DNA sequence collections by direct comparison of compressed text indexes. In: Raphael, B., Tang, J. (eds.) WABI 2012. LNCS, vol. 7534, pp. 214–224. Springer, Heidelberg (2012)
6. Ferragina, P., Gagie, T., Manzini, G.: Lightweight data indexing and compression in external memory. Algorithmica 63(3), 707–730 (2012)
7. Ferragina, P., Manzini, G.: Indexing compressed text. J. ACM 52(4), 552–581 (2005)
8. Lam, T., Li, R., Tam, A., Wong, S., Wu, E., Yiu, S.: High throughput short read alignment via bi-directional BWT. In: BIBM 2009, pp. 31–36 (2009)
9. Myers, E.: The fragment assembly string graph. Bioinformatics 21, ii79–ii85 (2005)
10. Peng, Y., Leung, H.C.M., Yiu, S.M., Chin, F.Y.L.: IDBA – A practical iterative de bruijn graph de novo assembler. In: Berger, B. (ed.) RECOMB 2010. LNCS, vol. 6044, pp. 426–440. Springer, Heidelberg (2010)
11. Salzberg, S.L., et al.: GAGE: A critical evaluation of genome assemblies and assembly algorithms. Genome Res. 22(3), 557–567 (2012)
12. Shi, F.: Suffix arrays for multiple strings: A method for on-line multiple string searches. In: Jaffar, J., Yap, R.H.C. (eds.) ASIAN 1996. LNCS, vol. 1179, pp. 11–22. Springer, Heidelberg (1996)
13. Simpson, J., Durbin, R.: Efficient construction of an assembly string graph using the FM-index. Bioinformatics 26(12), i367–i373 (2010)
14. Simpson, J., Durbin, R.: Efficient de novo assembly of large genomes using compressed data structures. Genome Res. 22, 549–556 (2012)
15. Simpson, J., Wong, K., Jackman, S., et al.: ABySS: a parallel assembler for short read sequence data. Genome Res. 19(6), 1117–1123 (2009)

Topology-Driven Trajectory Synthesis with an Example on Retinal Cell Motions

Chen Gu[1], Leonidas Guibas[2], and Michael Kerber[3]

[1] Google Inc., USA
guc@google.com
[2] Stanford University, USA
guibas@cs.stanford.edu
[3] Max-Planck-Institute for Informatics, Germany
mkerber@mpi-inf.mpg.de

Abstract. We design a probabilistic trajectory synthesis algorithm for generating time-varying sequences of geometric configuration data. The algorithm takes a set of observed samples (each may come from a different trajectory) and simulates the dynamic evolution of the patterns in $O(n^2 \log n)$ time. To synthesize geometric configurations with indistinct identities, we use the pair correlation function to summarize point distribution, and α-shapes to maintain topological shape features based on a fast persistence matching approach. We apply our method to build a computational model for the geometric transformation of the cone mosaic in retinitis pigmentosa — an inherited and currently untreatable retinal degeneration.

Keywords: trajectory, pair correlation function, alpha shapes, persistent homology, retinitis pigmentosa.

1 Introduction

The work presented in this paper is motivated from the investigation of a retinal disease called retinitis pigmentosa [18]. In this disease, a mutation kills the rod photoreceptors in the retina. A consequence of this death is that the geometry of the mosaic of cone photoreceptors deforms in an interesting way. Normally, cones form a relatively homogeneous distribution. But after the death of rods, the cones migrate to form an exquisitely regular array of holes.

Our central goal is to build a dynamic evolution model for the point distributions that arise from the cone mosaic in retinitis pigmentosa. In physics, the most classical method for modeling cell motions is to solve a system of differential equations from Newton's laws of motion with some predefined force field which specifies cell-to-cell interactions. However, in many cases it is difficult to understand how different types of cells (for example cones and rods) interact with each other. There are also mathematical models that do not presume much prior biological knowledge, such as flocking which has been widely used to simulate coordinated animal motions [16]. But as with all model-based approaches, the method is limited by the model chosen in the first place.

D. Brown and B. Morgenstern (Eds.): WABI 2014, LNBI 8701, pp. 326–339, 2014.

Instead of fitting a predefined model, we propose an alternative approach which relies only on geometric and topological multi-scale summaries. The input is a set of geometric configurations, each of which may come from a different trajectory of a cone migration. We design a probabilistic algorithm to synthesize trajectories from observed data. In short, we let the points move randomly and check whether the transformation brings them "closer" to our next observation. To define closeness, we combine two high-level distance measures: firstly, we employ the *pair correlation function* (PCF) which extracts pairwise correlations in the point cloud data by measuring how density varies as a function of distance from a reference point. The PCF is widely accepted as an informative statistical measure for point set analysis, and has been used for trajectory synthesis in previous work [14]. As major novelty, we propose to combine the PCF with a topological distance measure: we compare *persistence diagrams of alpha-shape filtrations* which capture the evolution of holes that arise when the points are thickened to disks with increasing radius. Persistence diagrams are currently a popular research topic with many theoretical and practical contributions; we refer to the surveys [1,6] for contemporary overviews. We demonstrate that the combination of PCF and persistence diagrams results in trajectories with a much cleaner hole structure than for trajectories obtained only by PCF (see Figure 6). We believe the problem of trajectory synthesis for very sparse data to be of more general importance in biological and medical contexts, and hope that our model-free methodology can be applied to other contexts. Moreover, our approach provides evidence that topological methods are useful in the analysis of point distributions which have been extensively studied in computer graphics and point processes [7,15,17].

2   Biological Background: Retinitis Pigmentosa

The retina is a light-sensitive layer of tissue that lines the inner surface of the eye. It contains photoreceptor cells that capture light rays and convert them into electrical impulses. These impulses travel along the optic nerve to the brain where they are turned into images of the visual world.

There are two types of photoreceptors in the retina: cones and rods. In adult humans, the entire retina contains about 6 million cones and 120 million rods. Cones are contained in the macula, the portion of the retina responsible for central vision. They are most densely packed within the fovea, the very center portion of the macula. Cones function best in bright light and support color perception. In contrast, rods are spread throughout the peripheral retina and function best in dim light. They are responsible for peripheral and night vision.

Retinitis pigmentosa is one of the most common forms of inherited retinal degeneration. This disorder is characterized by the progressive loss of photoreceptor cells and may lead to night blindness or tunnel vision. Typically, rods are affected earlier in the course of the disease, and cone deterioration occurs later. In the progressive degeneration of the retina, the peripheral vision slowly constricts and the central vision is usually retained until late in the disease.

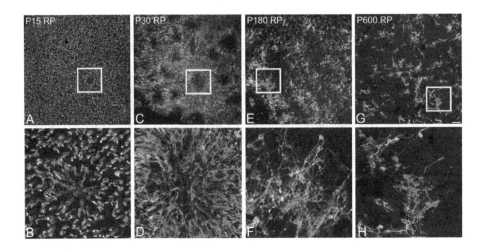

Fig. 1. Cone mosaic rearrangement in retinitis pigmentosa. The confocal micrographs in the top row show middle (red) and short (green) wavelength sensitive cones in whole mount retinas. Enlarged micrographs of marked regions are shown in the bottom row. This figure is taken from [8].

At present, there is no cure for retinitis pigmentosa. Researchers around the world are constantly working on development of treatments for this condition.

There have been some recent studies on the spatial rearrangement of the cone mosaic in retinitis pigmentosa [8,13]. These experiments are performed on a rat model in which a mutation in the retina triggers the cell death of rods, similar to those causing symptoms in humans. Figure 1 shows an example for the morphology and distribution of cones at postnatal days 15, 30, 180, and 600. In healthy retinas, the mosaic of cones exhibits a spatially homogeneous distribution. However, the death of rods causes cones to rearrange themselves into a mosaic comprising an orderly array of holes. These holes first begin to appear at random regions of the retina at day 15 and become ubiquitous throughout the entire tissue at day 30. Holes start to lose their form at day 180 and mostly disappear at day 600, at which time the cones are almost all dead.

Furthermore, it has been observed that both cones and rods follow the same retinal distribution. But the mechanisms of formations of holes of cones are different from those of rods. In fact, retinitis pigmentosa is caused by the initial loss of rods in the center of these holes, and then the death of rods tends to propagate as circular waves from the center of the holes outward. In contrast, the number of cones in normal and retinitis pigmentosa conditions do not show significant differences at stages as late as day 180. Therefore, holes of cones do not form by cell death at their centers, but by cell migration.

Since cones take a long time to die out, understanding whether and how these hole structures improve the survival of cones would provide scientific and clinical communities with better knowledge of how to preserve day and high acuity vision

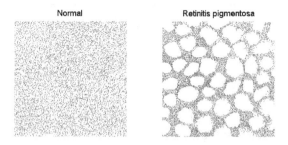

Fig. 2. Distribution of cones in normal (7523 cells in 1mm^2) and retinitis pigmentosa (6509 cells in 1mm^2) retinas at day 25

in retinitis pigmentosa. This motivated us to build a model for the geometric transformation of the cone mosaic in the retinal degeneration. The challenge in building such a model is that we only have access to one snapshot per rat, because the animals are killed before their retinas can be dissected. Therefore, we have very limited data and there is no correspondence between the cells in different snapshots.

3 Synthesis Algorithm

Suppose we are given a point set $X = \{x_1, x_2, \ldots, x_n\}$ at time t and we want to simulate the time evolution from t to $t + \Delta t$. Since we do not presume any biological knowledge about the system, in each step we simply move a point x_i to some random location x_i' within its neighborhood. We then compare both the old configuration $\{x_1, \ldots, x_i, \ldots, x_n\}$ and the new configuration $\{x_1, \ldots, x_i', \ldots, x_n\}$ after this point update to the real data at time $t + \Delta t$. If the new configuration is closer to the data than the old configuration, we accept this movement for x_i, otherwise we accept it with some probability which depends on their difference. We iteratively repeat this process for each point in X until the result converges.

The details of the trajectory synthesis algorithm are shown in Algorithm 1. It can be seen as a variant of the simulated annealing algorithm [9], in which the acceptance probability also depends on a temperature parameter to avoid local minima in optimization. There are two questions we have not addressed:

- how do we compare synthetic configurations to the real data at time $t + \Delta t$?
- what happens if we do not have observation at time $t + \Delta t$?

In fact, we have reduced the problem of motion modeling to quantifying some kind of distances between point sets in the synthetic and real data. Note that the number of points n in the synthesis algorithm is kept constant during simulation, but the point sets in the observed data may have different cardinalities (see Figure 2). Furthermore, since we only have access to one snapshot per animal, there is no correspondence between point sets in different snapshots. In the next

---

Algorithm 1. Trajectory synthesis

Input: sample point sets $\{X^{t_0}, X^{t_1}, \ldots, X^{t_M} \mid t_0 = 0 < t_1 < \ldots < t_{M-1} < t_M = 1\}$,
      number of frames N.
Output: synthesis point sets $\{Y^t \mid t = 0, 1/N, \ldots, (N-1)/N, 1\}$.
Procedure:
1: set time $t = 0$, point set $Y^0 = X^0$.
2: **while** $t < 1$ **do**
3:    set $t = t + 1/N$, initialize $Y^t = Y^{t-1/N}$.
4:    find time interval $t_i < t \leq t_{i+1}$.
5:    interpolate target pair correlation function g_{X^t} between $g_{X^{t_i}}$ and $g_{X^{t_{i+1}}}$.
6:    interpolate target persistence diagram P_{X^t} between $P_{X^{t_i}}$ and $P_{X^{t_{i+1}}}$.
7:    pick three random directions for persistence matching.
8:    set iteration $k = 0$, select initial temperature T_0.
9:    **repeat**
10:      set $k = k + 1$, $T = T_0/k$.
11:      **for** each point y in Y^t **do**
12:         replace y by a random neighbor y' to form a new point set Y'.
13:         compute distance d_1 between pair correlation functions g_{X^t} and g_{Y^t}.
14:         compute distance d_2 between persistence diagrams P_{X^t} and P_{Y^t} based on
             their three 1D projections.
15:         define distance between X^t and Y^t as $d = d_1 + \lambda d_2$.
16:         repeat lines 13–15 to compute distance d' between X^t and Y'.
17:         **if** $d' < d$ **then**
18:            accept new point set $Y^t = Y'$.
19:         **else**
20:            accept Y' with probability $\exp(\frac{d-d'}{T})$.
21:         **end if**
22:      **end for**
23:    **until** Y^t converges.
24: **end while**

---

two sections, we will describe how to define distances on configurations with indistinct identities and use them to interpolate missing data.

4 Geometry: Pair Correlation Function

Given a point set $X = \{x_1, x_2, \ldots, x_n\}$ in $\mathbb{R}^d$ with number density ρ, the pair correlation function is defined as

$$g_X(r) = \frac{1}{S_{d-1}(r)\rho n} \sum_{i=1}^{n} \sum_{j=1}^{n} G(\|x_i - x_j\| - r), \ \forall r \geq 0 \tag{1}$$

where $S_{d-1}(r)$ is the surface area of a ball of radius r in $\mathbb{R}^d$, and G is a 1D Gaussian kernel $G(x) = \frac{1}{\sqrt{2\pi}\sigma} \exp(-\frac{x^2}{2\sigma^2})$.

The PCF provides a compact representation for the characteristics of point distribution. Note that in (1) there are two normalization factors ρ and $S_{d-1}(r)$.

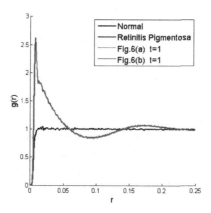

Fig. 3. Pair correlation function

The number density ρ is an intensive quantity to describe the degree of concentration of points in the space, and is typically defined as $\rho = n/V$ where V is the volume of the observation region. Since it is more likely to find two points with a given distance in a more dense system, this factor is used for comparing point sets with different cardinalities. The other inverse weighting factor $S_{d-1}(r)$ is the surface area of a ball of radius r in $\mathbb{R}^d$ (for example $S_1(r) = 2\pi r$). This accounts for the fact that as r gets larger, there will be naturally more points with the given distance from a reference point. After these normalizations, it can be shown that $\lim_{r\to\infty} g_X(r) = 1$ for any infinite point set X, and hence most information about the point set is contained in $g_X(r)$ for the lower values of r. For a finite point set, we can apply periodic boundary conditions to remove the window edge effects.

Figure 3 shows the PCF for the photoreceptor point sets in Figure 2 with $\sigma = 0.1r_{\max}$, where $r_{\max} = \sqrt{\frac{1}{2\sqrt{3}n}}$ is the maximum possible radius for n identical circles that can be packed into a unit square [12]. For the normal point set, we see that the density is almost 1 everywhere except for $r < 0.005$mm, which is about the diameter of cone cell bodies — such a pattern is called blue-noise where points are distributed randomly with a minimum distance between each pair. For healthy primate retinas, it is well-known that photoreceptor distributions may follow a blue-noise-like arrangement to yield good visual resolution [19]. In contrast, for the retinitis pigmentosa point set, the high densities at small distances show the clustering of cones in the sick retina, implying the cells become closer by migration. After we have computed the PCFs, it is natural to define their distance as

$$d(g_X, g_Y) = \left(\int_r (g_X(r) - g_Y(r))^2 \, \mathrm{d}r \right)^{1/2} \qquad (2)$$

It is obvious that computing the PCF for a point set $X = \{x_1, x_2, \ldots, x_n\}$ takes $O(n^2)$ time in (1). However, when we move a point x_i to x_i' in the synthesis algorithm, it only takes $O(n)$ time to update the PCF for the new point set X':

$$g_{X'}(r) = g_X(r) + \frac{2}{S_{d-1}(r)\rho n} \sum_{j \neq i} (G(\|x_i' - x_j\| - r) - G(\|x_i - x_j\| - r))$$

Of course since we compute the densities at different distances, the running time may also depend on the range and discretization of the distance r. But for Gaussian kernels we can set a cutoff threshold δ, so that for each pairwise distance $\|x_i - x_j\|$ we only need to update $g_X(r)$ at distance $\|x_i - x_j\| - \delta < r < \|x_i - x_j\| + \delta$, which contains $O(1)$ discretized values of r.

Data Interpolation. Now we answer the two questions proposed at the end of Section 3. Consider we have a set of observed samples $\{X^{t_0}, X^{t_1}, \ldots, X^{t_M}\}$. Without loss of generality, we can assume the observation time $t_0 = 0 < t_1 < \ldots < t_{M-1} < t_M = 1$. We start with $Y^0 = X^0$ as the initial point set, and run the synthesis algorithm to simulate its time evolution. By matching the PCFs of $\{X^{t_1}, X^{t_2}, \ldots, X^{t_M}\}$, we can obtain a sequence of point sets $\{Y^{t_1}, Y^{t_2}, \ldots, Y^{t_M}\}$ at all observation time. For each sample X^{t_i}, the goal is to minimize the distance between $g_{X^{t_i}}$ and $g_{Y^{t_i}}$ defined in (2). Furthermore, if there is more than one sample observed at time t_i, we can extend the objective function in standard ways, by taking the minimum or average distance from the synthetic point set to all samples at that time.

Note that in the above approach we can only synthesize point sets at the observation time $\{t_0, t_1, \ldots, t_M\}$. But how do we simulate during the time intervals between successive observations? Suppose we want to generate a point distribution at time $t_i < t < t_{i+1}$. Although there is no real data X^t, it is possible to approximate the PCF g_{X^t} by linear interpolation

$$g_{X^t} = \frac{t_{i+1} - t}{t_{i+1} - t_i} g_{X^{t_i}} + \frac{t - t_i}{t_{i+1} - t_i} g_{X^{t_{i+1}}}$$

It has been shown that such a simple linear interpolation can generate valid PCFs from which distributions can be synthesized [14]. Thus, we can use the synthesis algorithm to generate data at any time $t_0 \leq t \leq t_M$.

5   Topology: Distance of Persistence Diagrams

In Section 4, we have seen that the PCF can be used to characterize the distributions of photoreceptor point sets. However, this function only considers pairwise correlations and misses higher-order information in the data. As we will show in Section 6, there are point sets with almost same PCF but very different shape features. In this section, we present another way to summarize point distribution without correspondence from a topological perspective.

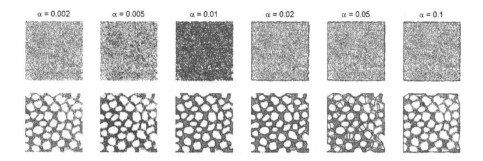

Fig. 4. Alpha shapes

Alpha Shapes. Suppose we are given a point set and we want to understand the shape formed by these points. Of course there are many possible interpretations for the notion of shape, the α-shape being one of them [4]. In geometry, α-shapes are widely used for shape reconstruction, as they give linear approximations of the original shape.

The concept of α-shapes is generally applicable to point sets in any Euclidean space $\mathbb{R}^d$, but for our application we will illustrate in the 2D case. Given a point set S in $\mathbb{R}^2$, the α-shape of S is a straight line graph whose vertices are points in S and whose edges connect pairs of points that can be touched by the boundary of an open disk of radius α containing no points in S. The parameter α controls the desired level of detail in shape reconstruction. For any value of α, the α-shape is a subgraph of the Delaunay triangulation, and thus it can be computed in $O(n \log n)$ time.

Figure 4 shows the α-shapes for the photoreceptor point sets in Figure 2 with different values of α. As α increases, we see that edges appear in the graph and some of them form cycles. For the normal point set, these edges and cycles disappear very quickly since there is no space for empty disks of large radius α. In contrast, for the retinitis pigmentosa point set, some cycles can stay for long time in the large empty regions. Therefore, α-shapes can successfully capture the hole structures formed by cone migration.

In Figure 4, we see that $\alpha = 0.02$mm gives a nice example to distinguish between the two photoreceptor point sets. However, in general how do we choose the right value of α? Indeed, what we are more interested in is to summarize information of α-shapes at different scale levels. So, we next turn to its topological definition — the α-complex. Given a point set in $\mathbb{R}^d$, the α-complex is a simplicial subcomplex of its Delaunay triangulation. For each simplex in the Delaunay triangulation, it appears in the α-complex $K(\alpha)$ if its circumsphere is empty and has a radius less than α, or it is a face of another simplex in $K(\alpha)$.

Although we can choose infinite numbers for α, there are only finite many α-complexes for a point set S. They are totally ordered by inclusion giving rise to filtration of the Delaunay triangulation $K_0 = \phi \subset K_1 = S \subset ... \subset K_m = \text{Del}(S)$. For a point set in $\mathbb{R}^2$, α-complexes consist of vertices, edges, and triangles.

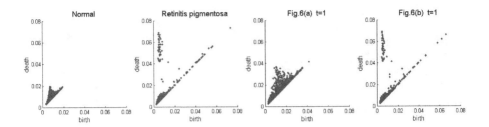

Fig. 5. Persistence diagram

The first non-empty complex K_1 is the point set S itself. As α increases, edges and triangles are added into $K(\alpha)$ until we eventually arrive at the Delaunay triangulation. The relation between α-shape and α-complex is that the edges in the α-shape make up the boundary of the α-complex.

Persistence. In Figure 4, we have seen that cycles appear and disappear in the α-complexes during the filtration. The cycles that stay for a while are important ones since they characterize major shape features of the data set. In algebraic topology, the cycles are defined based on homology groups: there is one group of cycles H_d per dimension d, and the rank of H_d is called the d-th Betti number β_d which can be considered as the number of d-dimensional holes in the space [5]. For example in the 2D case, β_0 is the number of connected components and β_1 is the number of holes in the plane. In the evolution from K_0 to K_m, adding an edge will create a new hole (except for $n-1$ edges in a spanning tree which change β_0 by merging connected components), while adding a triangle will fill a hole. The persistence of a hole is the difference between its birth time and death time which are paired by following the elder rule.

Given a point set S, the information about persistence of holes can be encoded into a two-dimensional persistence diagram P_S. As depicted in Figure 5, each point in the diagram represents a hole (or a class of cycles) during the filtration, where the x and y coordinates are the birth time and death time respectively. In the normal case all cycles have short persistence, while in the retinitis pigmentosa case some cycles have very long persistence and they capture the large hole features in the point set. Note that there are also some cycles with large birth time and very short persistence (the points near the diagonal). This is because the holes in the point set may not be perfectly round (such as ellipses), and thus some cycles can be split by adding long edges at large α. These cycles of short persistence can be considered as noise and ignored in the analysis of the data.

For a Delaunay triangulation with m simplicies, the persistence diagram can be computed using a matrix reduction in $O(m^3)$ time. In the 2D case, $m = O(n)$ and the running time can be reduced to $O(n\alpha(n))$ using the union-find data structure [5], where $\alpha(n)$ is the inverse of Ackermann function which grows very slowly with n. We also apply periodic boundary conditions by computing the periodic Delaunay triangulation of a point set [2,10].

There are two distances often used to measure the similarity between persistence diagrams: the bottleneck and Wasserstein distances [5]. Computing both distances reduces to the problem of finding the optimal matching in a bipartite graph. With the optimal matching, we can also interpolate between two persistence diagrams by linearly interpolating between the matched pairs of points. However, solving a minimum cost perfect matching problem in non-Euclidean spaces takes $O(n^3)$ time [11], so we should avoid recomputing this matching distance after each point update in the synthesis algorithm.

Matching. In this section, we present a faster algorithm to measure the similarity between persistence diagrams from their 1D projections. Instead of computing the optimal matching between 2D persistence diagrams, we take several directions and match their 1D projections in each direction independently. Given two persistence diagrams X, Y and k directions $w_1, w_2, \ldots, w_k$, we define the distance between X and Y as the sum of their 1D matching costs

$$d(X, Y) = \sum_{i=1}^{k} (\min_{f_i : X_{w_i} \to Y_{w_i}} \sum_{x \in X_{w_i}} |x - f_i(x)|) \tag{3}$$

where X_{w_i} is the projection of X onto direction w_i, and f_i is a bijection between X_{w_i} and Y_{w_i} (for simplicity we first assume that X and Y have same cardinality). It is easy to verify that the minimal matching cost over all bijections between X_{w_i} and Y_{w_i} can be computed in $O(n \log n)$ time by sorting X_{w_i} and Y_{w_i}, and matching pairs in ascending order. Furthermore, by randomly choosing three directions, we can uniquely reconstruct a point set from its three 1D projections with high probability.

Theorem 1. *Given a 2D multiset of points $P = \{(x_1, y_1), (x_2, y_2), \ldots, (x_n, y_n)\}$ in general position[1], the set of directions $x + cy$ such that P cannot be uniquely reconstructed from its 1D projections P_x, P_y, and P_{x+cy} has measure zero.*

Proof. Assuming there is another multiset of points $P' \neq P$ with the same three 1D projections. We take a point $p \in P' - P$ which consists of points that appear more times in P' than P. We first claim that $p \notin P$, otherwise since P is in general position there is no point other than p in P with the same y-coordinate, and thus the y-coordinate of p will appear more times in P'_y than P_y.

Let p be reconstructed from projection lines $x = x_i$, $y = y_j$, and $x + cy = x_k + cy_k$ where (x_i, y_i), (x_j, y_j), and (x_k, y_k) are all in P. So $x_i + cy_j = x_k + cy_k$. If $y_j = y_k$, then $x_i = x_k$. Thus $p = (x_i, y_j) = (x_k, y_k)$ is also in P — a contradiction. So $y_j \neq y_k$ and $c = (x_k - x_i)/(y_j - y_k)$. Therefore, there are at most $O(n^3)$ values of c without a unique reconstruction. □

---

[1] We define a 2D multiset of points $P = \{(x_1, y_1), (x_2, y_2), \ldots, (x_n, y_n)\}$ to be in general position if two points in P cannot have same y-coordinate unless they are coincident ($y_i = y_j \Rightarrow x_i = x_j$). If a multiset of points is not in general position, we can always rotate the point set by some angle ω clockwise to make it in general position. This is equivalent to reconstruct the original point set if we rotate directions x, y, and $x + cy$ by angle ω counter-clockwise.

To focus on major shape features of point sets, we choose the projections as follows: we randomly select three directions $x + cy$, where $c = \tan\theta$ and the angle θ is uniformly chosen from $(0, \pi/2) \cup (\pi/2, 3\pi/4)$. For each direction, we project all points (x, y) to $f_n(x, y) = ((x\cos\theta + y\sin\theta)^n - (x\cos\theta + x\sin\theta)^n)^{1/n}$, where n is a large positive number. It is easy to verify that $\lim\limits_{y \to x} f_n(x, y) =$

0 and $\lim\limits_{n \to \infty} f_n(x, y) = \begin{cases} x\cos\theta + y\sin\theta \text{ , if } x < y \\ 0 \qquad\qquad\quad \text{ , if } x = y \end{cases}$. As $n \to \infty$, the function

$f_n(x, y)$ captures the projection of persistence diagram onto direction $x + cy$, while it ignores noise near the diagonal. In practice, we find that $n = 8$ is usually good enough to serve as ∞. Finally, when comparing persistence diagrams with different cardinalities, we may assume that there exist infinitely many extra points on the diagonal — which all map to zero after projection.

6   Experimental Results: Cone Mosaic Rearrangement

We first test the performance of the synthesis algorithm using PCF only [14]. In this case, the distance between two point sets measures the difference between their pairwise correlations in (2). For point update, in each step we move a point to a random location within its neighborhood of radius $r_{\max}$ (as defined in Section 4). We generate $N = 16$ frames to simulate the cone mosaic rearrangement in retinitis pigmentosa. In Figure 6(a), we have labeled some points in red color to show their correspondences in different snapshots. By matching PCFs, we see that the algorithm creates several sparse regions in the point set. However, the synthetic point set $(t = 1)$ looks very different from the real data shown in Figure 2 — there are many outliers inside the sparse regions by the synthesis algorithm, while the holes of cones in retinitis pigmentosa seem to be very clean. If we compare the shape features for these two point sets, their PCFs are almost well-matched (see Figure 3). On the other hand, there is a big difference between their persistence diagrams because these outliers would significantly shorten the persistence of cycles in the α-complex (see Figure 5).

So, we next incorporate α-shapes to maintain the topological features. In this case, the distance function involves two parts: let d_1 be the distance between PCFs of two point sets in (2), and d_2 be the distance between their persistence diagrams in (3). We define the new distance as $d = d_1 + \lambda d_2$, where $\lambda > 0$ is a weight parameter and in our implementation we set the two parts to be equally weighted. The synthesis result using both PCF and α-shapes is shown in Figure 6(b). We see that holes appear in random positions and grow gradually in size as time increases. At the end of simulation, the points labeled in red color move close to the boundaries of holes. In Figures 3 and 5 we can see that the shape features for the synthetic point set match the targets very well. There are only some small differences between persistence diagrams near the diagonal, but they are considered as noise.

Figure 6(c) shows the simulation result in the reverse direction where we start with a retinitis pigmentosa distribution and move points towards a normal distribution. Actually the synthesis in this direction is much easier because we

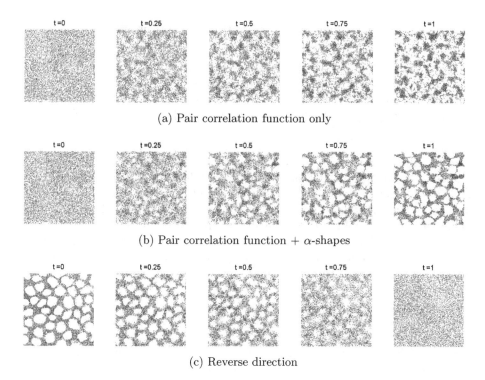

(a) Pair correlation function only

(b) Pair correlation function + α-shapes

(c) Reverse direction

Fig. 6. Simulation results

already know the hole positions. After filling the holes we end up with a blue-noise pattern. By reversing the sequence of snapshots in Figure 6(c), it gives another example on retinal cell motions in retinitis pigmentosa. Furthermore, we can start with a point set at any time t and run bidirectional simulations to synthesize trajectories for the time evolution of this sample.

Running Time. There are four main components of the trajectory synthesis algorithm (see Table 1). For PCF, we only need to compute it for the initial and target point sets, which takes $O(n^2)$ time. After that, it takes $O(n)$ time per point update. For α-shapes, it takes $O(n \log n)$ time for Delaunay triangulation and persistence matching, as well as $O(n\alpha(n))$ time for persistence diagram. Therefore, the running time for all these four parts is almost linear per point update, and hence the algorithm runs in $O(n^2 \log n)$ time per iteration.

We have also tested the real running time for each part of the synthesis algorithm on the photoreceptors data set. The experiment is performed on a computer with Intel® Core™ 2 Quad Processor Q6600 and 4GB Memory. In the current implementation, the periodic Delaunay triangulation is the slowest part which takes about half of the computation time. However, for each point update there is no need to recompute the whole Delaunay triangulation, and indeed it

Table 1. Running time of trajectory synthesis algorithm

Algorithm	Computation	Update	Real time
Pair correlation function	$O(n^2)$	$O(n)$	8 %
Delaunay triangulation (periodic)	$O(n \log n)$	$[O(\log n)]$	46 %
Persistence diagram	$O(n\alpha(n))$	—	27 %
Persistence matching	$O(n \log n)$	—	18 %

can be maintained in $O(\log n)$ expected time per point update [3]. So, by using dynamic Delaunay triangulation we can improve the real running time by almost a factor of 2, but theoretically the algorithm still takes $O(n \log n)$ time per point update — for persistence matching the input is the persistence diagram and it is not clear how to bound its change after we move a point.

Note that in the initialization part, we may need to interpolate the target persistence diagram at time t if we do not have the real data at that time. As mentioned in Section 5, this would take $O(n^3)$ time. Therefore, if we synthesize N frames and run L iterations per frame, the total running time is bounded by $O(N(n^3 + Ln^2 \log n))$. Although the initialization part has a larger theoretical cost $O(n^3)$, in practice the main synthesis part $O(Ln^2 \log n)$ may take longer time because its unit cost $O(1)$ is more expensive. For the simulation results shown in Figure 6(b–c), they take about 3450 seconds for initialization and 290 seconds per iteration, with $L = 20$ iterations for each frame. Furthermore, since the synthesis algorithm is probabilistic, we can use it to generate multiple trajectories from a data set, while the initialization can be considered as a preprocessing step and only needs to be computed once.

Acknowledgements. We wish to thank Dr. Norberto Grzywacz for providing us with the biological data for our experiments. This work was partially supported by the NSF grant DMS 0900700 and the Max Planck Center for Visual Computing and Communication.

References

1. Carlsson, G.E.: Topology and data. Bulletin of the American Mathematical Society 46, 255–308 (2009)
2. Caroli, M., Teillaud, M.: Computing 3D periodic triangulations. In: Proceedings of the European Symposium on Algorithms, pp. 59–70 (2009)
3. Devillers, O., Meiser, S., Teillaud, M.: Fully dynamic Delaunay triangulation in logarithmic expected time per operation. Computational Geometry: Theory and Applications 2(2), 55–80 (1992)
4. Edelsbrunner, H.: Alpha shapes — a survey. Tessellations in the Sciences (2011)
5. Edelsbrunner, H., Harer, J.L.: Computational Topology. An Introduction. American Mathematical Society (2010)
6. Edelsbrunner, H., Morozov, D.: Persistent homology: theory and practice. In: Proceedings of the European Congress of Mathematics, pp. 31–50 (2012)

7. Illian, J., Penttinen, A., Stoyan, H., Stoyan, D.: Statistical Analysis and Modelling of Spatial Point Patterns. Wiley Interscience (2008)
8. Ji, Y., Zhu, C.L., Grzywacz, N.M., Lee, E.-J.: Rearrangement of the cone mosaic in the retina of the rat model of retinitis pigmentosa. The Journal of Comparative Neurology 520(4), 874–888 (2012)
9. Kirkpatrick, S., Gelatt, C.D., Vecchi, M.P.: Optimization by simulated annealing. Science 220(4598), 671–680 (1983)
10. Kruithof, N.: 2D periodic triangulations. CGAL User and Reference Manual (2009)
11. Kuhn, H.W.: The Hungarian method for the assignment problem. Naval Research Logistics 2(1-2), 83–97 (1955)
12. Lagae, A., Dutre, P.: A comparison of methods for generating Poisson disk distributions. Computer Graphics Forum 27(1), 114–129 (2008)
13. Lee, E.-J., Ji, Y., Zhu, C.L., Grzywacz, N.M.: Role of muller cells in cone mosaic rearrangement in a rat model of retinitis pigmentosa. Glia 59(7), 1107–1117 (2011)
14. Oztireli, A.C., Gross, M.: Analysis and synthesis of point distributions based on pair correlation. Transactions on Graphics 31(6), 170 (2012)
15. Schlomer, T., Deussen, O.: Accurate spectral analysis of two-dimensional point sets. Journal of Graphics, GPU, and Game Tools 15(3), 152–160 (2011)
16. Vicsek, T., Czirok, A., Ben-Jacob, E., Cohen, I., Shochet, O.: Novel type of phase transition in a system of self-driven particles. Physical Review Letters 75(6), 1226–1229 (1995)
17. Wilkinson, L., Anand, A., Grossman, R.: Graph-theoretic scagnostics. In: Proceedings of the Symposium on Information Visualization, pp. 157–164 (2005)
18. Yanoff, M., Duker, J.S.: Ophthalmology. Elsevier (2009)
19. Yellott, J.I.: Spectral consequences of photoreceptor sampling in the rhesus retina. Science 221(4608), 382–385 (1983)

A Graph Modification Approach for Finding Core–Periphery Structures in Protein Interaction Networks

Sharon Bruckner[1], Falk Hüffner[2,*], and Christian Komusiewicz[2]

[1] Institut für Mathematik, Freie Universität Berlin, Germany
sharonb@mi.fu-berlin.de
[2] Institut für Softwaretechnik und Theoretische Informatik, TU Berlin, Germany
{falk.hueffner,christian.komusiewicz}@tu-berlin.de

Abstract. The core–periphery model for protein interaction (PPI) networks assumes that protein complexes in these networks consist of a dense core and a possibly sparse periphery that is adjacent to vertices in the core of the complex. In this work, we aim at uncovering a global core–periphery structure for a given PPI network. We propose two exact graph-theoretic formulations for this task, which aim to fit the input network to a hypothetical ground truth network by a minimum number of edge modifications. In one model each cluster has its own periphery, and in the other the periphery is shared. We first analyze both models from a theoretical point of view, showing their NP-hardness. Then, we devise efficient exact and heuristic algorithms for both models and finally perform an evaluation on subnetworks of the *S. cerevisiae* PPI network.

1 Introduction

A fundamental task in the analysis of PPI networks is the identification of protein complexes and functional modules. Herein, a basic assumption is that complexes in a PPI network are strongly connected among themselves and weakly connected to other complexes [22]. This assumption is usually too strict. To obtain a more realistic network model of protein complexes, several approaches incorporate the *core–attachment* model of protein complexes [12]: In this model, a complex is conjectured to consist of a stable core plus some attachment proteins, which only interact with the core temporally. In graph-theoretic terms, the core thus is a dense subnetwork of the PPI network. The attachment (or: periphery) is less dense, but has edges to one or more cores.

Current methods employing this type of modeling are based on *seed growing* [18, 19, 23]. Here, an initial set of promising small subgraphs is chosen as cores. Then, each core is separately greedily expanded into cores and attachments to satisfy some objective function. The aim of these approaches was to predict protein complexes [18, 23] or to reveal biological features that are correlated with topological properties of core–periphery structures in networks [19].

---

* Supported by DFG project ALEPH (HU 2139/1).

D. Brown and B. Morgenstern (Eds.): WABI 2014, LNBI 8701, pp. 340–351, 2014.

In this work, we use core–periphery modeling in a different context. Instead of searching for *local* core–periphery structures, we attempt to unravel a *global* core–periphery structure in PPI networks.

To this end, we hypothesize that the true network consists of several core–periphery structures. We propose two precise models to describe this. In the first model, the core–periphery structures are disjoint. In the second model, the peripheries may interact with different cores, but the cores are disjoint. Then, we fit the input data to each formal model and evaluate the results on several PPI networks.

Our approach. In spirit, our approach is related to the clique-corruption model of the CAST algorithm for gene expression data clustering [1]. In this model, the input is a similarity graph where edges between vertices indicate similarity. The hypothesis is that the objects corresponding to the vertices belong to disjoint biological groups of similar objects, the clusters. In the case of gene expression data, these are assumed to be groups of genes with the same function. Assuming perfect measurements, the similarity graph is a *cluster graph*, that is, a graph in which each connected component is a clique.

Because of stochastic measurement noise, the input graph is not a cluster graph. The task is to recover the underlying cluster graph from the input graph. Under the assumption that the errors are independent, the most likely cluster graph is one that disagrees with the input graph on a minimum number of edges. Such a graph can be found by a minimum number of edge modifications (that is, edge insertions or edge deletions). This paradigm directly leads to the optimization problem CLUSTER EDITING [3, 4, 21].

We now apply this approach to our hypothesis that there is a global core–periphery structure in the PPI networks. In both models detailed here, we assume that all proteins of the cores interact with each other; this implies that the cores are cliques. We also assume that the proteins in the periphery interact only with the cores but not with each other. Hence, the peripheries are independent sets.

In the first model, we assume that ideally the protein interactions give rise to *vertex-disjoint* core–periphery structures, that is, there are no interactions between different cores and no interactions between cores and peripheries of other cores. Then each connected component has at most one core which is a clique and at most one periphery which is an independent set. This is precisely the definition of a split graph.

Definition 1. *A graph $G = (V, E)$ is a* split graph *if V can be partitioned into V_1 and V_2 such that $G[V_1]$ is an independent set and $G[V_2]$ is a clique.*

The vertices in V_1 are called *periphery vertices* and the vertices in V_2 are called *core vertices.* Note that the partition for a split graph is not always unique. Split graphs have been previously used to model core–periphery structures in social networks [5]. There, however, the assumption is that the network contains exactly one core–periphery structure. We assume that each connected component is a split graph; we call graphs with this property *split cluster graphs.* Our fitting model is described by the following optimization problem.

SPLIT CLUSTER EDITING
Input: An undirected graph $G = (V, E)$.
Task: Transform G into a split cluster graph by applying a minimum number of edge modifications.

In our second model, we allow the vertices in the periphery to be attached to an arbitrary number of cores, thereby connecting the cores. In this model, we thus assume that the cores are disjoint cliques and the vertices of the periphery are an independent set. Such graphs are called *monopolar*.

Definition 2. *A graph is* monopolar *if its vertex set can be two-partitioned into V_1 and V_2 such that $G[V_1]$ is an independent set and $G[V_2]$ is a cluster graph. The partition (V_1, V_2) is called* monopolar partition.

Again, the vertices in V_1 are called periphery vertices and the vertices in V_2 are called core vertices. Our second fitting model now is the following.

MONOPOLAR EDITING
Input: An undirected graph $G = (V, E)$.
Task: Transform G into a monopolar graph by applying a minimum number of edge modifications and output a monopolar partition.

Clearly, both models are simplistic and cannot completely reflect biological reality. For example, subunits of protein complexes consisting of two proteins that first interact with each other and subsequently with the core of a protein complex are supported by neither of our models. Nevertheless, our models are less simplistic than pure clustering models that attempt to divide protein interaction networks into disjoint dense clusters. Furthermore, there is a clear trade-off between model complexity, algorithmic feasibility of models, and interpretability.

Further related work. The related optimization problem SPLIT EDITING asks to transform a graph into a (single) split graph by at most k edge modifications. SPLIT EDITING is, somewhat surprisingly, solvable in polynomial time [13]. Another approach of fitting PPI networks to specific graph classes was proposed by Zotenko et al. [25] who find for a given PPI network a close chordal graph, that is, a graph without induced cycles of length four or more. The modification operation is insertion of edges.

Preliminaries. We consider undirected simple graphs $G = (V, E)$ where $n := |V|$ denotes the number of vertices and $m := |E|$ denotes the number of edges. The *open neighborhood* of a vertex u is defined as $N(u) := \{v \mid \{u, v\} \in E\}$. We denote the *neighborhood of a set U* by $N(U) := \bigcup_{u \in U} N(u) \setminus U$. The *subgraph induced by a vertex set S* is defined as $G[S] := (S, \{\{u, v\} \in E \mid u, v \in S\})$.

2   Combinatorial Properties and Complexity

Before presenting concrete algorithmic approaches for the two optimization problems, we show some properties of split cluster graphs and monopolar graphs

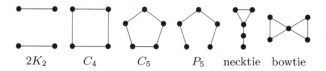

Fig. 1. The forbidden induced subgraphs for split graphs ($2K_2$, C_4, and C_5) and for split cluster graphs (C_4, C_5, P_5, necktie, and bowtie)

which will be useful in the various algorithms. Furthermore, we present computational hardness results for the problems which will justify the use of integer linear programming (ILP) and heuristic approaches.

Split Cluster Editing. Each connected component of the solution has to be a split graph. These graphs can be characterized by forbidden induced subgraphs (see Fig. 1).

Lemma 1 ([10]). *A graph G is a split graph if and only if G does not contain an induced subgraph that is a cycle of four or five edges or a pair of disjoint edges (that is, G is $(C_4, C_5, 2K_2)$-free).*

To obtain a characterization for split cluster graphs, we need to characterize the existence of $2K_2$'s within connected components.

Lemma 2. *If a connected graph contains a $2K_2$ as induced subgraph, then it contains a $2K_2 = (V', E')$ such that there is a vertex $v \notin V'$ that is adjacent to at least one vertex of each K_2 of (V', E').*

Proof. Let G contain the $2K_2$ $\{x_1, x_2\}, \{y_1, y_2\}$ as induced subgraph. Without loss of generality, let the shortest path between any x_i, y_j be $P = (x_1 = p_1, p_2, \ldots, p_k = y_1)$. Clearly, $k > 2$. If $k = 3$, then x_1 and y_1 are both adjacent to p_2. Otherwise, if $k = 4$, then $\{x_2, x_1 = p_1\}, \{p_3, p_4 = y_1\}$ is a $2K_2$ and x_1 and p_3 are both adjacent to p_2. Finally, if $k > 4$, then P contains a P_5 as induced subgraph. The four outer vertices of this P_5 induce a $2K_2$ whose K_2's each contain a neighbor of the middle vertex. $\square$

We can now provide a characterization of split cluster graphs.

Theorem 1. *A graph G is a split cluster graph if and only if G is a $(C_4, C_5, P_5, necktie, bowtie)$-free graph.*

Proof. Let G be a split cluster graph, that is, every connected component is a split graph. Clearly, G does not contain a C_4 or C_5. If a connected component of G contains a P_5, then omitting the middle vertex of the P_5 yields a $2K_2$, which contradicts that the connected component is a split graph. The same argument shows that the graph cannot contain a necktie or bowtie.

Conversely, let G be $(C_4, C_5, P_5, necktie, bowtie)$-free. Clearly, no connected component contains a C_4 or C_5. Assume for a contradiction that a connected component contains a $2K_2$ consisting of the K_2's $\{a, b\}$ and $\{c, d\}$. Then according to Lemma 2 there is a vertex v which is, without loss of generality, adjacent to a and c. If no other edges between the $2K_2$ and v exist, then $\{a, b, v, c, d\}$ is a P_5. Adding exactly one of $\{b, v\}$ or $\{d, v\}$ creates a necktie, and adding both edges results in a bowtie. No other edges are possible, since there are no edges between $\{a, b\}$ and $\{c, d\}$.   □

This leads to a linear-time algorithm for checking whether a graph is a split cluster graph.

Theorem 2. *A forbidden subgraph for a split cluster graph can be found in* $O(n + m)$ *time.*

Proof. For each connected component, we run an algorithm by Heggernes and Kratsch [14] that checks in linear time whether a graph is a split graph, and if not, produces a $2K_2$, C_4, or C_5. If the forbidden subgraph is a C_4 or C_5, we are done. If it is a $2K_2$, we can find in linear time a P_5, necktie, or bowtie, using the method described in the proof of Lemma 2.   □

In contrast, SPLIT CLUSTER EDITING is NP-hard even in restricted cases. We reduce from CLUSTER EDITING which has as input an undirected graph $G = (V, E)$ and an integer k, and asks whether G can be transformed into a cluster graph by applying at most k edge modifications. CLUSTER EDITING is NP-hard even if the maximum degree of the input graph is five [11] and it cannot be solved in $2^{o(k)} \cdot n^{O(1)}$ time assuming the so-called exponential-time hypothesis (ETH) [11, 17]. The reduction simply attaches to each vertex u an additional $\deg_G(v)$ many new degree-one vertices; we omit the correctness proof.

Theorem 3. SPLIT CLUSTER EDITING *is NP-hard even on graphs with maximum degree 10. Further, it cannot be solved in* $2^{o(k)} \cdot n^{O(1)}$ *or* $2^{o(n)} \cdot n^{O(1)}$ *time if the exponential-time hypothesis (ETH) [15] is true.*

This hardness result motivates the study of algorithmic approaches such as fixed-parameter algorithms or ILP-formulations. For example, SPLIT CLUSTER EDITING is fixed-parameter tractable for the parameter number of edge modifications k by the following search tree algorithm: Check whether the graph contains a forbidden subgraph. If this is the case, branch into the possibilities to destroy this subgraph. In each recursive branch, the number of allowed edge modifications decreases by one. Furthermore, since the largest forbidden subgraph has five vertices, at most ten possibilities for edge insertions or deletions have to be considered to destroy a forbidden subgraph. By Theorem 2, forbidden subgraphs can be found in $O(n + m)$ time. Altogether, this implies the following.

Theorem 4. SPLIT CLUSTER EDITING *can be solved in* $O(10^k \cdot (n + m))$ *time.*

This result is purely of theoretical interest. With further improvements of the search tree algorithm, practical running times might be achievable.

Monopolar Graphs. The class of monopolar graphs is hereditary, and thus it is characterized by forbidden induced subgraphs, but the set of minimal forbidden induced subgraphs is infinite [2]; for example among graphs with five or fewer vertices, only the wheel W_4 (⧈) is forbidden, but there are 34 minimal forbidden subgraphs with six vertices. In contrast to the recognition of split cluster graphs, which is possible in linear time by Theorem 2, deciding whether a graph is monopolar is NP-hard [9]. Thus MONOPOLAR EDITING is NP-hard already for $k = 0$ edge modifications.

3   Solution Approaches

Integer Linear Programming. We experimented with a formulation based directly on the forbidden subgraphs for split cluster graphs (Theorem 1). However, we found a formulation based on the following observation to be faster in practice, and moreover applicable also to MONOPOLAR EDITING: If we correctly guess the partition into clique and independent set vertices, we can get a simpler characterization of split cluster graphs by forbidden subgraphs.

Lemma 3. *Let $G = (V, E)$ be a graph and $C \dot\cup I = V$ a partition of the vertices. Then G is a split cluster graph with core vertices C and periphery vertices I if and only if it does not contain an edge with both endpoints in I, nor an induced P_3 with both endpoints in C.*

Proof. "⇒": Clearly, if there is an edge with both endpoints in I or an induced P_3 with both endpoints in C, then I is not an independent set or C does not form a clique in each connected component, respectively.

"⇐": We again use contraposition. If G is not a split cluster graph with core vertices C and periphery vertices I, then it must contain an edge with both endpoints in I, or $C \cap H$ does not induce a clique for some connected component H of G. In the first case we are done; in the second case, there are two vertices $u, v \in C$ in the same connected component with $\{u, v\} \notin E$. Consider a shortest path $u = p_1, \ldots, p_l = v$ from u to v. If it contains a periphery vertex $p_i \in I$, then p_{i-1}, p_i, p_{i+1} forms a forbidden subgraph. Otherwise, p_1, p_2, p_3 is one.   □

With a very similar proof, we can get a simpler set of forbidden subgraphs for annotated monopolar graphs.

Lemma 4. *Let $G = (V, E)$ be a graph and $C \dot\cup I = V$ a partition of the vertices. Then G is a monopolar graph with core vertices C and periphery vertices I if and only if it does not contain an edge with both endpoints in I, nor an induced P_3 whose vertices are contained in C.*

Proof. "⇒": Easy to see as in Lemma 3.

"⇐": If G is not monopolar with core vertices C and periphery vertices I, then it must contain an edge with both endpoints in I, or C does not induce a cluster graph. In the first case we are done; in the second case, there is a P_3 with all vertices in C, since that is the forbidden subgraph for cluster graphs.   □

From Lemma 3, we can directly derive an integer linear programming formulation for SPLIT CLUSTER EDITING. We introduce binary variables e_{uv} indicating whether the edge $\{u, v\}$ is present in the solution graph and binary variables c_u indicating whether a vertex u is part of the core. Defining $\bar{e}_{uv} := 1 - e_{uv}$ and $\bar{c}_u := 1 - c_u$, and fixing an arbitrary order on the vertices, we have

$$\text{minimize} \quad \sum_{\{u,v\} \in E} \bar{e}_{uv} + \sum_{\{u,v\} \notin E} e_{uv} \text{ subject to} \tag{1}$$

$$c_u + c_v + \bar{e}_{uv} \geq 1 \ \forall u, v \tag{2}$$

$$\bar{e}_{uv} + \bar{e}_{vw} + e_{uw} + \bar{c}_u + \bar{c}_w \geq 1 \ \forall u \neq v, v \neq w > u. \tag{3}$$

Herein, Eq. (2) forces that the periphery vertices are an independent set and Eq. (3) forces that core vertices in the same connected component form a clique. For MONOPOLAR EDITING, we can replace Eq. (3) by

$$\bar{e}_{uv} + \bar{e}_{vw} + e_{uw} + \bar{c}_u + \bar{c}_v + \bar{c}_w \geq 1 \ \forall u \neq v, v \neq w > u \tag{4}$$

which forces that the graph induced by the core vertices is a cluster graph.

Data Reduction. Data reduction (preprocessing) proved very effective for solving CLUSTER EDITING optimally [3, 4]. Indeed, any instance can be reduced to one of at most $2k$ vertices [7], where k is the number of edge modifications. Unfortunately, the data reduction rules we devised for SPLIT CLUSTER EDITING were not applicable to our real-world test instances. However, a simple observation allows us to fix the values of some variables of Eqs. (1) to (3) in the SPLIT CLUSTER EDITING ILP: if a vertex u has only one vertex v as neighbor and $\deg(v) > 1$, then set $c_u = 0$ and $e_{uw} = 0$ for all $w \neq v$. Since our instances have many degree-one vertices, this considerably reduces the size of the ILPs.

Heuristics. The integer linear programming approach is not able to solve the hardest of our instances. Thus, we employ the well-known *simulated annealing* heuristic, a local search method. For SPLIT CLUSTER EDITING, we start with a clustering where each vertex is a singleton. As random modification, we move a vertex to a cluster that contains one of its neighbors. Since this allows only a decrease in the number of clusters, we also allow moving a vertex into an empty cluster. For a fixed clustering, the optimal number of modifications can be computed in linear time by counting the edges between clusters and computing for each cluster a solution for SPLIT EDITING in linear time [13]. For MONOPOLAR EDITING, we also allow moving a vertex into the independent set. Here, the optimal number of modifications for a fixed clustering can also be calculated in linear time: all edges in the independent set are deleted, all edges between clusters are deleted, and all missing edges within clusters are added.

4   Experimental Results

We test exact algorithms and heuristics for SPLIT CLUSTER EDITING (SCE) and MONOPOLAR EDITING (ME) on several PPI networks, and perform a biological

Table 1. Network statistics. Here, n is the number of proteins, without singletons, and m is the number of interactions; n_{lcc} and m_{lcc} are the number of proteins and interactions in the largest connected component; C is the number of CYC2008 complexes with at least 50% and at least three proteins in the network, p is the number of network proteins that do not belong to these complexes, and A_C is the average complex size. Finally, i_g is the number of genetic interactions between proteins without physical interaction.

	n	m	n_{lcc}	m_{lcc}	C	p	A_C	i_g
cell cycle	196	797	192	795	7	148	21.8	1151
transcription	215	786	198	776	11	54	28.0	1479
translation	236	2352	186	2351	5	88	29.8	174

evaluation of the modules found. We use two known methods for comparison. The algorithm by Luo et al. [19] ("LUO" for short) produces clusters with core and periphery, like SCE, but the clusters may overlap and might not cover the whole graph. The SCAN algorithm [24], like ME, partitions the graph vertices into "clusters", which we interpret as cores, and "hubs" and "outliers", which we interpret as periphery.

4.1 Experimental Setup

Data. We perform all our experiments on subnetworks of the *S. cerevisiae* (yeast) PPI network from BioGRID [6], version 3.2.101. Our networks contain only physical interactions; we use genetic interactions only for the biological evaluation. From the complete BioGRID yeast network with 6377 vertices and 81549 edges, we extract three subnetworks, corresponding to three essential processes: cell cycle, translation, and transcription. These are important subnetworks known to contain complexes. To determine the protein subsets corresponding to each process, we select all yeast genes annotated with the relevant GO terms: GO:0007049 (cell cycle), GO:0006412 (translation), and GO:0006351 (DNA-templated transcription). Table 1 shows some properties of these networks.

Implementation details. The integer linear program and simulated annealing heuristic were implemented in C++ and compiled with the GNU g++ 4.7.2 compiler. As ILP solver, we used CPLEX 12.6.0. For the ILP, we use the heuristic solution found after one minute as MIP start, and initially add all independent set constraints (2). In a cutting plane callback, we add the 500 most violated constraints of type (3) or (4).

The test machine is a 4-core 3.6 GHz Intel Xeon E5-1620 (Sandy Bridge-E) with 10 MB L3 cache and 64 GB main memory, running Debian GNU/Linux 7.0.

Biological evaluation. We evaluate our results using the following measures. First, we examine the coherence of the GO terms in our modules using the semantic similarity score calculated by G-SESAME [8]. We use this score to test the hypothesis that the cores are more stable than the peripheries. If the hypothesis is true, then the pairwise similarity score within the core should be higher than in the periphery. We test only terms relating to process, not function, since proteins in the same complex play a role in the same biological process. Since MONOPOLAR EDITING and SCAN return multiple cores and only a single periphery, we assign to each cluster C its neighborhood $N(C)$ as periphery. We consider only clusters with at least two core vertices and one periphery vertex.

Next, we compare the resulting clusters with known protein complexes from the CYC2008 database [20]. Since the networks we analyze are subnetworks of the larger yeast network, we discard for each network the CYC2008 complexes that have less than 50% of their vertices in the current subnetwork, restrict them to proteins contained in the current subnetwork, and then discard those with fewer than three proteins. We expect that the cores mostly correspond to complexes and that the periphery may contain complex vertices plus further vertices.

Finally, we analyze the genetic interactions between and within modules. Ideally, we would obtain significantly more genetic interactions outside of cores than within them. This is supported by the *between pathways model* [16], which proposes that different complexes can back one another up, thus disabling one would not harm the cell, but disabling both complexes would reduce its fitness or kill it. Here, when counting genetic interactions, we are interested only in genetic interactions that occur between proteins that do not physically interact.

4.2   Results

Our results are summarized in Table 2. For SPLIT CLUSTER EDITING, the ILP approach failed to solve the cell cycle and transcription network, and for MONOPOLAR EDITING, it failed to solve the transcription network, with CPLEX running out of memory in each case. The fact that for the "harder" problem ME more instances were solved could be explained by the fact that the number k of necessary modifications is much lower, which could reduce the size of the branch-and-bound tree. For the three optimally solved instances, the heuristic also finds the optimal solution after one minute for two of them, but for the last one (ME transcription) only after several hours; after one minute, it is 2.9% too large. This indicates the heuristic gives good results, and in the following, we use the heuristic solution for the three instances not solvable by ILP.

Table 3 gives an overview of the results. We say that a cluster is *interesting* if it contains at least two vertices in the core and at least one in the periphery. In the cell cycle network (see Fig. 2), the SCE solution identifies ten interesting clusters, along with four clusters containing only cores, and some singletons. Only for one of the ten clusters is the GO term coherence higher in the periphery than in the core, as expected (for two more the scoring tool does not return a result).

Following our hypothesis, we say that a complex is *detected* by a cluster if at least 50% of the core belongs to the complex and at least 50% of the complex belongs to the cluster. Out of the seven complexes, three are detected without any error, and one is detected with an error of two additional proteins in the core that are not in the complex. The periphery contains between one and eight extra proteins that are not in the complex (which is allowed by our hypothesis).

Table 2. Experimental results. Here, K is the number of clusters with at least two vertices in the core and at least one in the periphery, p is the size of the periphery, k is the number of edge modifications, and c_t, c_c, and c_p is the average coherence within the cluster, core, and periphery, respectively.

	cell-cycle						transcription						translation					
	K	p	k	c_t	c_c	c_p	K	p	k	c_t	c_c	c_p	K	p	k	c_t	c_c	c_p
SCE	10	108	321	0.60	0.64	0.40	13	112	273	0.54	0.54	0.57	6	94	308	0.63	0.73	0.69
ME	24	75	126	0.46	0.58	0.39	26	78	106	0.55	0.61	0.54	11	129	240	0.52	0.58	0.53
SCAN	28	48	—	0.42	0.62	0.34	26	58	—	0.53	0.51	0.47	2	25	—	0.59	0.59	0.76
Luo	16	84	—	0.34	0.50	0.31	12	125	—	0.40	0.52	0.38	4	137	—	0.72	0.84	0.67

Table 3. Experimental results for the complex test. Here, D is the number of detected complexes ($\geq$ 50% of core contained in complex and $\geq$ 50% of complex contained in cluster), $\text{core}_\%$ is among the detected complexes the median percentage of core vertices that are in this complex and $\text{comp}_\%$ is the median percentage of complex proteins that are in the cluster.

	cell-cycle			transcription			translation		
	D	$\text{core}_\%$	$\text{comp}_\%$	D	$\text{core}_\%$	$\text{comp}_\%$	D	$\text{core}_\%$	$\text{comp}_\%$
SCE	4	100	100	7	89	100	4	100	96
ME	5	100	100	11	100	100	4	100	96
SCAN	4	91	100	8	84	100	0	—	—
Luo	5	81	100	6	87	100	4	92	96

The MONOPOLAR EDITING result contains more interesting clusters than SCE (24). Compared to SCE, clusters are on average smaller and have a smaller core, but about the same periphery size (recall that a periphery vertex may occur in more than one cluster). ME detects the same complexes as SCE, plus one additional complex.

SCAN identifies 7 hubs and 41 outliers, which then comprise the periphery. SCAN fails to detect one of the complexes ME finds. It also has slightly more errors, for example having three extra protein in the core for the anaphase-promoting complex plus one missing. Luo identifies only large clusters (this is true for all subnetworks we tested). It detects the same complexes as ME, but also finds more extra vertices in the cores.

In the transcription network, for GO-Term analysis, we see a similar pattern here that Luo has worse coherence, but all methods show less coherence in the peripheries than in the cores. The ME method comes out a clear winner here with detecting all 11 complexes and generally fewer errors.

In the translation network, SCE and ME find about the same number of interesting clusters (22 and 24) and detect the same four complexes. The SCAN algorithm does not seem to deal well with this network, since it finds only two interesting clusters and does not detect any complex. Luo finds only four interesting clusters, corresponding to the four complexes also detected by SCE and ME; this might also explain why it has the best coherence values here.

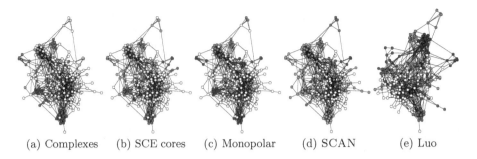

(a) Complexes   (b) SCE cores   (c) Monopolar   (d) SCAN   (e) Luo

Fig. 2. Results of the four algorithms on the cell-cycle network. The periphery is in white, remaining vertices are colored according to their clusters.

Counting genetic interactions. Since the identified clusters largely correspond to known protein complexes, it is not surprising that we identify a higher than expected number of genetic interactions between these complexes. For SCE, the binomial test to check whether the frequency of genetic interactions within the periphery is higher than the frequency in the entire network gives p-values lower than $4 \cdot 10^{-7}$ for all networks, thus the difference is significant.

Experiments conclusion. The coherence values for cores and peripheries indicate that a division of clusters into core and periphery makes sense. In detecting complexes, the ME method does best (20 detected), followed by SCE and LUO (15 each), and finally SCAN (12). This indicates that the model that peripheries are shared is superior. Note however that SCE is at a disadvantage in this evaluation, since it can use each protein as periphery only once, while having large peripheries makes it easier to count a complex as detected.

5   Outlook

There are many further variants of our models that could possibly yield better biological results or have algorithmic advantages. For instance, one could restrict the cores to have a certain minimum size. Also, instead of using split graphs as a core–periphery model, one could resort to dense split graphs [5] in which every periphery vertex is adjacent to all core vertices. Finally, one could allow some limited amount of interaction between periphery vertices.

References

[1] Ben-Dor, A., Shamir, R., Yakhini, Z.: Clustering gene expression patterns. Journal of Computational Biology 6(3-4), 281–297 (1999)

[2] Berger, A.J.: Minimal forbidden subgraphs of reducible graph properties. Discussiones Mathematicae Graph Theory 21(1), 111–117 (2001)

[3] Böcker, S., Baumbach, J.: Cluster editing. In: Bonizzoni, P., Brattka, V., Löwe, B. (eds.) CiE 2013. LNCS, vol. 7921, pp. 33–44. Springer, Heidelberg (2013)

[4] Böcker, S., Briesemeister, S., Klau, G.W.: Exact algorithms for cluster editing: Evaluation and experiments. Algorithmica 60(2), 316–334 (2011)

[5] Borgatti, S.P., Everett, M.G.: Models of core/periphery structures. Social Networks 21(4), 375–395 (1999)

[6] Chatr-aryamontri, A., et al.: The BioGRID interaction database: 2013 update. Nucleic Acids Research 41(D1), D816–D823 (2013)

[7] Chen, J., Meng, J.: A $2k$ kernel for the cluster editing problem. Journal of Computer and System Sciences 78(1), 211–220 (2012)

[8] Du, Z., Li, L., Chen, C.-F., Yu, P.S., Wang, J.Z.: G-SESAME: web tools for GO-term-based gene similarity analysis and knowledge discovery. Nucleic Acids Research 37(suppl. 2), W345–W349 (2009)

[9] Farrugia, A.: Vertex-partitioning into fixed additive induced-hereditary properties is NP-hard. The Electronic Journal of Combinatorics 11(1), R46 (2004)

[10] Foldes, S., Hammer, P.L.: Split graphs. Congressus Numerantium 19, 311–315 (1977)

[11] Fomin, F.V., Kratsch, S., Pilipczuk, M., Pilipczuk, M., Villanger, Y.: Subexponential fixed-parameter tractability of cluster editing. CoRR, abs/1112.4419 (2011)

[12] Gavin, A.-C., et al.: Proteome survey reveals modularity of the yeast cell machinery. Nature 440(7084), 631–636 (2006)

[13] Hammer, P.L., Simeone, B.: The splittance of a graph. Combinatorica 1(3), 275–284 (1981)

[14] Heggernes, P., Kratsch, D.: Linear-time certifying recognition algorithms and forbidden induced subgraphs. Nordic Journal of Computing 14(1-2), 87–108 (2007)

[15] Impagliazzo, R., Paturi, R., Zane, F.: Which problems have strongly exponential complexity? Journal of Computer and System Sciences 63(4), 512–530 (2001)

[16] Kelley, R., Ideker, T.: Systematic interpretation of genetic interactions using protein networks. Nature Biotechnology 23(5), 561–566 (2005)

[17] Komusiewicz, C., Uhlmann, J.: Cluster editing with locally bounded modifications. Discrete Applied Mathematics 160(15), 2259–2270 (2012)

[18] Leung, H.C., Xiang, Q., Yiu, S.-M., Chin, F.Y.: Predicting protein complexes from PPI data: a core-attachment approach. Journal of Computational Biology 16(2), 133–144 (2009)

[19] Luo, F., Li, B., Wan, X.-F., Scheuermann, R.: Core and periphery structures in protein interaction networks. BMC Bioinformatics (Suppl. 4), S8 (2009)

[20] Pu, S., Wong, J., Turner, B., Cho, E., Wodak, S.J.: Up-to-date catalogues of yeast protein complexes. Nucleic Acids Research 37(3), 825–831 (2009)

[21] Shamir, R., Sharan, R., Tsur, D.: Cluster graph modification problems. Discrete Applied Mathematics 144(1-2), 173–182 (2004)

[22] Spirin, V., Mirny, L.A.: Protein complexes and functional modules in molecular networks. PNAS 100(21), 12123–12128 (2003)

[23] Wu, M., Li, X., Kwoh, C.-K., Ng, S.-K.: A core-attachment based method to detect protein complexes in PPI networks. BMC Bioinformatics 10(1), 169 (2009)

[24] Xu, X., Yuruk, N., Feng, Z., Schweiger, T.A.J.: SCAN: a structural clustering algorithm for networks. In: Proc. 13th KDD, pp. 824–833. ACM (2007)

[25] Zotenko, E., Guimarães, K.S., Jothi, R., Przytycka, T.M.: Decomposition of overlapping protein complexes: a graph theoretical method for analyzing static and dynamic protein associations. Algorithms for Molecular Biology 1(7) (2006)

Interpretable Per Case Weighted Ensemble Method for Cancer Associations

Adrin Jalali[1,2,*] and Nico Pfeifer[1]

[1] Department of Computational Biology and Applied Algorithmics, Max Planck
Institute for Informatics, Campus E1 4, 66123 Saarbrücken, Germany
[2] Saarbrücken Graduate School of Computer Science, Saarland University,
Saarbrücken, Germany
ajalali@mpi-inf.mpg.de

Abstract. Over the past decades, biology has transformed into a high
throughput research field both in terms of the number of different measurement techniques as well as the amount of variables measured by
each technique (e.g., from Sanger sequencing to deep sequencing) and is
more and more targeted to individual cells [3]. This has led to an un-
precedented growth of biological information. Consequently, techniques
that can help researchers find the important insights of the data are becoming more and more important. Molecular measurements from cancer
patients such as gene expression and DNA methylation are usually very
noisy. Furthermore, cancer types can be very heterogeneous. Therefore,
one of the main assumptions for machine learning, that the underlying
unknown distribution is the same for all samples in training and test
data, might not be completely fulfilled.

In this work, we introduce a method that is aware of this potential bias
and utilizes an estimate of the differences during the generation of the
final prediction method. For this, we introduce a set of sparse classifiers
based on L1-SVMs [1], under the constraint of disjoint features used by
classifiers. Furthermore, for each feature chosen by one of the classifiers,
we introduce a regression model based on Gaussian process regression
that uses additional features. For a given test sample we can then use
these regression models to estimate for each classifier how well its features are predictable by the corresponding Gaussian process regression
model. This information is then used for a confidence-based weighting
of the classifiers for the test sample. Schapire and Singer showed that
incorporating confidences of classifiers can improve the performance of
an ensemble method [2]. However, in their setting confidences of classifiers are estimated using the training data and are thus fixed for all test
samples, whereas in our setting we estimate confidences of individual
classifiers per given test sample.

In our evaluation, the new method achieved state-of-the-art performance on many different cancer data sets with measured DNA methylation or gene expression. Moreover, we developed a method to visualize
our learned classifiers to find interesting associations with the target label. Applied to a leukemia data set we found several ribosomal proteins

---

* Corresponding author.

D. Brown and B. Morgenstern (Eds.): WABI 2014, LNBI 8701, pp. 352–353, 2014.
© Springer-Verlag Berlin Heidelberg 2014

associated with leukemia that might be interesting targets for follow-up studies and support the hypothesis that the ribosomes are a new frontier in gene regulation.

Keywords: machine learning, cancer biomarkers, supervised prediction, ensemble methods, support vector machines, Gaussian processes.

References

1. Bradley, P.S., Mangasarian, O.L.: Feature selection via concave minimization and support vector machines. In: ICML, vol. 98, pp. 82–90 (1998)
2. Schapire, R.E., Singer, Y.: Improved boosting algorithms using confidence-rated predictions. Machine Learning 37(3), 297–336 (1999)
3. Shapiro, E., Biezuner, T., Linnarsson, S.: Single-cell sequencing-based technologies will revolutionize whole-organism science. Nat. Rev. Genet. 14(9), 618–630 (2013)

Reconstructing Mutational History in Multiply Sampled Tumors Using Perfect Phylogeny Mixtures

Iman Hajirasouliha and Benjamin J. Raphael

Department of Computer Science and Center for Computational Molecular Biology,
Brown University, Providence, RI, 02906, USA
imanh@cs.brown.edu, braphael@brown.edu

Abstract. High-throughput sequencing of cancer genomes have motivated the problem of inferring the ancestral history of somatic mutations that accumulate in cells during cancer progression. While the somatic mutation process in cancer cells meets the requirements of the classic Perfect Phylogeny problem, nearly all cancer sequencing studies do not sequence single cancerous cells, but rather thousands-millions of cells in a tumor sample. In this paper, we formulate the Perfect Phylogeny Mixture problem of inferring a perfect phylogeny given somatic mutation data from multiple tumor samples, each of which is a superposition of cells, or "species." We prove that the Perfect Phylogeny Mixture problem is NP-hard, using a reduction from the graph coloring problem. Finally, we derive an algorithm to solve the problem.

Keywords: DNA sequencing, Cancer genomics, perfect phylogeny, Graph coloring.

1 Introduction

Cancer is an evolutionary process, where somatic mutations accumulate in a population of cells during the lifetime of an individual. The clonal theory of cancer posits that all cells in a tumor are descended from a single founder cell, and that selection for advantageous mutations and clonal expansions of cells containing these mutations leads to uncontrolled growth of a tumor [16]. Traditional genome-wide profiling, using comparative genomic hybridisation (CGH), initially shed light on cancer progression. Using CGH for the purpose of analysis, mathematical models such as oncogenetic tree models [2], were developed to describe the pathways and the order of somatic alterations in cancer genomes. In recent years, high-throughput DNA sequencing technologies has enabled large-scale measurement of somatic mutations in many cancer genomes [4,12,14,22]. This new data has led to much interest in modeling the mutational process within a tumor, and reconstructing the history of somatic mutations [10,20].

At first glance, the problem of reconstructing the history of somatic mutations is a phylogenetic problem, where the "species" are the individual cells in the

D. Brown and B. Morgenstern (Eds.): WABI 2014, LNBI 8701, pp. 354–367, 2014.

tumor, and the "characters" are the somatic mutations. Since the number of single-nucleotide mutations in a tumor is a very small percentage of the number of positions in the genome, one may readily assume that somatic mutations follow the infinite sites assumption whereby a mutation occurs at a particular locus only once during the course of tumor evolution. Moreover, we may assume that mutations have one of two possible states: 0 = normal, and 1 = mutated. Under these assumptions the mutational process in a tumor follows a *perfect phylogeny* (Figure 1(a)). Reconstructing perfect phylogenies has been extensively studied, and many algorithms to construct perfect phylogenies from data have been developed. See [6] for a survey.

Nearly all cancer sequencing studies to date do not measure somatic mutations in single cells, but rather sequence DNA from a tumor sample containing thousands-millions of tumor cells. This is because of technological and cost issues in single cell sequencing [23,11,5]. Reconstructing the history of somatic mutations from a *single* sample is very different from a traditional phylogenetic problem, as the data is not from individual species, but rather from a mixture of all species. Thus, researchers have instead focused on identifying subpopulations of tumor cells that share somatic mutations by clustering mutations according to their inferred frequencies within the tumor [3,15,19,10]. In contrast with the traditional perfect phylogeny problems, these works instead solve a *deconvolution* problem.

A few recent studies have sequenced multiple spatially distinct samples from the same tumor [7,18]. This data presents an interesting intermediate between the perfect phylogeny problem – where characters are measured in individual species – and the single sample problem – where the goal is to deconvolve a mixture. Other related recent studies include computational methods to infer tumor phylogenies, with the goal of improving single nucleotide variant (SNV) calling [17].

In this paper, we formulate a hybrid problem, the *Perfect Phylogeny Mixture problem*. In this problem, we are given a collection of samples, each of which is a superposition of the characters from a subset of species. The problem is to reconstruct the state of each character in each species so that the resulting species satisfies a perfect phylogeny. Note that, similiar to some of the studies we mentioned above (e.g. [3,15,19,10]), we also restrict our problem formulation to single nucleotide variant (SNV) markers. Although consideration of additional somatic events such as copy number variations (CNVs) might be informative, the perfect phylogeny assumptions do not apply to CNVs. We demonstrate that one instance of this problem, using a cost function based on parsimony, is NP-hard, by using a reduction from the problem of finding the minimum vertex coloring of a graph. Finally, we develop an algorithm to solve the problem.

2   Preliminaries and Problem Definition

In this section, we first define the Perfect Phylogeny Mixture problem, and then formulate the *Minimum Split Row problem* as a specific optimization version of

Perfect Phylogeny Mixture. We assume that we measure the state (mutated or not mutated) at each of n positions in the genome for m tumor samples. Thus, the sequencing data from the m tumor samples is represented by an $m \times n$ binary $(0/1)$ matrix M, where each row represents a sample and each column represents a mutation. The value 1 in the entry (i, j) of the matrix shows that the mutation j is present in the sample i. Without loss of generality, we assume there is no duplicated columns in the matrix.

Our goal is to reconstruct the phylogenetic tree that describes the relationship between these samples. If each sample consists of a single tumor cell, then by the infinite sites assumption (a mutation occurs at a position at most once) the samples satisfy a *perfect phylogeny*. Because we also know that the normal, unmutated genome (all characters are 0) is the ancestral state, we can describe the relationship between the samples using a rooted perfect phylogeny tree [8]: each cell corresponds to a leaf in the tree, and each edge is labeled by the character(s) that change state from $0 \rightarrow 1$ on the edge of the tree. Figure 1(a) shows a perfect phylogeny tree (on the left), where three mutations m_1, m_2 and m_3 occurred on its branches. The leaves, each of them assigned with a binary representation of their mutation sets, correspond to distinct subpopulations.

Now, because each sample from the tumor consists of more than one tumor cell, we do not measure the sequences at individual leaves of the tree. Rather, each sample is a mixture of cells, some of which share common mutations. Thus, each sample corresponds to a superposition, or union, of mutations in the subset of leaves from the perfect phylogeny tree of the cells. Figure 1(b) shows an example of four different samples, shown in different colors. Each sample contains a mixture of cells and the we measure the *union* of the mutation sets belonging to the cells in the sample.

Recall by the Perfect Phylogeny Theorem [9] that a $(0/1)$ matrix M exhibits a perfect phylogeny if and only if no pair of columns conflict, according to the following definition.

Definition 1. *Columns C_i and C_j in M **conflict** if and only if there exists three rows $\mathbf{r}_1$, $\mathbf{r}_2$ and $\mathbf{r}_3$ in M such that their (i, j) positions are $(1, 1), (0, 1), (1, 0)$, respectively.*

For example, consider a $(0/1)$ matrix M with 4 rows corresponding to each sample in Figure 1, and 3 columns corresponding to each of the mutations m_1, m_2 and m_3. For each entry (i, j) of this matrix, we place a 1, if the sample i contains the mutation j. It is easy to see that a perfect phylogeny does not exists for matrix M since m_2 and m_3 *conflict*.

Since we assume that the cells in the tumor exhibit a perfect phylogeny, then any conflicts in the matrix M result from either errors in the data or from the fact that each row represents the mutations in a multiple cells in the tumor. Thus, assuming no errors we formulate the **Perfect Phylogeny Mixture** problem as *deconvolving* each row in the matrix M into one or more rows (representing individual cells in the tumor) such that the resulting matrix has no conflicts.

Before formally defining this problem, we recall a few more facts about the perfect phylogeny models.

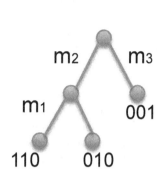

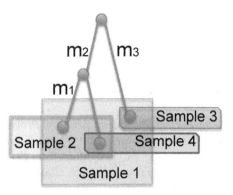

(a) A perfect phylogeny with 3 mutations m_1, m_2 and m_3. The leaves are assigned binary representations 110, 010, 001, representing mutation sets $\{m_1, m_2\}$, $\{m_2\}$, $\{m_3\}$, respectively

(b) Taking 4 different samples from the tumor phylogeny. The binary representation of the mutations sets in samples 1,2,3 and 4 will be $111, 110, 010$ and 001, respectively.

Fig. 1. This figure shows a tumor phylogeny with 3 mutations. Multiple samples are taken from the tumor which overlap and may create conflicts in observed mutations.

Definition 2. *A mutation matrix M is **conflict-free** if and only if it has no pairs of conflicting columns.*

Note that following the Perfect Phylogeny Theorem [8] with the assumption that an all-zero ancestral sequence is present at the root, a conflict-free mutation matrix corresponds to a perfect phylogeny tree [8].

Also, recall the following alternative statement of the Perfect Phylogeny Theorem [8], which was proved in [9]:

Theorem 1. *[9] There is a perfect-phylogeny for a mutation matrix M if and only if for every pair of columns, either one column is contained in the other (i.e. the mutation set corresponding to one column is a subset of the mutations of the other column) or the intersection of the mutation sets corresponding to the columns is empty.*

Using the above theorem, it is solvable in polynomial time to check whether a $(0/1)$ mutation matrix has a perfect phylogeny. However, the more general perfect phylogeny problem where characters may have arbitrary integer values as their states, is NP-hard. The general perfect phylogeny problem reduces to the graph-theoretic problem of finding a chordal completion of a colored graph (also known as the triangulating colored graphs problem) [1]. See [21] for a survey on combinatorial optimization problems related to perfect phylogeny.

We formally define the *deconvolution* of a row (a sample containing two or more subpopulations of tumor cells), as the following operation on the row, and then define the Perfect Phylogeny Mixture problem.

Definition 3. *Given a row* **r** *of a mutation matrix* M, *a* **split-row operation** $S_{\mathbf{r}}$ *on* M *is the following transformation: replace* **r** *by* k *rows* $\hat{r}_1, \ldots, \hat{r}_k$ *such that and for every* i, *where the* i^{th} *position in* **r** *is equal to* 0, *all rows* $\hat{r}_i$ *has a* 0 *in that position. In other words,* **r** *equals the bitwise OR of the rows* $\hat{r}_1, \ldots, \hat{r}_k$.

The Perfect Phylogeny Mixture Problem Given a binary matrix M with m rows and n columns, find a binary matrix M' with m' rows and n columns such that: (1) M' is conflict-free. (2) For every row **r** in M, there exists k rows $\hat{r}_1, \ldots, \hat{r}_k$ in M' such that **r** can be replaced by $\hat{r}_1, \ldots, \hat{r}_k$, using an Split-Row operation.

Note that for any given binary matrix M with m rows and n columns, the identity matrix I_n is always a trivial solution for the Perfect Phylogeny Mixture problem. In what follows, we consider optimization versions of the Perfect Phylogeny Mixture problem and define the Minimum-Split-Row problem.

We use Split-Row operations, to distinguish distinct leaves in a Perfect Phylogeny model whose corresponding subpopulations were mixed with each other in one sample. This mixture may cause conflicts in the input mutation matrix, and we ask to perform Split-Row operations to convert the input mutation matrix to a conflict-free one. For a split row operation $S_{\mathbf{r}}$, we define the cost function $\gamma(M, \mathbf{r}) = k - 1$, the number of additional rows to M. An alternative cost function $\eta(M, \mathbf{r})$ is the number of additional *unique* rows that were not identical [1] to any of the original rows in M, after the Split-Row operation on $S_{\mathbf{r}}$. Note that, we only discuss the problem under the cost function γ. We leave the study of the problem under the cost function η to a future work.

The following matrix on the left shows an example where a conflict exists between columns m_2 and m_3. After performing an Split-Row operation on the first row of the matrix (shown in blue), two new rows are created (shown in red). The resulting matrix on the right is conflict-free and thus corresponds to a perfect phylogeny tree. In this example $\gamma(M, 1) = 1$, while $\eta(M, 1) = 0$.

$$
\text{Split-Row operation}
\begin{array}{c}
m_1 \quad m_2 \quad m_3 \\
\begin{pmatrix}
1 & 1 & 1 \\
1 & 1 & 0 \\
0 & 0 & 1 \\
0 & 1 & 0
\end{pmatrix}
\end{array}
\Rightarrow
\begin{array}{c}
\text{First new row} \\
\text{Second new row}
\end{array}
\begin{array}{c}
m_1 \quad m_2 \quad m_3 \\
\begin{pmatrix}
1 & 1 & 0 \\
0 & 0 & 1 \\
1 & 1 & 0 \\
0 & 0 & 1 \\
0 & 1 & 0
\end{pmatrix}
\end{array}
$$

In order to transform a matrix with conflicts to a conflict-free matrix using Split-Row operations, one has to be very careful:

---

[1] Two rows in the mutation matrix are identical, if their entries are identical at every position.

Observation 2. *Performing the Split-Row operation on a row may introduce new conflicts.*

Proof. See the following example in which the Split-Row operation on the first row creates new conflicts between the first and second columns, if $(1,1,1,1)$ is replaced by $(1,0,1,0)$ and $(0,1,0,1)$. In this case, to avoid creating new conflicts, the Split-Row operation must replace $(1,1,1,1)$ with $(0,0,1,1)$ and $(1,1,0,0)$.

$$M = \begin{pmatrix} 1\ 1\ 1\ 1 \\ 1\ 1\ 0\ 0 \\ 0\ 0\ 1\ 1 \\ 1\ 1\ 1\ 0 \\ 0\ 1\ 1\ 1 \end{pmatrix}$$

Given a matrix mutation M, our aim is to find a series of Split-Row operations that transforms M into a *conflict-free* matrix. Note that since a *conflict-free* matrix gives a perfect phylogeny tree, the Split-Row operations help distinguish tumor sub-populations in each sample. In an ideal case, if rows correspond to individual leaves of the tree, the resulting mutation matrix will be *conflict-free*.

Following the principle of maximum parsimony, we consider an optimization problem:

The Minimum-Split-Row Problem Given a binary matrix mutation M, perform split-row operations on a subset S its rows, such that the resulting matrix is *conflict-free* and $\sum_{r \in S} \gamma(M, r)$ is minimized.

Given a mutation matrix M with n rows and m columns, there is an elegant way of obtaining a *conflict-free* mutation matrix from M, using Split-Row operations on each row (we will discuss further). We conclude this section with a definition and a lemma statement. Corresponding to each row r of M, we construct a graph $\mathcal{G}_{M,r}$ as follows: For each entry 1 in r, we put a node in $\mathcal{G}_{M,r}$ and two nodes in $\mathcal{G}_{M,r}$ are connected with an edge if and only their corresponding columns in M are *in conflict* (See the definition of Conflicting Columns above).

It is clear that in order to transform a matrix M to a conflict-free matrix with a series of Split-Row operations, we need to perform an independent Split-Row operation on every row r which has a corresponding non-empty graph $\mathcal{G}_{M,r}$ (i.e. a graph with at least one edge). Moreover the following lemma shows each row r must be replaced by at least $\chi(\mathcal{G}_{M,r})$ rows, where χ is the chromatic number[2] of the graph.

Lemma 1. *Given a mutation matrix M with n rows and m columns, in a series of Split-Row operations that transfer M to a conflict-free mutation matrix M',*

---

[2] The smallest number of colors needed to color nodes of a graph G such that adjacent nodes are assigned distinct colors, is called its chromatic number.

each row **r** *in* M *must be (independently) replaced by at least* $\chi(\mathcal{G}_{M,\mathbf{r}})$ *rows, where* $\mathcal{G}$ *is the corresponding graph for row* **r** *as defined above.*

Proof. First note that, in the given mutation matrix, if the corresponding graph for a particular row **r** is not empty, then we have to perform the Split-Row operation on **r**. If two columns of **r** are in conflict, then performing the Split-Row operation on other rows does not affect the conflicts in **r**. Moreover, if **r** is replaced by less than $\chi(\mathcal{G}_{M,\mathbf{r}})$ rows, then due to the pigeonhole principle, there would still be a row with conflicting columns.

3   Complexity of the Minimum-Split-Row Problem

In this section, we prove the following theorem.

Theorem 3. *The Minimum-Split-Row problem is NP-Compete.*

The proof of this theorem is by a reduction from the minimum vertex coloring problem. To prove Theorem 3, we first show that for any simple graph G, there exists a mutation matrix M whose first row is all 1's and such that conflict graph of the first row is exactly G. Using this construction and Lemma 1, we show that it is NP-hard to find the minimum number of new rows in a series of Split-Row operations that lead to a conflict-free matrix.

Theorem 4. *Given a simple graph* G *with* n *vertices and* m *edges, there exists a binary matrix* M *with* n *columns and* $2m + 1$ *rows, such that all entries of its first row are equal to* 1 *and* $\mathcal{G}_{M,1} = G$.

Proof. Let $V(G) = \{v_1, \cdots, v_n\}$ be the vertex set of G, and $E(G) = \{e_1, \cdots, e_m\}$ be the edge set of G. Assume for any $1 \leq k \leq m, e_k = (x_k, y_k)$. That is, the edge e_k connects vertices x_k and y_k (See Figure 2(a)) for an example of such a conflict graphs with 5 vertices and 4 edges). We construct a mutation matrix with n columns and $2m + 1$ rows such that each column corresponds a vertex in G. All entries in the first row are equal to 1, and corresponding to each edge $e_k (1 \leq k \leq m)$, we have *two* rows (i.e. rows $2k$ and $2k + 1$) in M. For each $k (1 \leq k \leq m)$, the entries of rows $2k$ and $2k + 1$ are determined as follows. For row $2k$ of the matrix, we place a 0 at entry x_k and we place a 1 at entry y_k, while for row $2k + 1$, we place a 1 at entry x_k and we place a 0 at entry y_k. Since all the entries of the first row are equal to 1, this configuration leads to a conflict between any pairs of columns x_k and y_k that correspond to an edge e_k in G. We name the above as Step 1 of our construction.

Now, we fill the rest of the matrix with 0's and 1's such that we do not create new conflicts among the columns that were not originally in conflict. In order to guide filling of the other entries of the matrix so that no new conflict is created, we maintain a principle that if two columns are not in conflict (i.e. there is no edge between the corresponding vertices in G) then the column on the right must *contain* the column on the left. That is, for each row, the entry of the column on the right is greater or equal to the entry of the column on the left. In other

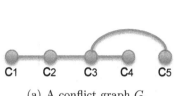

(a) A conflict graph G.

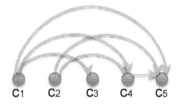

(b) An underlying containment graph G' is shown. All edges of G' are oriented from left to right.

Fig. 2. This figure shows a conflict graph G (on the left) and an underlying containment graph G' (on the right) corresponding to a row of a mutation matrix that is known to be all 1's. Hence, the edge set of G' is the complement of the edge set of G.

words, if we consider an underlying *containment graph* G' on the columns of the matrix M, we fill the entries of the matrix such that the order of columns from left to right implies a topologically sorted containment graph.

Corresponding to each column, there is a vertex in G', and the column C_i is connected to the column C_j with a directed edge, if and only if C_j contains C_i. Note that since all entries of the first row of M are equal to 1, the edge set of G' is the complement of the edge set of G. An example of an underlying containment graph is shown in Figure 2 (b).

We fill the empty cells of the matrix column by column, and from the leftmost column to the right-most one. During this process and while at column $\ell (1 \leq \ell \leq n)$, we fill each entry (ℓ, p) of the matrix, which was not originally filled due to a conflict, as follows We fill the entry (ℓ, p) with a 1, if there exists an edge in the underlying conflict graph of the matrix from any column $\ell' < \ell$ and the value of the entry (ℓ', p) is already set to 1. Otherwise, we fill the entry (ℓ, p) with a 0.

We call the above process, Step 2 of the construction. Figure 3 provides an example.

Note that the above construction does not lead to new conflicts. For the purpose of our argument, we distinguish the entries which received their values due to a connected pair of vertices in the conflict graph (shown in black color in the example matrix) with those which received their values in Step 2 of the algorithm (shown in color red in the example).

Assume after filling the matrix, there exists a pair of *conflicting* columns whose their corresponding vertices are not connected in the conflict graph. Let i and $j (i < j)$ be a new conflicting pairs of columns, in which j is the smallest index among all the new conflicting pairs of conflicting columns. Since columns i and j are now in conflict, there exists a row r in the matrix in which the entries $(r, i) = 1$ and $(r, j) = 0$, respectively. The entry (r, j) could not be empty after we initially planted $(0/1)$ in the matrix based on the original conflicts (i.e. Step 1 of the algorithm). In other words, the zero at (r, j) must have been planted similar

$$
M = \quad \text{Step 1} \begin{array}{ccccc} c_1 & c_2 & c_3 & c_4 & c_5 \\ \end{array} \begin{pmatrix} 1 & 1 & 1 & 1 & 1 \\ 1 & 0 & & & \\ 0 & 1 & & & \\ & 1 & 0 & & \\ & 0 & 1 & & \\ & & 1 & 0 & \\ & & 0 & 1 & \\ & & & 1 & 0 \\ & & & 0 & 1 \end{pmatrix} \Rightarrow \quad \text{Step 2} \begin{array}{ccccc} c_1 & c_2 & c_3 & c_4 & c_5 \\ \end{array} \begin{pmatrix} 1 & 1 & 1 & 1 & 1 \\ 1 & 0 & 1 & 1 & 1 \\ 0 & 1 & 0 & 1 & 1 \\ 0 & 1 & 0 & 1 & 1 \\ 0 & 0 & 1 & 0 & 0 \\ 0 & 0 & 1 & 0 & 0 \\ 0 & 0 & 0 & 1 & 1 \\ 0 & 0 & 1 & 0 & 0 \\ 0 & 0 & 0 & 0 & 1 \end{pmatrix}
$$

Fig. 3. The partially-filled matrix on the left shows the mutation matrix after Step 1 of the construction described in the proof of Theorem 4, while the matrix on the right shows the mutation matrix after Step 2

to the (0/1) entries shown in black. Because the entry (r, i) value is equal to 1 and vertices corresponding to i and j are not connected in the conflict graph (and thus connected in the underlying containment graph, oriented from i to j), filling (r, j) with a 0 would be a contradiction. Now, if the 0 was originally planted in the entry(r, j), the value 1 at the entry (r, j) cannot also be an originally planted 1 in the row r. Because in this scenario the pair must have been already a conflict in the graph. Thus the entry (r, j) must have been filled with 1 during Step 2 of the algorithm, essentially from a chain of *containment* relationships edges which starts from an entry on row r which has an originally planted 1. Note that the containment relationships are transitive. This is also a contradiction, and thus no new conflicting pairs of columns exists in the filled mutation matrix.   □

Using this result, we now prove Theorem 3.

Proof (of Theorem 3). We use a reduction from the minimum vertex coloring problem (one of Karp's 21 NP-hard problems [13]). Given a simple graph G, by Theorem 4 there exists a mutation matrix M whose conflict graph of the first row is G and whose size is polynomially bounded. If we solve the Minimum-Split-Row problem for M, then by Lemma 1, the first row of M is replaced by at least $\chi(G)$ rows, each of which defines a color for a vertex of G. Thus, we obtain a vertex coloring for G. Furthermore, given an instance of the Minimum-Split-Row problem, we can show that minimum vertex colorings of the conflict graphs corresponding to each row, will lead to a solution for the Minimum-Split-Row problem.

4   A Graph-Theoretic Algorithm

In this section, we provide an algorithm for the Minimum-Split-Row problem. Our algorithm achieves the lower bound given in Lemma 1: i.e. for a mutation matrix M, we replace each row $\mathbf{r}$ by exactly $\chi(\mathcal{G}_{M,\mathbf{r}})$ rows, where χ is the chromatic number of the graph $\chi(\mathcal{G}_{M,\mathbf{r}})$.

To state our algorithm, we first make the following definitions.

Definition 4. *Given a binary matrix M, we say column i **contains** j (or j is smaller than i), if and only if, for every row k, $M_{k,i} \geq M_{k,j}$.*

Definition 5. *Given a binary matrix M, the **containment graph** H_M is the directed graph whose vertices are the columns of M and such that there is an directed edge $i \to j$ if and only if column j contains column i.*

Note that H_M is a directed acyclic graph (DAG). Further note that if column i contains column j then there is no conflict between i and j in the original mutation matrix [9].

As we discussed earlier (Observation 2), performing a Split-Row operation on a row may introduce new conflicts. Informally, a Split-Row operation may affect pairs of columns i and j, where j contains i, in a way that a new conflict arises. Thus, any series of Split-Row operations that aim to make the mutation matrix M conflict free, must be aware of the underlying containment graph of M. In what follows, we describe an algorithm that uses what we call *Containment-Aware* series of Split-Row operations and achieves the lower bound of $\chi(\mathcal{G}_{M,\mathbf{r}})$ additional rows.

Algorithm. Without loss of generality, we permute the columns of the mutation matrix M such that the corresponding vertices of the underlying containment graph of M are topologically sorted. We remind the reader that the underlying containment graph is always a DAG. Let $k = \chi(\mathcal{G}_{M,\mathbf{r}})$. We assign k colors (from the set $\{1, \ldots, k\}$) to the vertices of $\mathcal{G}_{M,\mathbf{r}}$ such that two vertices whose corresponding columns are in conflict receive different colors. We now replace the row $\mathbf{r}$ of the matrix M with exactly k rows as follows. In the ith row among these k rows ($1 \leq i \leq k$), we place 1 in all corresponding columns of the vertices which received i as their color in the proper coloring. We place 0 in all other entries. It is easy to observe that, after this Split-Row operation, conflicting pairs of the columns corresponding to vertices in $\mathcal{G}_{M,\mathbf{r}}$ are no longer in conflict.

However, placing 0's in the new rows may affect the containment edges and create new conflicts elsewhere in the matrix. We avoid this problem as follows. Without loss of generality, consider the first new row after the split-Row operation and call that row $\mathbf{r}'_1$. If there exists a pair of columns c and d such that corresponding vertex v_c in the underlying containment graph is connected to v_d, but the new entry at $\mathbf{r}'_1, c$ is 1 and the new entry at $\mathbf{r}'_1, d$ is 0, then a new conflict may arise. To prevent this situation, we perform a Depth-first search (DFS), starting from the leftmost column, on the underlying containment graph and if we traverse an edge which connects 1 to 0 in corresponding entries in the matrix, we change the 0 to 1.

Note that fliping some of the 0 entires in the new rows to 1, as described above, guarantees that no new conflict will be created. Because, after the DFS and fliping the zeros accordingly, every containment relationship will be maintaied and thus no new conflict would arise.

Since the above algorithm depends on proper vertex coloring in graphs, in the worst case the runtime of the algorithm is exponential. If the optimal vertex coloring of $\mathcal{G}_{M,\mathbf{r}}$ is given, for each row $\mathbf{r}$ of the matrix, the rest of the algorithm is very efficient and can be executed in $O(n \ (n + m))$, where n is the number of rows and m is the number of columns. Note that once every row $\mathbf{r}$, with a non-empty $\mathcal{G}_{M,\mathbf{r}}$ is split, the DFS step will be done for the whole matrix. Thus a fast vertex graph coloring heuristic leads to an efficient algorithm, in practice.

5   Discussion and Future Work

In this paper, we provide a rigorous formulation of the Perfect Phylogeny Mixture problem and an optimization version of the problem. Given a collection of samples taken from the same tumor and the set of somatic mutations present in each of those samples, the Perfect Phylogeny Mixture problem asks for a deconvolution of the samples such that the mutational history can be reconstructed, using a perfect phylogeny model. Following the maximum parsimony principle, we introduce a novel cost function and discuss the NP-hardness of the Perfect Phylogeny Mixture problem under the cost function. We also present an algorithm to solve an optimization version of the Perfect Phylogeny Mixture problem using graph coloring to resolve conflicts in an appropriately sorted mutation matrix.

The novel problem formulated in this paper provides additional areas for further investigation, both theoretical and applied. One immediate open problem is to consider the Perfect Phylogeny Mixture problem under alternative cost functions that are independent of the Split-Row operations. One alternative is to use the cost function $\eta(M, \mathbf{r})$ that does not penalize the creation of rows that are identical to a current row of the matrix (i.e. contain the exact same mutations). In other words, given a mutation matrix M, the problem under this cost function asks for construction of a conflict-free matrix M' with the **minimum** number of rows (irrespective of the Split-Row operations) such that for each row in M, the set of mutations of the row is identical to the union of the mutations in a subset of rows in M'.

In the applied direction, it will be interesting to apply our algorithm for the Perfect Phylogeny Mixture problem to real cancer sequencing data, and to see how often deconvolution of a mixture will resolve conflicts in the data. As more datasets of genome sequencing from multiple samples of a tumor become available, there will be increasing need for computational models to infer mutational history of the data. Handling errors in sequencing data, while using our model, is another important future direction. The Perfect Phylogeny Mixture problem ignores the complications due to errors in real data sets. Finally, as single cell sequencing technologies continue to improve [5], new opportunities to develop and validate models based on the Perfect Phylogeny Mixture problem will arise.

Acknowledgments. The authors would like to thank Ahmad Mahmoody for fruitful discussions on the problem. This work was supported by National Science Foundation CAREER Award (CCF-1053753 to B.J.R.) and the National Institutes of Health (R01HG5690 to B.J.R.). B.J.R. is also supported by a Career Award at the Scientific Interface from the Burroughs Wellcome Fund, an Alfred Sloan Research Fellowship. I.H. is also supported by a Natural Sciences and Engineering Research Council of Canada (NSERC) Postdoctoral Fellowship.

Conflict of Interest: None declared.

References

1. Buneman, P.: A characterization of rigid circuit graphs. Discrete Mathematics 9, 205–212 (1974)
2. Desper, R., Jiang, F., Kallioniemi, O.-P., Moch, H., Papadimitriou, C.H., Schäffer, A.A.: Inferring tree models for oncogenesis from comparative genome hybridization data. Journal of Computational Biology 6(1), 37–51 (1999)
3. Ding, L., Ley, T.J., Larson, D.E., Miller, C.A., Koboldt, D.C., Welch, J.S., Ritchey, J.K., Young, M.A., Lamprecht, T., McLellan, M.D., McMichael, J.F., Wallis, J.W., Lu, C., Shen, D., Harris, C.C., Dooling, D.J., Fulton, R.S., Fulton, L.L., Chen, K., Schmidt, H., Kalicki-Veizer, J., Magrini, V.J., Cook, L., McGrath, S.D., Vickery, T.L., Wendl, M.C., Heath, S., Watson, M.A., Link, D.C., Tomasson, M.H., Shannon, W.D., Payton, J.E., Kulkarni, S., Westervelt, P., Walter, M.J., Graubert, T.A., Mardis, E.R., Wilson, R.K., DiPersio, J.F.: Clonal evolution in relapsed acute myeloid leukaemia revealed by whole-genome sequencing. Nature 481(7382), 506–510 (2012)
4. Ding, L., Raphael, B.J., Chen, F., Wendl, M.C.: Advances for studying clonal evolution in cancer. Cancer Lett. (January 2013)
5. Eberwine, J., Sul, J.-Y., Bartfai, T., Kim, J.: The promise of single-cell sequencing. Nat. Methods 11(1), 25–27 (2014)
6. Fernandez-Baca, D.: The Perfect Phylogeny Problem (retrieved September 30, 2012)
7. Gerlinger, M., Rowan, A.J., Horswell, S., Larkin, J., Endesfelder, D., Gronroos, E., Martinez, P., Matthews, N., Stewart, A., Tarpey, P., Varela, I., Phillimore, B., Begum, S., McDonald, N.Q., Butler, A., Jones, D., Raine, K., Latimer, C., Santos, C.R., Nohadani, M., Eklund, A.C., Spencer-Dene, B., Clark, G., Pickering, L., Stamp, G., Gore, M., Szallasi, Z., Downward, J., Futreal, P.A., Swanton, C.: Intratumor heterogeneity and branched evolution revealed by multiregion sequencing. N. Engl. J. Med. 366(10), 883–892 (2012)
8. Gusfield, D.: Efficient algorithms for inferring evolutionary trees. Networks 21, 19–28 (1991)
9. Gusfield, D.: Algorithms on Strings, Trees and Sequences: Computer Science and Computational Biology (1997)
10. Hajirasouliha, I., Mahmoody, A., Raphael, B.J.: A combinatorial approach for analyzing intra-tumor heterogeneity from high-throughput sequencing data. Bioinformatics 30(12), 78–86 (2014)
11. Hou, Y., Song, L., Zhu, P., Zhang, B., Tao, Y., Xu, X., Li, F., Wu, K., Liang, J., Shao, D., Wu, H., Ye, X., Ye, C., Wu, R., Jian, M., Chen, Y., Xie, W., Zhang, R.,

Chen, L., Liu, X., Yao, X., Zheng, H., Yu, C., Li, Q., Gong, Z., Mao, M., Yang, X., Yang, L., Li, J., Wang, W., Lu, Z., Gu, N., Laurie, G., Bolund, L., Kristiansen, K., Wang, J., Yang, H., Li, Y., Zhang, X., Wang, J.: Single-cell exome sequencing and monoclonal evolution of a JAK2-negative myeloproliferative neoplasm. Cell 148(5), 873–885 (2012)

12. Kandoth, C., McLellan, M.D., Vandin, F., Ye, K., Niu, B., Lu, C., Xie, M., Zhang, Q., McMichael, J.F., Wyczalkowski, M.A., Leiserson, M.M., Miller, C.A., Welch, J.S., Walter, M.J., Wendl, M.C., Ley, T.J., Wilson, R.K., Raphael, B.J., Ding, L.: Mutational landscape and significance across 12 major cancer types. Nature 502(7471), 333–339 (2013)

13. Karp, R.M.: Reducibility among combinatorial problems. In: Complexity of Computer Computations, pp. 85–103 (1972)

14. Lawrence, M.S., Stojanov, P., Polak, P., Kryukov, G.V., Cibulskis, K., Sivachenko, A., Carter, S.L., Stewart, C., Mermel, C.H., Roberts, S.A., Kiezun, A., Hammerman, P.S., McKenna, A., Drier, Y., Zou, L., Ramos, A.H., Pugh, T.J., Stransky, N., Helman, E., Kim, J., Sougnez, C., Ambrogio, L., Nickerson, E., Shefler, E., Cortés, M.L., Auclair, D., Saksena, G., Voet, D., Noble, M., DiCara, D., Lin, P., Lichtenstein, L., Heiman, D.I., Fennell, T., Imielinski, M., Hernandez, B., Hodis, E., Baca, S., Dulak, A.M., Lohr, J., Landau, D.-A., Wu, C.J., Melendez-Zajgla, J., Hidalgo-Miranda, A., Koren, A., McCarroll, S.A., Mora, J., Lee, R.S., Crompton, B., Onofrio, R., Parkin, M., Winckler, W., Ardlie, K., Gabriel, S.B., Roberts, C.M., Biegel, J.A., Stegmaier, K., Bass, A.J., Garraway, L.A., Meyerson, M., Golub, T.R., Gordenin, D.A., Sunyaev, S., Lander, E.S., Getz, G.: Mutational heterogeneity in cancer and the search for new cancer-associated genes. Nature 499(7457), 214–218 (2013)

15. Nik-Zainal, S., Van Loo, P., Wedge, D.C., Alexandrov, L.B., Greenman, C.D., Lau, K.W., Raine, K., Jones, D., Marshall, J., Ramakrishna, M., Shlien, A., Cooke, S.L., Hinton, J., Menzies, A., Stebbings, L.A., Leroy, C., Jia, M., Rance, R., Mudie, L.J., Gamble, S.J., Stephens, P.J., McLaren, S., Tarpey, P.S., Papaemmanuil, E., Davies, H.R., Varela, I., McBride, D.J., Bignell, G.R., Leung, K., Butler, A.P., Teague, J.W., Martin, S., Jonsson, G., Mariani, O., Boyault, S., Miron, P., Fatima, A., Langerod, A., Aparicio, S.A., Tutt, A., Sieuwerts, A.M., Borg, A., Thomas, G., Salomon, A.V., Richardson, A.L., Borresen-Dale, A.L., Futreal, P.A., Stratton, M.R., Campbell, P.J.: The life history of 21 breast cancers. Cell 149(5), 994–1007 (2012)

16. Nowell, P.C.: The clonal evolution of tumor cell populations. Science 194(4260), 23–28 (1976)

17. Salari, R., Saleh, S.S., Kashef-Haghighi, D., Khavari, D., Newburger, D.E., West, R.B., Sidow, A., Batzoglou, S.: Inference of tumor phylogenies with improved somatic mutation discovery. In: Deng, M., Jiang, R., Sun, F., Zhang, X. (eds.) RECOMB 2013. LNCS, vol. 7821, pp. 249–263. Springer, Heidelberg (2013)

18. Schuh, A., Becq, J., Humphray, S., Alexa, A., Burns, A., Clifford, R., Feller, S.M., Grocock, R., Henderson, S., Khrebtukova, I., Kingsbury, Z., Luo, S., McBride, D., Murray, L., Menju, T., Timbs, A., Ross, M., Taylor, J., Bentley, D.: Monitoring chronic lymphocytic leukemia progression by whole genome sequencing reveals heterogeneous clonal evolution patterns. Blood 120(20), 4191–4196 (2012)

19. Shah, S.P., Roth, A., Goya, R., Oloumi, A., Ha, G., Zhao, Y., Turashvili, G., Ding, J., Tse, K., Haffari, G., Bashashati, A., Prentice, L.M., Khattra, J., Burleigh, A., Yap, D., Bernard, V., McPherson, A., Shumansky, K., Crisan, A., Giuliany, R., Heravi-Moussavi, A., Rosner, J., Lai, D., Birol, I., Varhol, R., Tam, A., Dhalla, N., Zeng, T., Ma, K., Chan, S.K., Griffith, M., Moradian, A., Cheng, S.W., Morin, G.B., Watson, P., Gelmon, K., Chia, S., Chin, S.F., Curtis, C., Rueda, O.M., Pharoah, P.D., Damaraju, S., Mackey, J., Hoon, K., Harkins, T., Tadigotla, V., Sigaroudinia, M., Gascard, P., Tlsty, T., Costello, J.F., Meyer, I.M., Eaves, C.J., Wasserman, W.W., Jones, S., Huntsman, D., Hirst, M., Caldas, C., Marra, M.A., Aparicio, S.: The clonal and mutational evolution spectrum of primary triple-negative breast cancers. Nature 486(7403), 395–399 (2012)
20. Strino, F., Parisi, F., Micsinai, M., Kluger, Y.: TrAp: a tree approach for finger-printing subclonal tumor composition. Nucleic Acids Res. 41(17), e165 (2013)
21. Warnow, T.: Some combinatorial problems in phylogenetics. In: Invited paper, Proceedings of the International Colloquium on Combinatorics and Graph Theory, Balatonlelle, Hungary (1999)
22. Vogelstein, B., Papadopoulos, N., Velculescu, V.E., Zhou, S., Diaz Jr., L.A., Kinzler, K.W.: Cancer genome landscapes. Science 339(6127), 1546–1558 (2013)
23. Xu, X., Hou, Y., Yin, X., Bao, L., Tang, A., Song, L., Li, F., Tsang, S., Wu, K., Wu, H., He, W., Zeng, L., Xing, M., Wu, R., Jiang, H., Liu, X., Cao, D., Guo, G., Hu, X., Gui, Y., Li, Z., Xie, W., Sun, X., Shi, M., Cai, Z., Wang, B., Zhong, M., Li, J., Lu, Z., Gu, N., Zhang, X., Goodman, L., Bolund, L., Wang, J., Yang, H., Kristiansen, K., Dean, M., Li, Y., Wang, J.: Single-cell exome sequencing reveals single-nucleotide mutation characteristics of a kidney tumor. Cell 148(5), 886–895 (2012)

Author Index